Advancing Global Bioethics

Volume 7

Series editors
Henk A.M.J. ten Have
Pittsburgh, USA

Bert Gordijn
Dublin, Ireland

The book series Global Bioethics provides a forum for normative analysis of a vast range of important new issues in bioethics from a truly global perspective and with a cross-cultural approach. The issues covered by the series include among other things sponsorship of research and education, scientific misconduct and research integrity, exploitation of research participants in resource-poor settings, brain drain and migration of healthcare workers, organ trafficking and transplant tourism, indigenous medicine, biodiversity, commodification of human tissue, benefit sharing, bio-industry and food, malnutrition and hunger, human rights, and climate change.

More information about this series at http://www.springer.com/series/10420

Evaristus Chiedu Obi

Post-Trial Access to Drugs in Developing Nations

Global Health Justice

Springer

Evaristus Chiedu Obi
Duquesne University
Pittsburgh, PA, USA

ISSN 2212-652X ISSN 2212-6538 (electronic)
Advancing Global Bioethics
ISBN 978-3-319-86762-5 ISBN 978-3-319-60028-4 (eBook)
DOI 10.1007/978-3-319-60028-4

Printed on acid-free paper

This Springer imprint is published by Springer Nature
The registered company is Springer International Publishing AG
The registered company address is: Gewerbestrasse 11, 6330 Cham, Switzerland

The book is dedicated to my beloved mother, the late Mrs. Elizabeth Obi, for championing the cause of justice for the less privileged and the voiceless all through her life and to all those who continue to champion the cause of justice for the global poor afflicted with various diseases.

Abstract

The pandemic nature of HIV/AIDS in developing countries engenders global health emergency which establishes the urgent need to address the issue of affordable access to anti-retroviral drugs in developing countries. The book discusses an ethical justification for posttrial access to drugs for participants and host populations in developing countries within the context of global justice, stressing the combination of national and global responsibilities in realizing this objective. Drawing on the strengths of Rawls's statist and Pogge's cosmopolitan theories and on the International Human Rights Law, the book proposes a paradigm of global health justice involving a sliding scale of national and global responsibilities for the realization of the right to health in general and access to drugs in particular.

Every nation has the primary responsibility for realizing the right to health, including affordable access to drugs for its citizens. However, when poor developing countries demonstrate their best efforts by spending at least 3% of their average gross domestic product (GDP) on national health and still are not able to realize the right to health, including affordable access to drugs for its citizens, developed countries should intervene to exercise their global responsibilities in realizing this objective. The proposed paradigm emphasizes that global responsibility supplements rather than replaces national responsibility for health. Thus, health is considered a shared responsibility in the contemporary era.

International human rights law was argued as providing a theoretical framework for national and global responsibilities for realizing the core obligations that stem from socioeconomic rights and for addressing global health inequalities. The obligation to provide international assistance in realizing the minimum essential level of the right to health which includes access to drugs was argued as imperative.

The book proposes an international agency such as the Global Health Fund for the distribution of health-related goods that would rectify the injustice stemming from the current global system. Expanding the mandate of the current Global Fund to fight AIDS, tuberculosis, and malaria to include interventions for other diseases and addressing the weakness of the public health systems in developing countries were proposed as a good start for establishing the Global Health Fund. An effective Global Health Fund rooted in the concept of financial sustainability would significantly enhance the realization of the right to health and posttrial access to drugs for participants and host populations in developing countries.

Acknowledgments

Gratitude they say is the distinctive mark of a gentle person. The completion of a monumental task such as a book would inevitably involve the significant contributions of many individuals. First and foremost, my boundless gratitude goes to God for His unfathomable benevolence and sustenance. I am highly indebted and grateful to my mentor Dr. Gerard Magill for his assistance in completing this work. I really can't thank him enough for his unparalleled support, direction, generosity, good will, and motivation. His wealth of knowledge, mastery, insight, and commitment to academic excellence were critical for the diligent completion of this book. I am also grateful to Dr. Henk ten Have for his outstanding dedication and commitment to medicine, global health, ethics, and research and for his powerful insights. My gratitude also goes to Dr. David Kelly, the founding director of the Center for Healthcare Ethics, Duquesne University, who laid the strong foundation of my healthcare ethics education. I immensely thank current and past faculty, staff, and students of the Center for Healthcare Ethics for their various contributions toward my growth in healthcare ethics education.

I am also exceedingly grateful to my deceased parents, Mr. James Obi and Mrs. Elizabeth Obi, for teaching, nurturing, and cultivating in me important moral values of justice, integrity, honesty, compassion, personal responsibility, and excellence. I am as well immeasurably grateful to my siblings, especially Ego Appolonia Ike for all her selfless sacrifice for me and for the entire family and for significantly contributing to whatever each person in the family has become.

Finally, I thank my deceased former bishop, Most Rev. Albert K. Obiefuna, for his support and for giving me the opportunity to come to the United States to study, and the current bishop, Most Rev. Valerian Okeke, for his support.

Contents

Chapter 1
Global Health Inequalities and the Quest for Justice

1.1 Introduction: Contextualizing the Debate

The pandemic nature of HIV/AIDS in developing countries results in global health emergency which creates the urgent need to address the issue of affordable access to anti-retroviral drugs in developing countries. The work focuses on the ethical justification of post-trial access to drugs for participants and host populations in developing countries by combining national and international responsibilities in realizing this objective.

However, I specifically discussed anti-retroviral drugs to illustrate for the general analysis but the argument applies to post-trial access to any drug. The choice of using HIV/AIDS to illustrate post-trial access to drugs may be contentious as there has been significant progress over the years in access to antiretroviral drugs in developing countries, but the author justifies it for several reasons. Free access to antiretroviral drugs in developing countries may have significantly improved but there are other barriers to access identified in the literature. Posse et al. articulate, "The cost of drugs seems not to be the most important constraint of access to (anti-retroviral therapy) ART, as many of the drugs are provided free of cost to eligible patients through combined efforts by countries and bilateral and multilateral partners.".[1]

A wide range of other factors impede access to ART in developing countries. For example, at the population level, several barriers were cited, "lack of information about ART, perceived high costs for ART and stigma."[2] At the health system level, barriers cited were "long distance from home to the health facility, lack of

[1] Mariana Posse et al., "Barriers to Access to Antiretroviral Treatment in Developing Countries: A Review", *Tropical Medicine and International Health* 13, no. 7 (July 2008): 904.

[2] Posse et al., "Barriers to Access to Antiretroviral Treatment in Developing Countries: A Review", 904.

E.C. Obi, *Post-Trial Access to Drugs in Developing Nations*, Advancing Global Bioethics 7, DOI 10.1007/978-3-319-60028-4_1

co-ordination across services and limited involvement of the community in the programme planning process."[3]

Several authors also identified economic, sociocultural, and behavioral barriers to access ART in resource-limited countries for people living with HIV/AIDS (PLWHA).[4] Economic obstacles to access focus on loss of wages due to regular clinic visits for treatment, problem with affording expenses for transportation to hospital for regular appointments, inadequate HIV/AIDS training for service providers, and lack of equipped facilities and resources.[5] In some situations, even though ART is available for free or significantly discounted, costs of laboratory testing are unaffordable for patients which is a prerequisite for free treatment.[6]

Some authors also predict what they call a "treatment timebomb" that would lead to formidable challenges after the significant progress of the past decade in scaling up access to antiretroviral (ARV) drugs.[7] First, the cost of treatment will increase due to emergence of new ARVs likely to be under patent in developing countries and therefore exorbitant. Second, an expanded drug formulary would be immediately needed because more people would need access to new version ARVs. Furthermore, roughly two-thirds of people in need of ARV could not access first-line medicines. Third, research progresses on newer drugs and combinations should be widely available.[8]

The book argues for a paradigm involving a sliding scale of national and global responsibilities for the right to health in general and access to drugs in particular. The implication is that every nation has the primary responsibility for providing affordable access to drugs for its citizens. However, when developing countries exhaust their domestic resources and still are not able to provide affordable access to drugs for its citizens, developed countries should intervene to exercise their global responsibilities in realizing this objective. Post-trial access to drugs in developing countries is defended in this book within the context of realizing the minimum essential level of the right to health which includes access to drugs.

---

[3] Posse et al., "Barriers to Access to Antiretroviral Treatment in Developing Countries: A Review", 904.

[4] John A. Bartlett, Ashwini Shewade, and Rukmini Rajagopalan, "Obstacles and Proposed Solutions to Effective Antiretroviral Therapy in Resource-Limited Settings", *Journal of the International Association of Physicians in AIDS Care* 8, no. 4 (July/August 2009): 253.

[5] Bartlett, Shewade, and Rajagopalan, "Obstacles and Proposed Solutions to Effective Antiretroviral Therapy in Resource-Limited Settings", 255–256.

[6] Bartlett, Shewade, and Rajagopalan, "Obstacles and Proposed Solutions to Effective Antiretroviral Therapy in Resource-Limited Settings", 255–256.

[7] Ellen 't Hoen et al., "Driving a Decade of Change: HIV/AIDS, Patents and Access to Medicines for All", *Journal of the International AIDS Society* 14, no. 15 (March 2011): https://msfaccess.org/sites/default/files/MSF_assets/HIV_AIDS/Docs/HIV_MedJourn_JournIntnlAIDS_Decadeofchange_ENG_2011.pdf (Accessed April 2, 2017).

[8] Ellen 't Hoen et al., "Driving a Decade of Change: HIV/AIDS, Patents and Access to Medicines for All", *Journal of the International AIDS Society* 14, no. 15 (March 2011): https://msfaccess.org/sites/default/files/MSF_assets/HIV_AIDS/Docs/HIV_MedJourn_JournIntnlAIDS_Decadeofchange_ENG_2011.pdf (Accessed April 2, 2017).

Global inequality in health and wealth between developed and developing countries continue to be a challenge for the global community. No place is this inequality felt more than in the catastrophic impact of HIV/AIDS in developing countries, where those living with HIV/AIDS lacked access to anti-retroviral drugs. Global health crisis created by HIV/AIDS results in many clinical trials in developing countries in search of anti-retroviral drugs for combating the disease. Some of the clinical trials conducted in developing countries were marred by allegations of violation of rights of research participants and host populations.

A related issue discussed by some scholars was the availability of successful products, e.g. anti-retroviral drugs at the end of clinical trials.[9] Although, the industrialized world shoulders very few research burdens, it enjoys most research benefits because, unlike the developing world it can buy a proven intervention. On the other hand, poor inhabitants of developing countries who serve as research subjects assume the vast burdens of research but rarely share economic reach of both research subjects and their governments.[10] The problem of affordable access to drugs in developing countries has been exacerbated by Trade-Related Aspects of Intellectual Property Rights Agreement (TRIPS) and TRIPS-plus measures. TRIPS agreement and TRIPS-plus measures ensure strong patent protection which drive up the price of drugs through monopoly pricing system and block generic alternatives.[11] Forman notes, "The price impact of excluding access to generic medicines is particularly acute, since generic competition is a critical factor in reducing drug prices."[12] Trade agreements currently being negotiated by United States and other western governments may severely limit production of generic drugs which is considered the primary source of affordable medications in developing countries.[13]

A related issue regarding the problem of affordable access to anti-retroviral drugs in developing countries was discussed by some authors as the demand factors that influence access to HIV/AIDS drugs in Africa.[14] Schuklenk and Ashcroft observe that on the demand side that the health care infrastructure needed to make

[9] Christie Grady, "Ethics of International Research: What Does Responsiveness Mean?", *Virtual Mentor* 8, no. 4 (April 2006): 237.

[10] Alice K. Page, "Prior Agreements in International Clinical Trials: Ensuring the Benefits of Research to Developing Countries", *Health Policy, Law and Ethics* 3, no.1 (Winter 2002): 35–64.

[11] Jillian Clare Cohen and Patricia Illingworth, "The Dilemma of Intellectual Property Rights for Pharmaceuticals: The Tension between Ensuring Access of the Poor to Medicines and Committing to International Agreements", *Developing World Bioethics* 3, no. 1 (November 2003): 27–48.

[12] Lisa Forman, "Trade Rules, Intellectual Property, and the Right to Health", *Ethics & International Affairs* 21, no. 3 (Fall 2007): 340.

[13] Michael Westerhaus and Arachu Castro, "How Do Intellectual Property Law and International Trade Agreements Affect Access to Antiretroviral Therapy?, *Plos Medicine* 3, no. 8 (August 2006): 1234.

[14] Udo Schuklenk and Richard Ashcroft, "Affordable Access to Essential Medication in Developing Countries: Conflicts between Ethical and Economic Imperatives", in *Ethics & AIDS in Africa: The Challenge to Our Thinking* ed. Anton A. Van Niekerk and Loretta M. Kopelman (Walnut Creek, California: Left Coast Press Inc., 2005), 128.

use of these drugs effectively is lacking in developing countries.[15] Three factors were identified as impeding affordable access to anti-retroviral drugs in developing countries, highly strong legal protection of patents, lack of or slow third world government focus on the crisis and economic programs that have largely reduced funding for public health.[16]

However, some scholars advocate for the social responsibility of pharmaceutical companies as a way of dealing with the affordable access issue.[17] Private donation of drugs, differential pricing, price reductions, prior agreements, public-private partnerships and manufacture of generic copies of patented drugs and compulsory licensing were highlighted as strategies for dealing with affordable access to anti-retroviral drugs in developing countries.[18] Private donations, differential pricing and price reductions are improvised solutions that merely rely on the generosity of pharmaceutical companies and they do not go far enough in dealing with affordable access to anti-retroviral drugs in developing countries. These strategies are considered substandard in both preventing avoidable death and in relation to the sustainability of the policy and the recognition of the social responsibilities of companies and states.[19] Compulsory licensing is considered more effective because it helps the countries to fulfill the duty of meeting the health needs of their populations in national health emergencies.

The debate for post-trial access to drugs in developing countries has been dominated by two approaches, John Rawls's statist approach and Thomas Pogge's cosmopolitan approach. Rawls's statist approach relies on humanitarian assistance from the perspective of global justice to provide post-trial access to drugs in developing countries. In contrast, Pogge's cosmopolitan approach adopts a much more international perspective to global justice to justify access to drugs in developing countries. The book introduces a paradigm which combines Rawls's national responsibilities and Pogge's emphasis on international responsibilities for global justice to address the ethical problem of post-trial access to drugs in developing countries. The discussion on ethical justification for post-trial access to drugs for participants and host populations in developing countries begins with the issue of global health inequalities.

[15] Udo Schuklenk and Richard Ashcroft, "Affordable Access to Essential Medication in Developing Countries: Conflicts between Ethical and Economic Imperatives", 128.

[16] James T. Gathii, "Third World Perspectives on Global Pharmaceutical Industry", in *Ethics and the Pharmaceutical Industry* ed. Michael A. Santoro and Thomas M. Gorrie (New York: Cambridge University Press, 2005): 336.

[17] Cohen and Illingworth, "The Dilemma of Intellectual Property Rights for Pharmaceuticals: The Tension between Ensuring Access of the Poor to Medicines and Committing to International Agreements", 27–48.

[18] Ruth Macklin, *Double Standards in Medical Research in Developing Countries* (Cambridge: Cambridge University Press, 2004), 163–192.

[19] Schuklenk and Ashcroft, "Affordable Access to Essential Medication in Developing Countries: Conflicts between Ethical and Economic Imperatives", 135.

1.2 Global Health Inequalities

1.2.1 Developed and Developing Countries

The global inequality in health is morally alarming. The gap between developed and developing countries is increasingly widening. A child's birth either in a developed country or a developing country determines the life's chances and opportunities of the child.[20] Ruger acknowledges wide and growing global health inequalities in relation to life expectancies and child mortality rates between developed and developing countries. For example, a child born today in Afghanistan is 75 times as likely to die by age 5 years as a child born in Singapore. Furthermore, in Africa the number of children at risk of dying is 35% higher today than it was 10 years ago.[21] The global health inequalities exist between developed and developing countries in several areas, health outcomes, supply of health care services and funding of such services by public and private agents.[22] We will examine some aspects of global health inequalities, inequalities of life and death between countries and causes of death and disability.

1.2.1.1 Inequalities of Life and Death

The global burden of disease is disproportionately borne by developing countries, which has resulted in significant different health outcomes between developed and developing countries.[23] An abundance of data exists that shows vast disparities in life expectancy, child mortality, adult mortality and maternal mortality among rich and poor countries. Average life expectancy in Africa is nearly 30 years less than in the Americas or Europe.[24] World Health Organization reports vast disparity in average life expectancy between inhabitants of Africa, America and Europe. The disparity is also significant between the rich and the poor relative to the number of years of healthy life.[25] Life expectancy in Zimbabwe or Swaziland is less than half that in

[20] Lawrence O. Gostin, "Meeting Basic Survival Needs of the World's Least Healthy People: Toward a Framework Convention on Global Health", *The Georgetown Law Journal* 96, no. 2 (2008): 336.

[21] J. P. Ruger, "Ethics and Governance of Global Health Inequalities", *Journal of Epidemiology and Community Health* 60 (May 2006); 998.

[22] The Department of Economic and Social Affairs of the United Nations, *Committee for Development Policy, Policy Note: Implementing the Millennium Development Goals: health Inequality and the Role of Global Health Partnerships*, (New York: United Nations, 2009), 8.

[23] Gostin, "Meeting Basic Survival Needs of the World's Least Healthy People: Toward a Framework Convention on Global Health", 336–339.

[24] World Health Organization (WHO), *World Health Statistics* 30 (2007), http://who.int/whosis/whostat2007.pdf (Accessed May 3, 2013).

[25] WHO, *World Health Statistics* 30 (2007), http://who.int/whosis/whostat2007.pdf (Accessed May 3, 2013).

Japan.[26] WHO reports that "A person born in Zimbabwe can hope to live only to age thirty-four for men or thirty-three for women, whereas a person born in Japan is expected to live to age seventy-nine for men or eighty-six for women."[27] A child born in Angola is 65 times more likely to die in the first few years of life than a child born in Norway.[28] A woman giving birth in sub-Saharan Africa is 100 times more likely to die in labor than a woman in a rich country.[29] Although life expectancy has increased in developed countries in the past five decades, it has been decreasing in developing countries.[30] Infectious disease epidemics, especially HIV/AIDS which kills over 5800 African, but only 49 North Americans, each day,[31] and increased chronic disease have been instrumental in the decline of life expectancy in developing countries.[32]

Other key health status indicators that are used to measure the global health inequalities are child mortality rate or under five mortality rate and infant mortality rate. Child mortality rate or under five mortality is "the probability that a newborn baby will die before reaching age five, expressed as a number per 1000 live births."[33] This rate varies to a great extent with the wealth of a country. In developed countries the rate is about 20 per 1000 live births, while, in developing countries the rate can be as high as 170 per 1000 live births, as in the African region of the World Health Organization (WHO).[34] In many developing countries, child mortality rates can be twenty-five to thirty times higher than the rate in developed countries.[35] Statistics available indicate that of the 10.8 million children under five who die each year, 10 million are from low-income countries more- than twice the number of children born annually in the United States and Canada combined.[36] Infant mortality rate which is another health status indicator is the "the number of deaths of infants under

---

[26] WHO, *World Health Statistics* 24, 30 (2007), http://who.int/whosis/whostat2007.pdf (Accessed May 3, 2013).

[27] WHO, *World Health Statistics* 24, 30 (2007), http://who.int/whosis/whostat2007.pdf (Accessed May 3, 2013).

[28] UNICEF, *The State of the World's Children 2007* (2006), http://www.unicef.org/sowc07/docs/sowc07.pdf. (Accessed May 3, 2013).

[29] WHO/UNICEF/UNFPA, *Maternal Mortality in 2000* (2006), http://childinfo.org/areas/maternalmortality/countrydata.php (Accessed May 3, 2013).

[30] Ruger, "Ethics and Governance of Global Health Inequalities", 998.

[31] Joint U. N. Programme on HIV/AIDS, *Global Facts and Figures* (2006).

[32] Joint U. N. Programme on HIV/AIDS, *Report on the Global AIDS Epidemic* (2004).

[33] A. Haupt and T. T. Kane, *Population Handbook* (Washington, D.C.: Population Reference Bureau, 2004), quoted in Richard Skolnik, *Essentials of Global Health* (Sadbury, Massachusetts: Jones and Bartlett Publishers, 2008), 20.

[34] WHO, *Health Status Statistics: Mortality*, http://www.who.int/healthinfo/statistic/indneonatalmortality/en/(Accessed June 25, 2006).

[35] Global Health Council, *Child Mortality*, http://www.globalhealth.ord/child_health/child_mortality (Accessed October 15, 2007).

[36] Global Health Council, *The Importance of Child Health*, http://www.globalhealth.org/childhealth (Accessed August 21, 2007).

age 1 per 1000 live births in a given year."[37] The infant mortality rate varies largely with the income status of a country.[38] Some of the poor countries, such as Niger, have infant mortality rates as high as 150 infants deaths for every 1000 live births, whereas Sweden has only about 3 infants for every 1000 live births.[39] Data provided by the World Bank on the global health gap between the rich and the poor indicates that in one year alone, 14 million of the poorest people in the world died, while only 4 million would have died if this population had the same death rate as the global rich.[40]

The health gap between developed and developing countries is consistently increasing. In richer nations, the population is increasingly healthy and living longer, while in the least developed countries, the population is getting sicker and dying younger.[41] In countries with the highest child and adult mortality rates, people suffer multiple deprivations when compared with their low-mortality counterparts. They are four times more likely to live on less than one dollar per day, have twice the female illiteracy rate, and a 20 fold for adults or 65 fold for children difference in per capita health spending.[42]

1.2.1.2 Causes of Death and Disability

The causes of death and disability differ significantly between developing and developed countries. Leading causes of death and disability in developed countries are chronic, non-communicable diseases because they have technologies to prevent and treat most communicable diseases.[43] Gwatkin and Guillot write, "Among the global rich, all of the top five causes of death and of disability-adjusted life year (DALY) losses are non-communicable diseases, with ischaemic heart disease and malignant neoplasms at or near the top."[44] The DALY is defined as a measure of premature deaths and losses due to illness and disabilities in a population. A DALY

---

[37] Richard Skolnik, *Essentials of Global Health* (Sadbury, Massachusetts: Jones and Bartlett Publishers, 2008), 20.

[38] Skolnik, *Essentials of Global Health*, 20.

[39] World Bank, *WDI Data Query*, http://devdata.worldbank.org/data-query/ (Accessed July1, 2006).

[40] Davison R. Gwatkin and Michel, Guillot, *The Burden of Disease Among the Global Poor: Current Situation, Future Trends, and Implications for Strategy* (Washington, D.C.: The World Bank, 2000), 12.

[41] Jennifer P. Ruger and Hak-Ju Kim, "Global Health Inequalities: An International Comparison," *Journal of Epidemiology and Community Health* 60, no. 11 (November 2006): 928, 935.

[42] Ruger and Kim, "Global Health Inequalities: An International Comparison," 929.

[43] Abdallah Daar et al., "Top Ten Biotechnologies for Improving Health in Developing Countries," *Nature Genetics* 32, no. 2 (October 2002): 229.

[44] Gwatkin and Guillot, *The Burden of Disease Among the Global Poor: Current Situation, Future Trends, and Implications for Strategy*, 9.

measures the number of healthy years of life lost between the population being measured and the healthiest possible population.[45]

On the other hand, communicable diseases are the leading causes of death and disability in developing countries.[46] Infections defy geographical boundaries especially in an age of advanced innovation in transportation which makes for easy transmission of infectious agents.[47] In developing countries, chronic diseases are also becoming more widespread, thus, producing a double burden of disease.[48] In poor developing countries, communicable diseases account for majority of deaths (58.6%) and DALY loss (63.6%).[49] Similarly, non-communicable diseases are responsible for more than half the disease burden in low and middle income countries.[50] However, in rich developed countries, communicable diseases are responsible for 7.7% of all deaths and 10.9% of DALY loss.[51] We will examine the preventable and treatable diseases prevalent in developing countries usually classified as diseases of poverty.

1.2.1.3 Diseases of Poverty

Infectious diseases of poverty are defined as "an umbrella term used to describe a number of diseases which are known to be more prevalent among poorer populations, rather than a definitive group of diseases."[52] It is a concept that acknowledges the need to focus on the poor and vulnerable, which have less power to intervene. A good number of such diseases are regarded as "neglected tropical diseases", as defined by WHO.[53] Griffiths and Zhou acknowledge that many other infectious diseases have not been high on the global agenda except tuberculosis, malaria and HIV/AIDS.[54] Diseases such as diarrhea, elephantiasis, guinea worm, malaria, measles, river blindness, schistosomiasis, and trachoma are largely unheard of in rich

[45] Skolnik, *Essentials of Global Health*, 24–25.

[46] Gostin, "Meeting Basic Survival Needs of the World's Least Healthy People: Toward a Framework Convention on Global Health", 338.

[47] Solomon R. Benatar, Abdallah S. Daar and Peter A. Singer, "Global Health Ethics: The Rationale for Mutual Caring," *International Affairs* 79, no. 1(2003): 111.

[48] Benatar, Daar and Singer, "Global Health Ethics: The Rational for Mutual Caring," 111.

[49] Gwatkin and Guillot, *The Burden of Disease Among the Global Poor: Current Situation, Future Trends, and Implications for Strategy,* 8.

[50] Gostin, "Meeting Basic Survival Needs of the World's Least Healthy People: Toward a Framework Convention on Global Health", 338.

[51] Gwatkin and Guillot, *The Burden of Disease Among the Global Poor: Current Situation, Future Trends, and Implications for Strategy*, 8.

[52] Sian Griffiths and Xiao-Nong Zhou, Why Research Infectious Diseases of Poverty?, *in Global Report for Research on Infectious Diseases of Poverty, World Health Organization* (Geneva, Switzerland: World Health Organization, 2012), 13.

[53] Griffiths and Zhou, Why Research Infectious Diseases of Poverty?, 13.

[54] Griffiths and Zhou, Why Research Infectious Diseases of Poverty?, 13.

countries, but in contrast are leading causes of sickness and death in poor countries.[55]

Data shows that diseases of poverty are responsible for 54% of deaths in high-mortality poor countries, compared with 6.2% of deaths in high-income countries.[56] Diseases of poverty are as well leading causes of child mortality in poor countries.[57] Eighty-five percent of the 2.1 million deaths each year from diarrheal disease are in low-income countries, primarily among infants.[58]

A large proportion of diseases in low-income countries are preventable and treatable with current medicines or interventions.[59] A clear majority of the disease burden in low- income countries is rooted in the consequences of poverty, such as poor nutrition, indoor air pollution and lack of access to proper sanitation and health education.[60] Malaria can be prevented through various interventions, spraying dwellings with DDT, using insecticide treated mosquito nets and taking prophylactic medicines such as mefloquine, doxycycline and malorone.[61] Tuberculosis can be prevented with improved nutrition and treatment with DOTS therapy. About 95% of infectious patients in poor countries can be detected and cured.[62] Education is necessary for the prevention of HIV/AIDS and this requires involvement of civil society. Combining anti-retroviral drugs and good nutrition can help in controlling the viral load and suppressing the symptoms of HIV/AIDS.[63] Macklin highlights the staggering figures of HIV infection in Africa and India, but, with very limited access to anti-retroviral drugs for a very few population in those countries and most other developing countries with exception of Brazil. On the other hand, most HIV infected individuals in United States and other industrialized countries are treated.[64] Vast inequalities between poor and rich countries relative to access to health care and essential drugs are clear indications of increasing global health inequalities.

Apart from affecting morbidity and premature mortality of populations in poor regions of the world, the diseases of poverty also result in physical anguish.[65] People who suffer from these diseases are often stigmatized and ostracized from the

[55] Gostin, "Meeting Basic Survival Needs of the World's Least Healthy People: Toward a Framework Convention on Global Health", 338.

[56] Philip Stevens, *Diseases of Poverty and the 10/90 Gap*, (London: International Policy Network, 2004), 5.

[57] Peter Hotez, "A New Voice for the Poor," *PLos Neglected Tropical Diseases* 1, no. 1(October 2007): e77.

[58] Umesh D. Parashar et al., "Global Illness and Deaths Caused by Rotavirus Disease in Children," *Emerging Infectious Diseases* 9, no.5 (May 2003): 565, 568.

[59] Stevens, *Diseases of Poverty and the 10/90 Gap*, 4.

[60] Stevens, *Diseases of Poverty and the 10/90 Gap*, 4.

[61] Stevens, *Diseases of Poverty and the 10/90 Gap*, 4.

[62] Stevens, *Diseases of Poverty and the 10/90 Gap*, 4.

[63] Stevens, *Diseases of Poverty and the 10/90 Gap*, 4.

[64] Macklin, *Double Standards in Medical Research in Developing Countries*, 72.

[65] Gostin, "Meeting Basic Survival Needs of the World's Least Healthy People: Toward a Framework Convention on Global Health", 339.

society.[66] Diseases of poverty impose a heavy health and economic burdens on poor populations in Africa, Asia and Latin America.[67] For example, malaria which is one of the leading causes of mortality in children under five years of age in Africa accounts for 40% of public health expenditure in areas with high malaria transmission.[68] These diseases perpetuate the cycle of poverty by decreasing earning ability and economic productivity.[69] Griffiths and Zhou articulate, "Lost labor time due to illness often means a reduction in household capacity to earn income, particularly at a time when the household needs additional money to pay for treatment."[70] The social and economic conditions of poor populations support poverty which can affect health status and health outcomes either directly or indirectly.

1.2.2 Social Determinants of Health

The commission on social determinants of health recognizes that the global burden of disease and the major causes of health inequalities, which are found in all countries, stem from the conditions in which people are born, grow, live, work and age.[71] Daniels also articulates, "Health is produced not just by having access to medical prevention and treatment but also, to a measurably great extent by the cumulative experience of social conditions across the life course."[72] These conditions are generally referred to as social determinants of health including early year's experiences, education, economic status, employment and decent work, housing and environment, and effective systems of preventing and treating ill health.[73]

Social disadvantages are linked with inequalities in socio-economic status, gender, ethnicity and geographical area.[74] On the other hand, social advantages result from socio-economic development, and are associated with cultural, political and historical factors. They are natural and "built in" environments as well as public policies.[75] In a nutshell, the structural determinants and conditions of daily life form

---

[66] Griffiths and Zhou, Why Research Infectious Diseases of Poverty?, 14.

[67] Griffiths and Zhou, Why Research Infectious Diseases of Poverty?, 14.

[68] Griffiths and Zhou, Why Research Infectious Diseases of Poverty?, 14.

[69] Gostin, "Meeting Basic Survival Needs of the World's Least Healthy People: Toward a Framework Convention on Global Health ", 339.

[70] Griffiths and Zhou, Why Research Infectious Diseases of Poverty?, 14.

[71] World health Organization, *Meeting Report: World Conference on Social Determinants of Health*, 2011(Geneva:WHO, 2012), IX.

[72] Norman Daniels, *Just Health: Meeting Health Needs Fairly, Fairly* (Cambridge: Cambridge University Press, 2008), 79.

[73] World health Organization, *Meeting Report: World Conference on Social Determinants of Health*, 48.

[74] The Department of Economic and Social Affairs of the United Nations, *Committee for Development Policy, Policy Note: Implementing the Millennium Development Goals: health Inequality and the Role of Global Health Partnerships*, 21.

[75] The Department of Economic and Social Affairs of the United Nations, *Committee for Development Policy, Policy Note: Implementing the Millennium Development Goals: health Inequality and the Role of Global Health Partnerships*, 21.

the social determinants of health as well as account for a substantial part of health inequalities between and within countries.[76]

1.2.2.1 Health and Socio-economic Status

Many studies have documented a strong correlation between health outcomes and socio-economic status. Daniels describes each increment up the socioeconomic hierarchy as associated with improved health outcomes.[77] Gostin also articulates, "If residence in a poor country significantly increases a person's risk of illness and premature death, it is only more disadvantageous to be a member of a low-income, low-status population in that country."[78] Poverty and ill-health are regarded as interwoven. Poor countries are usually predicted to have worse health outcomes than better-off countries. There is also a prediction that within countries, poor people have worse health outcomes than better-off people.[79] However, the wealth of a country does not solely determine the health outcomes of the particular country. There are other mediating factors certainly involved. For example, Costa Rica's life expectancy is nearly the same as that of the United States, despite about $21,000 large difference between the Gross Domestic Product per capita (GDPpc) for Costa Rica and the United States.[80] Daniels et al. contend that factors such as culture, social organization, and government policies contribute significantly to the determination of population health, and that variations in these factors go a great length to explain the differences in health outcomes between nations.[81]

Strong epidemiological evidence shows that individuals of low socioeconomic status live much shorter and less healthy lives than individuals of high socioeconomic status.[82] There is a consensus among numerous authors that the level of economic inequality in a society adversely affects population health.[83] The

[76] The Department of Economic and Social Affairs of the United Nations, *Committee for Development Policy, Policy Note: Implementing the Millennium Development Goals: health Inequality and the Role of Global Health Partnerships*, 21.

[77] Douglas Black et al., *Inequalities in Health: The Black Report: The Health Divide* (London: Penguin Group, 1988), quoted in Daniels, *Just Health: Meeting Health Needs Fairly*,84.

[78] Gostin, "Meeting Basic Survival Needs of the World's Least Healthy People: Toward a Framework Convention on Global Health", 339.

[79] Adam Wagstaff, "Poverty and Health Sector Inequalities," *Bulletin of the World Health Organization* 80, no. 2 (January 2002): 97.

[80] Daniels, *Just Health: Meeting Health Needs Fairly*, 83–84.

[81] Norman Daniels, Bruce P. Kennedy and Ichiro Kawachi, "Why Justice is Good for Our Health: The Social Determinants of Health Inequalities," *Daedalus* 128, no. 4 (Fall 1999): 220.

[82] Gostin, "Meeting Basic Survival Needs of the World's Least Healthy People: Toward a Framework Convention on Global Health", 339.

[83] Richard Wilkinson, *Unhealthy Societies: The Afflictions of Inequality*, (London and New York: Routledge, 1996), 3–4.

implication is that societies with wide inequalities between rich and poor tend to have worse health status than societies with smaller inequalities after controlling for per capita income.[84] Scholars who give credence to this line of thought argue that social justice is good for our health.[85]

Disparities in socioeconomic status show differently in both developed and developing countries. For example, in United States, "Persons of poverty, non-white race, and or menial position are more likely to experience significant health problems decades before their more privileged fellow citizens."[86] In developing countries, health and longevity of individuals are severely affected if they are poorer, less valued and less powerful.[87]

Furthermore, vulnerable populations consisting of women, children and indigenous persons in developing countries are less healthy than their counterparts. Women have very limited control over their social, political and economic lives, which are best indicators of poor health.[88] Such living conditions result in worse health outcomes for women. For example, in Angola, the maternal mortality ratio is 1700 deaths per 100,000 live births, compared with 7 deaths in Switzerland.[89] HIV infection rates in sub-Saharan Africa are 5–16 times higher among young girls than boys.[90] Women's physical health and mental health are also affected by gender-based violence such as rapes, forced pregnancies and sexual assaults.[91] In developing countries, children experience worse health outcomes than in developed countries. For example, in developing countries, 33 out of 1000 infants die during the neonatal period, compared with 4 out of 1000 in developed countries.[92] Studies have also shown that "indigenous groups historically have faced worse health

[84] Lynch J. Smith et al., "Is Income Inequality a Determinant of Population Health? Part 1. A Systematic Review," *Milbank Q* 82, no. 1 (March 2004): 5.

[85] Norman Daniels, Bruce P. Kennedy and Ichiro Kawachi, "Justice is Good for Our Health," *Boston Review* 25, no. 1 (February–March 2000), http://bostonreview.net/BR25.1/daniels.htm (accessed 26 April 2013).

[86] Nicole Lurie and Tamara Dubowitz, "Health Disparities and Access to Health," *JAMA* 297, no. 10 (March 2007): 1118–21.

[87] Angus Deaton, "Health Inequality and economic Development," *Journal of Economic Literature* XLI (41), no. 1 (March 2003): 113.

[88] Myra Buvinic et al., *Gender Differentials in Health, in Disease Control Priorities in Developing Countries* ed. Dean T. Jamison et al. (Washington DC: World Bank, 2006), 195, 202, http://files.dcp2.org/pdf/DCP10.pdf (accessed April 26 2013).

[89] Gostin, "Meeting Basic Survival Needs of the World's Least Healthy People: Toward a Framework Convention on Global Health", 341.

[90] Myra Buvinic et al., *Gender Differentials in Health, in Disease Control Priorities in Developing Countries*, 195, 199.

[91] Gita Sen, Piroska Ostlin and Asha George, *Unequal, Unfair, Ineffective and Inefficient Gender Inequity in Health: Why It Exists and How We Can Change It, Women and Gender Equity Knowledge Network*, (September 2007): http:www.who.int/social_determinants/resources/csdh_media/wgekn_ final_report_07.pdf (Accessed April 27, 2013).

[92] Joy E. Lawn et al., *Newborn Survival, in Disease Control Priorities in Developing Countries* ed. Dean T Jamison et al. (Washington DC: World Bank, 2006), 531, 532, http:files.dcp2.org/pdf/DCP27.pdf.(Accessed April 27, 2013).

outcomes due to low socioeconomic status (SES) and marginalization within states and communities.".[93]

Discrimination against various racial, ethnic or religious groups in both developing and developed countries usually has severe social and health consequences. For example, in Bulgaria, the life expectancy of the Roma people at any age is five to six years below the rest of the population, while their infant mortality rate is six times the national average.[94] Among the non-Tagalog speaking population of the Philippines, child mortality rates were 33% above those of Tagalog speakers. Child mortality rates for non-Christians were 47% above those of Christians.[95] In Latin America, the prevalence of child diarrhea and maternal mortality is much higher among indigenous people than among non-indigenous.[96] In United States, "the life expectancy of African-Americans in the District of Columbia is 63 years, compared with 80 years for whites in neighboring Montgomery county Maryland.[97] We will further examine the components of Socio-Economic Status (SES) including income, education and social class, and how they affect health outcomes differently in both developing and developed countries.

Relative Income

Studies have shown that there is a correlation between the income distribution in a society and the level of health achievement of its members.[98] Succinctly put, "It is not just the size of the economic pie but how the pie is shared that matters for population health.[99] Variations in life expectancy between countries were associated with income distributions. Countries with more equal income distributions have higher life expectancies than countries with lesser income distributions. For example, Japan and Sweden have higher life expectancies than U.S. that has higher GDP per

---

[93] Stephen J. Kunitz, "Globalization, States and the Health of Indigenous Peoples," *Am. J. Public Health* 90, no. 10 (October 2000): 1531–32.

[94] The Department of Economic and Social Affairs of the United Nations, *Committee for Development Policy, Policy Note: Implementing the Millennium Development Goals: health Inequality and the Role of Global Health Partnerships,* 22.

[95] The Department of Economic and Social Affairs of the United Nations, *Committee for Development Policy, Policy Note: Implementing the Millennium Development Goals: health Inequality and the Role of Global Health Partnerships*, 22–23.

[96] The Department of Economic and Social Affairs of the United Nations, *Committee for Development Policy, Policy Note: Implementing the Millennium Development Goals: health Inequality and the Role of Global Health Partnerships,* 23.

[97] The Department of Economic and Social Affairs of the United Nations, *Committee for Development Policy, Policy Note: Implementing the Millennium Development Goals: health Inequality and the Role of Global Health Partnerships*, 23.

[98] Richard G. Wilkinson, "Income Distribution and Life Expectancy," *British Medical Journal* 304, no. 6820 (January 1992): 165–168.

[99] Daniels, Kennedy and Kawachi, "Why Justice is Good for Our Health: The Social Determinants of Health Inequalities," 221.

capita.[100] Countries with lower GDP per capita, such as Costa Rica also enjoy high life expectancy due to a more equitable income distribution.[101]

Individual mortality rates are also associated with income distributions.[102] Wilkinson describes far reaching impact of income on mortality.[103] For example, a study "estimated that a move from household income of $20,000–$30,000 to a household income greater than $70,000 was associated with a halving of the odds of adult mortality."[104]

Parental income has strong effects on children's health. Family income is clearly associated with various measures of child health.[105] The association continues to be large after controlling for household composition, race, parental education, and parental labor force status.[106] Health, on the other hand has an effect on income and wealth. "Health improves one's ability to participate in the labor market and earn a decent wage."[107] Illness increases spending on health and consequently reduces wealth. The negative effect of poor health on income and on wealth accounts for the correlation between financial resources and health especially among adults.[108]

Education

Education plays a central role in any discussion and analysis of the SES-health gradient. Educational attainment has been used as the primary indicator of SES.[109] Education is a significant determinant of health for several reasons. "First, it brings

---

[100] Daniels, Kennedy and Kawachi, "Why Justice is Good for Our Health: The Social Determinants of Health Inequalities," 222.

[101] Daniels, Kennedy and Kawachi, "Why Justice is Good for Our Health: The Social Determinants of Health Inequalities," 222.

[102] David M. Cutler, Adriana Lleras-Muney and Tom Vogl, *Socioeconomic Status and Health: Dimensions and Mechanisms*, National Bureau of Economic Research working Paper, no.14333, (September 2008): 18, http://www.nber.org/papers/w14333 (Accessed April 26, 2013).

[103] Richard G. Wilkinson, "Income Distribution and Mortality: A Natural Experiment," *Sociology of Health and Illness* 12, no. 4 (December 1990): 412.

[104] P. McDonough et al., "Income Dynamics and Adult Mortality in the United States, 1972 through 1989," *American Journal of Public Health* 87, no. 9 (September 1997): 1476–1483.

[105] Anne Case, Darren Lubotsky, and Christina Paxson, "Economic Status and Health in Childhood: The Origins of the Gradient," *American Economic Review* 92, no.5 (December 2002): 1308–1334.

[106] Cutler, Lleras-Muney and Vogl, *Socioeconomic Status and Health: Dimensions and Mechanisms*, 20.

[107] Cutler, Lleras-Muney and Vogl, *Socioeconomic Status and Health: Dimensions and Mechanisms*, 18.

[108] Cutler, Lleras-Muney and Vogl, *Socioeconomic Status and Health: Dimensions and Mechanisms*, 18.

[109] Evelyn M. Kitagawa and Philip M. Hauser, *Differential Mortality in the United States: A Study in Socioeconomic Epidemiology*, (Cambridge, Mass: Havard University Press, 1973), quoted in Cutler, Lleras-Muney and Vogl, *Socioeconomic Status and Health: Dimensions and Mechanisms*, 10.

with it knowledge of good health practices. Second, it provides opportunities for gaining skills, getting better employment, raising one's income, and enhancing one's social status, all of which are also related to health."[110]

Studies show that the best predictor of the birth weight of a baby is the mother's level of educational attainment.[111] There is a strong and positive correlation between the level of education and all key health indicators.[112] For example, in United States more educated individuals report better health and suffer lower mortality risk.[113] Birn points out that people who are better educated have lower levels of infectious diseases and non-communicable diseases such as hypertension, emphysema, diabetes, anxiety and depression. They also show improved physical and mental functioning; health behaviors such as lower rates of smoking, heavy drinking and drug use. They as well have higher rates of exercise and better management of stress and chronic health conditions.[114]

The implication is that more educated individuals engage in more healthy behaviors because they are more informed. They are more inclined to comply with prescribed therapies and to utilize modern medical technologies to deal with their health problems.[115] Furthermore, studies have shown that education affects cognition, which invariably affects the ability to process information relative to healthy behaviors.[116] Studies also support that health affects education.[117] For example, in Britain, adolescents who were born with low birth weight or suffered health insults in childhood have worse schooling outcomes.[118] In developing countries, children with poor health have worse schooling outcomes.

Access to education and quality of education are significantly compromised in developing countries. Millions of children are excluded from formal education due to costly fees involved. Investing on education for girls is endangered when families prefer to have them focus more on household work rather than pay their school fees. The perception that boys benefit more from education also constitutes an impediment

---

[110] Skolnik, *Essentials of Global Health*, 18.

[111] John Hobcraft, "Women's Education, Child Welfare and Child Survival: A Review of the Evidence," *Health Transition Review* 3, no. 2 (October 1993): 159–173.

[112] Skolnik, *Essentials of Global Health*, 18.

[113] Cutler, Lleras-Muney and Vogl, *Socioeconomic Status and Health: Dimensions and Mechanisms*, 11.

[114] Anne-Emmanuelle Birn, Addresssing the Societal Determinants of Health: The Key Global Health Ethics Imperatives of Our Times, in *Global Health and global Health Ethics* ed. Solomon Benatar and Gillian Brock (Cambridge: Cambridge University Press, 2011), 43.

[115] Cutler, Lleras-Muney and Vogl, *Socioeconomic Status and Health: Dimensions and Mechanisms*, 16.

[116] David M. Cutler and Adriana Lleras-Muney, *Understanding Differences in Health Behaviors by Education* (Mimeo, Princeton University, 2007), quoted in Cutler, Lleras-Muney and Vogl, *Socioeconomic Status and Health: Dimensions and Mechanisms*, 16.

[117] Cutler, Lleras-Muney and Vogl, *Socioeconomic Status and Health: Dimensions and Mechanisms*, 13.

[118] Anne Case, Angela Fertig and Christina Paxson, "The Lasting Impact of Childhood Health and Circumstance," *Journal of Health Economics* 24, no.2 (March 2005); 365–389.

in educating girls. In many countries, deregulation and decrease in social sector spending result in a marked decline in the quality of education. Furthermore, factors such as high rates of migration to developed countries, civil conflict and HIV/AIDS have destroyed in recent decades the ranks of educators in some settings.[119] There are indications that those developing countries that invest heavily in human capital, for example in education, have better health outcomes. In developing countries, adult literacy is one of the best predictors of life expectancy.[120]

Social Class

Individuals attain social class through their educational attainment and their levels of income and wealth. Cutler et al., articulate, "Education, income, and wealth characterize individuals who are separated from the society in which they live."[121] There is a positive correlation between an individual's position in a social hierarchy and better health outcomes. Individuals of greater wealth and education enjoy better health for a greater extent, because of the individual's position in a social hierarchy not because of some process affecting the individuals in isolation.[122]

The Whitehall studies of British civil servants report that civil servants with lower prestige jobs experience higher rates of mortality from cardiovascular causes.[123] The studies report variations in behavior patterns based on the rank of the civil servants. For example, higher ranking officials show a lower obesity rate, a lower tendency to smoke and higher propensities to exercise and eat fruits and vegetables.[124] The rank or position in employment is positively associated with a sense of control over one's health and one's work, job satisfaction, social support, and the absence of stressful life events.[125] Apart from sources of international health inequalities that focus on disparities in global disease burden between developing and developed countries and marked variations in status, we will further explore other factors such as international practices that undermine the population health of developing nations.

---

[119] Birn, Addresssing the Societal Determinants of Health: The Key Global Health Ethics Imperatives of Our Times, 43.

[120] Daniels, *Just Health: Meeting Health Needs Fairly*, 88.

[121] Cutler, Lleras-Muney and Vogl, *Socioeconomic Status and Health: Dimensions and Mechanisms*, 22.

[122] Cutler, Lleras-Muney and Vogl, *Socioeconomic Status and Health: Dimensions and Mechanisms*, 22.

[123] Michael G. Marmot et al., "Employment Grade and Coronary Heart disease in British Civil Servants," *Journal of Epidemiology and Community Health* 32, no.4 (December 1978): 244–249, quoted in Cutler, Lleras-Muney and Vogl, *Socioeconomic Status and Health: Dimensions and Mechanisms*, 23.

[124] Cutler, Lleras-Muney and Vogl, *Socioeconomic Status and Health: Dimensions and Mechanisms*, 23.

[125] Cutler, Lleras-Muney and Vogl, *Socioeconomic Status and Health: Dimensions and Mechanisms*, 23.

1.2.2.2 International Practices and Developing Nations

Benatar identifies increasing poverty in most parts of the world as a primary factor preventing sustainable control of population growth, which invariably threatens physical and mental health as the necessary conditions for a decent human life and global survival.[126] He further traces the current global inequalities in health status and access to health care to poverty in developing countries resulting from the world expenditures on military goods and services.[127] As at 1990, "the world spends almost $1 trillion a year on military goods and services."[128] Developed countries spend an average of 5.4% of their Gross National Product (GNP) on the military and a meager 0.3% on aid to developing countries.[129] Similarly, United States spend 0.15% of its GNP on defense support for Egypt, Israel, Turkey, Parkistan and the Philippines.[130] Militarization and the associated militarism have been identified as compromising the health of individuals and nations in a variety of ways including killing, maiming, torture, refugeeism, destruction of livelihoods, diversion of resources, crime, terrorism, black markets, poverty, starvation, environmental damage, and destabilization within developing countries.[131]

Modern international economic policies and market driven health care also contributed to the poverty in developing countries, which undermines the population health. Some scholars describe the impact of such economic policies as removal of great quantities of material and human resources from poor developing countries to rich industrialized nations.[132] The debt burden of developing countries threatens the health of those nations. In 1990, total developing country world debt was $1.3 trillion, that is, double the level in 1980, and it had grown further to $1.9 trillion by 1995.[133] International lenders such as World Bank (WB) and International Monetary Fund (IMF) imposed structural adjustment programs on developing countries as a condition for lending money to them. The implication is that these poor countries were forced to embark on severe cut back on their publicly funded health systems and to take other necessary steps to cut spending deficit.[134] For example, in the 1990's Cameroon adopted Structural Adjustment Programs (SAP) measures which

[126] Soloman R. Benatar, "Global Disparities in Health and Human Rights: A Critical Commentary," *American Journal of Public Health* 88, no.2 (February 1998): 296.

[127] Benatar, "Global Disparities in Health and Human Rights: A Critical Commentary," 296.

[128] C. W. Kiefer, "Militarism and World Health," *Social Science Medicine* 34, no. 7 (April 1992): 719–724.

[129] Benatar, "*Global Disparities in Health and Human Rights: A Critical Commentary*," 296.

[130] Benatar, "*Global Disparities in Health and Human Rights: A Critical Commentary*," 296.

[131] Benatar, "Global Disparities in Health and Human Rights: A Critical Commentary," 296.

[132] A. Gilbert, *An Unequal World: The Links Between Rich and Poor Nations,* 2nd ed. (London, England: Nelson, 1992), quoted in Benatar, "Global Disparities in Health and Human Rights: A Critical Commentary," 296.

[133] Benatar, "Global Disparities in Health and Human Rights: A Critical Commentary," 296.

[134] Daniels, *Just Health: Meeting Health Needs Fairly*, 338.

include suspension of health worker recruitment, mandatory retirement at age fifty or fifty five, suspension of promotions, and reduction of benefits.[135]

The brain drain of health care workers from developing to developed countries also harms the population health. Daniel argues, "Rich countries have harmed health in poorer ones by solving their own labor shortages of trained health care personnel by actively and passively attracting immigrants from poorer countries."[136] Data shows that 23–34% of physicians in developed countries such as New Zealand, The United Kingdom, the United States, Australia and Canada are foreign trained.[137] WHO reports that, "Over 60 percent of the doctors trained in Ghana in the 1980s emigrated overseas."[138] International efforts to reduce poverty, lower mortality rates and treat HIV/AIDS patients as articulated in the Millenium Development Goals (MDG) are jeopardized by the loss of health personnel in sub-Saharan Africa.[139]

The global enforcement of intellectual property rights which resulted in impediment to access to essential drugs threatened the health of poor developing countries. The globalized patent regime raised the prices of essential drugs that poor patients in developing countries could not afford them. Furthermore, bilateral free trade agreements negotiated currently by United States made the problem of access to essential drugs worse, with the extension of 20 year patent and the suppression of generic production of drugs. In the wake of HIV/AIDS crisis in developing countries, millions of people have died due to the suppression of manufacture and trading of generic drugs.[140]

1.3 Global Health Justice

1.3.1 Developing Countries and HIV/AIDS

1.3.1.1 The Scope of HIV/AIDS Epidemics

Statistics on the spread of HIV/AIDS across the countries shows an upward trend. The Joint United Nations Programme on HIV/AIDS (UNAIDS) and WHO reported that, an estimate of 39.4 million (35.9 million- 44.3 million) people were living with HIV at the end of 2004. About 4.9 million (4.3 million–5.4 million) people were infected in 2004. An estimate of 3.1 million (2.8 million–3.5 million) people died of

[135] Daniels, *Just Health: Meeting Health Needs Fairly*, 338.

[136] Daniels, *Just Health: Meeting Health Needs Fairly*, 337–378.

[137] Daniels, *Just Health: Meeting Health Needs Fairly*, 338.

[138] WHO, *Recruitment of Health Workers from Developing World* (Geneva: WHO, 2004), quoted in Daniels, *Just Health: Meeting Health Needs Fairly*, 338.

[139] Daniels, *Just Health: Meeting Health Needs Fairly*, 338.

[140] Thomas Pogge, *World Poverty and Human Rights: Cosmopolitan Responsibilities and Reforms*, 2nd ed. (Cambridge: Polity Press, 2008), 222–225.

HIV/AIDS in the past year.[141] The data on the magnitude of HIV/AIDS shows that Africa has been hardest hit by the epidemics. Sub-Saharan Africa is the worst devastated region, with 25.4 million at the end of 2004, which is an increase of one million since 2002. Sixty four percent of all people living with HIV, that is, about two thirds are in sub-Saharan Africa.[142] Most severely affected regions are Southern and Eastern Africa. Seven countries have an estimate of adult (15–49) HIV prevalence of 20% or greater: Botswana, Lesotho, Namibia, South Africa, Swaziland, Zambia and Zimbabwe.[143] Other countries such as Burkina Faso, Cameroon, Central African Republic, Kenya, Malawi and Mozambique, have adult HIV prevalence levels higher than 10%.[144]

The onslaught of the HIV/AIDS epidemics was underestimated initially, mostly in South Africa due to the debate on the reliability of the data. The scope and the scale of the epidemics were undermined. Malan, a South African journalist argues that in as much as AIDS is a serious issue for Africa, the size of the problem and its long-term effects on society and economy have been blown out of proportion.[145] The denial is made worse when the former president of South Africa Mbeki used Malan's perspective to argue that, "AIDS is not a serious problem as we think."[146]

The data and data collection on the full extent of HIV/AIDS are plagued by many problems, high refusal rates, inadequate testing and reporting facilities, poor access to individuals who were selected and lax use of numbers by the press and AIDS activists. There were high refusal rates of people who would not be interviewed, provide specimens or who were not contactable in both Kenya and South Africa. The data of the prevalence of HIV in Swaziland for 2002, which stood at 38.6%, was loosely presented as if 38% of the adult were living with HIV.[147] There are indications of limitations relative to data and data collections but the overwhelming impact of the HIV/AIDS should not be undermined. Whiteside acknowledges the debate about the exact number of people currently living with HIV/AIDS in South Africa, but highlights increasing number of people being infected, continuing high prevalence and no sign of downturn. There is also a prediction of an astronomical rise of the number of people falling ill and dying, the number of orphans growing

---

[141] UNAIDS and WHO, *Global AIDS Epidemics*, (Geneva: UNAIDS, 2004), quoted in Niekerk and Kopelman, ed., *Ethics & AIDS in Africa: The Challenge to Our Thinking*, 1.

[142] Alan Whiteside, "AIDS in Africa: Facts, Figures and the Extent of the Problem," in *Ethics & AIDS in Africa: The Challenge to Our Thinking*, ed. Niekerk and Kopelman, 1.

[143] Commission on HIV/AIDS and Governance in Africa, *Africa: The Socio-Economic Impact of HIV/AIDS*, (Addis Ababa, Ethiopia: Economic Commission for Africa), 3, http://www.fsg.afre.msu.edu/adult_death/socio_ECO_IMPACT.pdf. (Accessed April 27, 2013.

[144] Commission on HIV/AIDS and Governance in Africa, *Africa: The Socio-Economic Impact of HIV/AIDS*, 3.

[145] R. Malan, "Africa isn't Dying of AIDS," *Spectator*, 18 December 2004, quoted in Niekerk and Kopelman, ed., *Ethics & AIDS in Africa: The Challenge to Our Thinking*, 2.

[146] Whiteside, "AIDS in Africa: Facts, Figures and the Extent of the Problem," 2.

[147] Whiteside, "AIDS in Africa: Facts, Figures and the Extent of the Problem," 4.

and a greater impact of the disease yet to be fully felt.[148] The devastating impact of HIV/AIDS has been felt in two areas, demographic and economic.

1.3.1.2 The Demographic Effects

The population dynamics is generally altered by unusual levels of death. Demography focuses on populations and their dynamics. It deals with the numbers, growth rates and structure of populations. It evaluates and predicts size and growth rates, structure by gender and age, and important indicators like birth, death and fertility rates, life expectancy and infant and child mortality. The demographic data is derived from two sources, the census and vital registration statistics.[149] Censuses are generally conducted in most countries every ten years. A census is defined as the total process of collecting, compiling, evaluating and publishing or otherwise disseminating demographic, economic and social data pertaining at a specific time to all persons in a country or a well delimitated part of a country.[150] On the other hand, vital registration deals with information about births, deaths and marriages. In developed countries, registration of these events is compulsory, while, in poor developing countries these statistics may not be recorded or collected.[151]

Exploring the demographic impact of HIV/AIDS presents some problems. There is a concern that what is finally recorded regarding the impact of HIV/AIDS is an event such as the death and its effects on household composition and dependency ratios rather than a process. The impacts of the events culminating to the death and stemming from it are not recorded.[152] The demographic impact of HIV/AIDS does not account for the process. Whiteside et al. write, "The impact of AIDS is felt as a process: a person begins to feel unwell and so, perhaps, does not grow as much food, the family has less to sell and can't afford to send children to school. When the person dies, the household composition changes."[153] Another problem is that demographic indicators may focus on nations, provinces or areas, while the impact of the epidemics may be felt more in households and among specific groups.[154] Another concern deals with demographic changes which are measured only after several years, as in the case of census which is usually done every ten years. The implication is that the impact of a new disease may not be tracked in most of the national and international statistics for a very long time.[155]

---

148 Whiteside, "AIDS in Africa: Facts, Figures and the Extent of the Problem," 3.

149 Whiteside et al., "Through a Glass, Darkly: Data and Uncertainty in the AIDS Debate," 25.

150 W. Petersen and R. Petersen, *Dictionary of Demography: Multilingual Glossary* (Westport: Greenwood Press, 1985), quoted in Niekerk and Kopelman, ed., *Ethics & AIDS in Africa: The Challenge to Our Thinking,* 25.

151 Whiteside et al., "Through a Glass, Darkly: Data and Uncertainty in the AIDS Debate," 25.

152 Whiteside et al., "Through a Glass, Darkly: Data and Uncertainty in the AIDS Debate," 25.

153 Whiteside et al., "Through a Glass, Darkly: Data and Uncertainty in the AIDS Debate," 25.

154 Whiteside et al., "Through a Glass, Darkly: Data and Uncertainty in the AIDS Debate," 25–26.

155 Whiteside et al., "Through a Glass, Darkly: Data and Uncertainty in the AIDS Debate," 26.

The demographic consequences of HIV/AIDS are experienced through increased mortality and decreased fertility in developing countries severely affected by the epidemics.[156] Furthermore, mortality increases among infected adults and those infants infected through mother to child transmission and adults.[157] The significant increase in the death toll from HIV/AIDS in developing countries has resulted in changes to the population structure. In South Africa, mortality of young adult women between 25 and 29 year range increased sharply by 3.5 times higher in 1999/2000 than in 1985. Mortality of young men between 30 to 39 year range increased nearly twice in 1999/2000 when compared to the 1985 rate.[158] The rapid increase in adult mortality and child mortality in developing countries has been attributed to HIV/AIDS epidemics. The South African data prove convincingly that there is increased mortality in the country. AIDS is blamed in the absence of any other reasonable explanation. In 2000, about 40% of adult deaths aged 15–49, were attributed to HIV/AIDS. An estimate of about 25% of all deaths was due to AIDS, making it single biggest cause of death.[159]

The decreased fertility among childbearing age women due to HIV/AIDS contributes to population changes. Women who are infected with HIV may have difficulty getting pregnant and carrying a child to term, resulting in premature mortality. There will be fewer childbearing age women, which significantly affect fertility.[160] For example, in Uganda the number of births was decreased by about 700,000, which is about 5.9% of all births that would have occurred during the last two decades.[161]

The astronomical increase in the number of orphans constitutes a demographic impact as well as social and economic consequences.[162] In 2003, The United Nations Children's Fund (UNICEF) projected that by 2010 an estimated 20 million children in Africa will have lost one or both parents to HIV/AIDS. HIV/AIDS accounts for over 80% of orphans in the worst affected countries. These children suffer severe stress and they are more likely to drop out from school and to be exploited. They may likely experience premature mortality and as well have a more pessimistic view of life.[163]

---

[156] Whiteside, "AIDS in Africa: Facts, Figures and the Extent of the Problem," 9.

[157] Whiteside, "AIDS in Africa: Facts, Figures and the Extent of the Problem," 9.

[158] R. Dorrington et al., "Some Implications of HIV/AIDS on Adult Mortality in South Africa," *AIDS Analysis Africa* 12, no. 5, quoted in Niekerk and Kopelman, ed., *Ethics & AIDS in Africa: The Challenge to Our Thinking*, 27.

[159] Whiteside et al., "Through a Glass, Darkly: Data and Uncertainty in the AIDS Debate," 28.

[160] Whiteside, "AIDS in Africa: Facts, Figures and the Extent of the Problem," 10.

[161] J.J. C. Lewis et al., *The Population Impact of HIV on Fertility in Sub-Saharan Africa*, quoted in Niekerk and Kopelman, ed., *Ethics & AIDS in Africa: The Challenge to Our Thinking*, 10.

[162] Whiteside, "AIDS in Africa: Facts, Figures and the Extent of the Problem," 10.

[163] UNICEF, *Africa's Orphaned Generation* (New York: UNICEF, 2003), quoted in Niekerk and Kopelman, ed., *Ethics & AIDS in Africa: The Challenge to Our Thinking*, 10.

1.3.1.3 The Economic Effects

A research presented at a meeting of economists at the XIII International Conference on AIDS indicates that, "HIV/AIDS is already starting to have immense impact on the economies of hard-hit countries, hurting not only individuals, families and firms, but also significantly slowing economic growth and worsening poverty."[164] An Increasing evidence shows that national wealth of the hardest hit countries of South Africa will be reduced by 15–20% over the next ten years as a result of HIV/AIDS.[165] UNAIDS and World Bank press release note, "Lower economic growth and increased poverty threaten to form a vicious cycle, in which HIV/AIDS drives many families into deepening poverty, and at the same time poverty accelerates the spread of HIV.".[166]

There is an enormous economic impact of HIV/AIDS on households in countries severely affected by the epidemics. The impact on the household begins to be felt whenever a member of the household begins to suffer from HIV-related illness. The first major impact is the loss of income of the patient who is likely the main breadwinner. This may be followed by substantial increase of household expenditures for medical expenses. Another ripple effect indicated is that daughters and wives of the patient may miss school or work less in order to care for the sick person. Finally, death results in a permanent loss of income, from less labor on the farm or from lower remittances, funeral and mourning costs and the removal of children from school so as to save on educational expenses and increase household labor, resulting in a severe loss of future earning potential.[167] Studies in African countries decimated by the epidemic highlight the significant burden of loss of income, large health care expenditures and draining of savings to pay for funeral and mourning costs.[168] For example, in Ethiopia, a study of 25 AIDS-afflicted rural families discover that the average cost of treatment, funeral and mourning expenses equaled to several times the average household income.[169] Steinberg et al., present the findings of a survey of household in South Africa on the impoverishing nature of the HIV/AIDS epidemic: "two thirds reported loss of income as a consequence of HIV/AIDS; half reported

---

[164] UNAIDS and World Bank, AIDS Hindering Economic Growth, Worsening Poverty in Hard Hit Countries," *The Body*, 11 July 2000, http:www.thebody.com/content/art641.html?ts=pf (Accessed 5 Feb.2013).

[165] UNAIDS and World Bank, AIDS Hindering Economic Growth, Worsening Poverty in Hard Hit Countries," *The Body*, 11 July 2000, http:www.thebody.com/content/art641.html?ts=pf (Accessed 5 Feb.2013).

[166] UNAIDS and World Bank, AIDS Hindering Economic Growth, Worsening Poverty in Hard Hit Countries," *The Body*, 11 July 2000, http:www.thebody.com/content/art641.html?ts=pf (Accessed 5 Feb.2013).

[167] John Stover and Lori Bollinger, *The Policy Project: The Economic Impact of AIDS*. (The Futures Group International, 1999), 3.

[168] Stover and Bollinger, *The Policy Project: The Economic Impact of AIDS*, 3.

[169] M. Demeka, "The Potential Impact of HIV/AIDS on the Rural Sector of Ethiopia," *Unpublished Manuscript*, Jan. 1993, quoted in Stover and Bollinger, *The Policy Project: The Economic Impact of AIDS*, 4.

not having enough food and that their children were going hungry; and almost a quarter of all children under age 15 had already lost at least one parent.".[170]

HIV/AIDS has a significant impact on firms. AIDS-related illnesses and deaths to employees affect a firm in two different ways, increasing expenditures and reducing revenues. Factors that lead to increased expenditure include, health care costs, burial fees and training and recruitment. On the other hand, factors that lead to decreased revenue consist of absenteeism due to illness, time-off to attend funerals, time spent on training and labor turnover. Labor turnover may result in a less experienced labor force that is less productive.[171] A study that examines several firms in Botswana and Kenya reveals that major factors in increased labor costs were absenteeism due to HIV/AIDS and increased burial costs.[172] The increased mortality and morbidity as a result of HIV/AIDS epidemics reduces labor supply in key sectors of the labor market. For example, in South Africa about 60% of the mining workforce with age range of 30 and 40 years is predicted to fall to 10% in 15 years.[173] Labor productivity also reduces as a result of a long period of illness associated with HIV/AIDS. The annual costs associated with sickness and reduced productivity as a result of HIV/AIDS per employee ranged from \$17 to \$300.[174] These costs adversely affect competitiveness and profits.

The situation results in a decline to government income. Dixon et al. articulate, "Government incomes also decline, as tax revenues fall, and governments are pressured to increase their spending to deal with the rising prevalence of AIDS, thereby creating the potential for fiscal crises."[175] The economic effects of HIV/AIDS which lead to lower domestic productivity reduce exports and increase imports of expensive health care goods. The significant decrease in export earnings and the increase in export expenditure may result in budget crises for the government. This could result in the government defaulting on debt repayments as well as requiring economic assistance from the international community.[176]

The macroeconomic impact of HIV/AIDS has been documented by several studies. Studies in Tanzania, Cameroon, Zambia, Swaziland, Kenya and other sub-Saharan African countries discover that the rate of economic growth could be

---

[170] M. Steinberg et al., *Hitting Home: How Household Cope with the Impact of the HIV/AIDS Epidemic. A Survey of Household Affected by HIV/AIDS in South Africa* (Washington, DC: Health Systems Trust and The Kaiser Family Foundation, 2002), quoted in Niekerk and Kopelman, ed., *Ethics & AIDS in Africa: The Challenge to Our Thinking*, 10.

[171] Stover and Bollinger, *The Policy Project: The Economic Impact of AIDS*, 4.

[172] Matthew Roberts and Bill Rau, *African Workplace Profiles: Private Sector AIDS Policy* (Arlington, VA: AIDSCAP, 1997, quoted in Stover and Bollinger, *The Policy Project: The Economic Impact of AIDS*, 6.

[173] Simon Dixon, Scott McDonald and Jennifer Roberts, "The Impact of HIV and AIDS on Africa's Economic Development," *BMJ* 324, no. 7331 (January 2002): 232.

[174] Stover and Bollinger, *The Policy Project: The Economic Impact of AIDS*, 7.

[175] Dixon, McDonald and Roberts, "The Impact of HIV and AIDS on Africa's Economic Development," 233.

[176] Dixon, McDonald and Roberts, "The Impact of HIV and AIDS on Africa's Economic Development," 233.

decreased by about 25% over a 20 year period.[177] The macroeconomic impact of HIV/AIDS varies across countries. The impact of HIV/AIDS on the macro-economy is felt in several areas, loss of experienced workers, reduced productivity, higher domestic production costs, and loss of international competitiveness, lower government revenues, reduced private savings and slower employment creation. Initially, the overall impact of AIDS on the macro-economy is minimal, but increases significantly over time.[178] Several studies show significant effects of HIV/AIDS in some African countries. A study focusing on the macro-economic impacts of AIDS in Zambia discover that as a result of HIV/AIDS epidemics, the GDP would be 5–10% lower by 2000.[179] The macroeconomic impact of AIDS in Tanzania assessed in 1991 shows that total GDP will decline by 15–25% in 2010 due to the impact of AIDS.[180] A study conducted on the impact of AIDS on the Kenyan economy projects that GDP will decrease by 14% in 2005 than it would have been without AIDS.[181] The impact of AIDS creates a global health emergency, since anti-retroviral drugs are not accessible to the majority of infected individuals in developing countries.

1.3.2 Distributive Justice and Global Research

Addressing the catastrophic impact of the HIV/AIDs in developing countries has given rise to many clinical trials, resulting in the discovery of anti-retroviral drugs. There are on-going debates regarding what justice requires when developed countries sponsor or conduct research in developing countries. Some scholars argue that global health inequalities and the devastating impact of HIV/AIDS can be redressed through biomedical research by providing access to anti-retroviral drugs resulting from clinical trials.[182] Concerns about justice in international clinical trials have shifted from focusing on the exploitation of research subjects or entire population in developing countries in the process of recruiting subjects and conducting the study to providing beneficial products at the end of the trials.[183] The Belmont report emphasizing the need for justice in research writes, "Research should not unduly

[177] Stover and Bollinger, *The Policy Project: The Economic Impact of AIDS*, 9.

[178] Stover and Bollinger, *The Policy Project: The Economic Impact of AIDS*, 10.

[179] Larry Forgy, "*Mitigating AIDS: The Economic Impacts of AIDS in Zambia and Measures to Counter Them*," REDSO/ESA, (Feb. 1993), quoted in Stover and Bollinger, *The Policy Project: The Economic Impact of AIDS*, 11.

[180] John T. Cuddington, "Modelling the Macroeconomic Effects of AIDS, with an Application to Tanzania," *World Bank Economic Review* 7, no. 2 (May 1993): 172–189.

[181] John Hancock et al., *The Macroeconomic Impacts of AIDS* (Washington, DC: AIDS in Kenya, Family Health International, 1996), quoted in Stover and Bollinger, *The Policy Project: The Economic Impact of AIDS*, 11.

[182] Macklin, *Double Standards in Medical Research in Developing Countries*, 68.

[183] Macklin, *Double Standards in Medical Research in Developing Countries*, 68.

involve persons from groups unlikely to be among the beneficiaries of subsequent applications of the research.".[184]

1.3.2.1 The Concept of Distributive Justice

Research in developing countries needs to fulfill the requirement of distributive justice which mandates a fair distribution of the benefits and burdens of research.[185] The two ethical concerns in any research are identified as an imposition of undue burdens and the absence of expected benefits. Despite the lasting concern about risks to research subjects, a major shortcoming in research sponsored by industrialized countries and conducted in resource-poor countries that has been recognized is the failure to share in the benefits of research when successful products or contributions to knowledge result.[186]

There have been debates regarding the use of the term distributive justice for the interactions between countries. A perspective on the issue argues that the scope of distributive justice lies within a single country.[187] The implication of this perspective is that the fair distribution of social benefits and burdens applies only among individuals living together in a society. A related perspective argues that distributive justice applies to group of collaborators in international research, regardless of any geographical distance that separates the countries involved.[188] A report published by the Institute of Medicine of the National Academy of Sciences writes, "Beneficiaries of the research outcomes must include people in the developing countries where the research is conducted, as well as in the developed country that sponsors the research.".[189]

Researchers in developed country usually establish a relationship with their collaborators in developing country and with the research subjects in the country where the research is conducted. The multinational clinical research is not regarded as interaction between countries. The research may be sponsored by several stakeholders, governmental agencies such as the U.S. National Institute of Health (NIH), U.K. Medical Research Council, private industry, private foundation, and international organizations such as WHO.[190]

---

[184] National Commission for the Protection of Human Subjects of Biomedical and Behavioral Research, *The Belmont Report: Ethical Principles and Guidelines for the Protection of Human Subjects of Research* (Washington DC: U.S. Government Printing Office, 1979).

[185] Macklin, *Double Standards in Medical Research in Developing Countries*, 69.

[186] Macklin, *Double Standards in Medical Research in Developing Countries*, 69.

[187] Macklin, *Double Standards in Medical Research in Developing Countries*, 69.

[188] Macklin, *Double Standards in Medical Research in Developing Countries*, 69–70.

[189] Anna C. Mastroianni, Ruth Faden and Daniel Federman, ed., *Women and Health Research: Ethical and Legal Issues of Including Women in Clinical Studies*, Vol. 1, (Washington DC: National Academy Press, 1994), 78.

[190] Macklin, *Double Standards in Medical Research in Developing Countries*, 70.

The concept of distributive justice is broad, and it covers not only social benefits and burdens but also other benefits such as health benefits, financial benefits etc. all of which may be regarded as social benefits.[191] Macklin argues, "There is nothing inherent in the concept of distributive justice that requires those benefits and burdens to result and arise from a group of people living together in a society."[192] The criteria for a fair distribution in the concept of distributive justice vary according to the context.[193] For example, equity is the core concept of fair distribution in the context of research involving human subjects.[194] "Equity requires that no one group-gender, racial, ethnic, geographic, or socio-economic receive disproportionate benefits or bear disproportionate burdens."[195] A glaring example of global inequity is seen in the disparity between the distribution of the global disease burden and the allocation of research funding. Less than 10% of global expenditures for health research by private and public sectors is devoted to addressing 90% of the world's burden of disease shouldered by developing countries. This is usually called "the 10/90 gap."[196] Resnik argues that 10/90 gap exists because of two reasons: (1) multinational pharmaceutical and biotechnology companies do not view research and development (R&D) investments on the health problems of developing nations to be economically advantageous; and (2) government biomedical research agencies encounter limited pressure to allocate funds for the problems of developing nations.[197] He argues further that developed nations have a moral obligation to address the disparities in connection to biomedical research funding. He proposes that developed countries should establish a trust fund in the form of Global AIDS Fund to sponsor research on the health problems of developing nations.[198]

Documented evidence shows that people in developing countries disproportionately shoulder the burden of research risks without enjoying corresponding benefits that may arise from it. Macklin articulating a similar view writes, "Residents of developing countries lack access to the products of research carried out in their countries if the medications are too expensive for individuals or the ministries of health to afford."[199] Some scholars stipulate two conditions for fulfilling the requirements of distributive justice in international research: (1) applying the same standards used in a research conducted in the sponsoring country in the evaluation of the design and determination of acceptable risk-benefit ratios; (2) beneficiaries of

[191] Macklin, *Double Standards in Medical Research in Developing Countries*, 70.

[192] Macklin, *Double Standards in Medical Research in Developing Countries*, 70.

[193] Mastroianni, Faden and Federman, ed., *Women and Health Research: Ethical and Legal Issues of Including Women in Clinical Studies*, 76.

[194] Macklin, *Double Standards in Medical Research in Developing Countries*, 70.

[195] Mastroianni, Faden and Federman, ed., *Women and Health Research: Ethical and Legal Issues of Including Women in Clinical Studies*, 76.

[196] David B. Resnik, "The Distribution of Biomedical Research Resources and International Justice," *Developing World Bioethics* 4, no.1 (May 2004): 42.

[197] Resnik, "The Distribution of Biomedical Research Resources and International Justice," 42.

[198] Resnik, "The Distribution of Biomedical Research Resources and International Justice," 42.

[199] Macklin, *Double Standards in Medical Research in Developing Countries*, 71.

research results must consist of people in the developing countries where the research is carried out, as well as in the United States.[200] These conditions imply that determining equity in international clinical trials requires that not only that the research participants bear the burdens and benefits from the research, but also the larger community needs to have opportunity to enjoy the successful products. Many of the HIV/AIDS clinical trials in developing countries have encountered a similar problem, because the participants were allowed to disproportionately bear the burden of the research without sharing equally in the benefits. In order to resolve this issue, with emphasis on satisfying the requirement of distributive justice, a solution has been proposed for developed countries to make a commitment to provide an affordable access to anti-retroviral drugs to host developing countries at the end of international clinical trials.[201] The discussion of the distributive justice principle which focuses on the fair distribution of burdens and benefits of research for participants and the larger community draws from two different interpretations of the distributive justice principle.

1.3.2.2 Theoretical Approaches to Distributive Justice

Cooley refutes critics who attacked HIV clinical trials in developing countries on the ground that they were not based on a distributive justice principle found in the commentary of the Council of International Organization of Medical Sciences (CIOMS) guidelines for international research on human subjects.[202] The Guideline stipulates that the sponsoring agency should guarantee that at the end of the trial, any successful product will be made reasonably available to inhabitants of the underdeveloped community in which the research was conducted. All parties involved in any research should justify and agree to exceptions to this general requirement before engaging in the research.[203] Critics of HIV clinical trials in developing countries argue that some researchers have violated the distributive justice principle and consequently exploited the poor in developing countries, because they did not ensure that successful products developed in the trials were made reasonably available to the community where the trials were conducted.[204]

---

200 Mastroianni, Faden and Federman, ed., *Women and Health Research: Ethical and Legal Issues of Including Women in Clinical Studies*, 78.

201 Macklin, *Double Standards in Medical Research in Developing Countries*, 71.

202 Dennis R. Cooley, "Distributive Justice and Clinical Trials in the Third World," *Theoretical Medicine and Bioethics* 22, no. 3 (June 2001): 151–167.

203 Council for International Organizations of Medical Sciences, *International Ethical Guidelines for Biomedical Research Involving Human Subjects* (Geneva: CIOMS, 1993), quoted in Cooley, "Distributive Justice and Clinical Trials in the Third World," 151.

204 Robert A. Crouch and John D. Arras, "AZT Trials and Tribulations," *Hastings Center Report* 28, no. 6 (November 1998): 26–34.

Scholars offer two different interpretations of distributive justice principle with their foundations on the opposing theories of Capitalism and Marxism.[205] Cooley explains that distributive justice requires that each social member receives a just distribution of the benefits and burdens of society, i.e., what he or she deserves from being a member of the society. The implication is that members of the community who did not participate in the trial and consequently did not shoulder any burden will also receive a huge benefit at no real cost to themselves. This interpretation cannot be based on capitalism, for capitalist justice requires that only those who have contributed to realizing their group's goals may receive benefits.[206] Cooley argues that an idea of distributive justice that would require that successful products be made reasonably available to the community or developing country that hosted the trial is "more closely aligned with Marxism, which requires that people work according to their abilities, while they receive according to their needs, than it does with capitalism."[207] He argues further that critics of HIV clinical trials in developing countries employed a notion of distributive justice that requires that goods be distributed according to need, but leaves out the part that emphasizes that abilities should determine contributions.[208] He contends that the principle is more extreme than Marxism, because it requires no contribution from those receiving benefits, regardless of whether or not they are able to contribute.[209] He argues that, "It is not clear that people in the country who do not receive any benefits are exploited when they did not share the burdens to obtain the benefits, especially since both capitalistic and Marxist justice require that they contribute in some way."[210] He argues further that human subjects were exploited for the benefit of non-participants who did not contribute anything to deserve the benefits. He points out that those who do not contribute to a research have no reasonable expectation of a benefit. They will require a justification for such a desert in order to have a reasonable expectation of a benefit. Simply indicating that you are a member of the community in which the research is conducted is not adequate to enable one to get benefits.[211] Cooley argues that need alone does not justify a desert but claims that justifying a desert requires individuals to make some contributions to the effort.[212]

An opposing view in the UNAIDS Guidance Document for preventive HIV/AIDS vaccine research argues that dire human need can justify the requirement of distributive justice. UNAIDS Guidance Document indicates that the severity of the epidemic makes it crucial that adequate incentives exist, both through financial rewards in the marketplace and through public subsidies to promote development of effective vaccines while also guaranteeing that vaccines are produced and distributed

[205] Macklin, *Double Standards in Medical Research in Developing Countries*, 78.

[206] Cooley, "Distributive Justice and Clinical Trials in the Third World," 157.

[207] Cooley, "Distributive Justice and Clinical Trials in the Third World," 157–158.

[208] Cooley, "Distributive Justice and Clinical Trials in the Third World," 158.

[209] Cooley, "Distributive Justice and Clinical Trials in the Third World," 158.

[210] Cooley, "Distributive Justice and Clinical Trials in the Third World," 158.

[211] Cooley, "Distributive Justice and Clinical Trials in the Third World," 158.

[212] Cooley, "Distributive Justice and Clinical Trials in the Third World," 158–159.

in a way that really makes them available to the population at greatest risk.[213] Cooley does not defend clinical trials conducted in developing countries on grounds of distributive justice principle which he considers too vague and ambiguous. Rather, he defends them based on utilitarian considerations, Kantian principle to treat people as ends in themselves and autonomy of the subjects to choose to participate in the trials.[214] Cooley argues that, "If a medical experiment in the Third World intended to help the citizens of those areas fails either to serve utility or not treat anyone as a mere means, it is unethical."[215]

Benatar disagrees with Cooley's utilitarian approach to clinical research in developing countries that disregards the principle of distributive justice in CIOMS guidelines and describes it as superficial for several reasons.[216] First, it is not proper to overlook the extent of injustice in health and health care research. Second, the belief that the principle of justice can be applied in a simple deductive manner, indicates lack of acknowledgment for the nature of principles and the need for interpreting how these should be applied in specific contexts. Third, there is a failure that the limited form of utility proposed has been linked with exploitative practices with far reaching effects on health and disease, and that such exploitation is perpetuated by some trials in developing countries. Fourth, a hidden form of paternalism underlies the author's proposals. Finally the important notion of informed consent is applied naively to clinical trials in developing countries.[217] Benatar argues further that, "These shortcomings serve only to entrench further a neo-liberal economic mindset deeply inimical to the progress required to rectify some of the widening disparities in wealth and health that characterize an increasingly unstable world.".[218]

A further discussion on distributive justice for clinical trials in developing countries explores other candidates for principle of justice or equity. A perspective termed equity as maximization describes a notion of justice or equity that focuses on maximizing health benefits for a population. In this context, some people regard it as self-evident that health policy should aim to produce as much health as possible for a given population.[219] Some scholars argue for maximizing principles as candidates for principles of justice or equity. Marchand et al. point out that the basic moral assumption for maximizing principles focuses on expressing equal respect for each person by according the same importance to every person's interests. The principle of equality emphasizes that each person's interests, in this case their health

[213] Joint United Nations Programme on HIV/AIDS, (UNAIDS), *Ethical Considerations in HIV Preventive Vaccine Research*, (Geneva:UNAIDS, 2000), 13, http://data.unaids.org/Publication/1RC-pub01/jc072-ethicalcons_en.pdf (Accessed May 10, 2013).

[214] Cooley, "Distributive Justice and Clinical Trials in the Third World," 151–162.

[215] Cooley, "Distributive Justice and Clinical Trials in the Third World," 162.

[216] Solomon R. Benatar, "Distributive Justice and Clinical Trials in the Third World," *Theoretical Medicine and Bioethics* 22, no. 3 (June 2001): 169.

[217] Benatar, "Distributive Justice and Clinical Trials in the Third World," 169.

[218] Benatar, "Distributive Justice and Clinical Trials in the Third World," 169.

[219] Sarah Marchand, Daniel Wikler, and Bruce Landesman, "Class, Health, and Justice," *The Milbank Quarterly* 76, no. 3 (September 1998): 456.

is considered a priority just as much and no more than anyone else's.[220] The implication is that "an improvement in health for the well-off is just as valuable and carries the same moral weight as an improvement in health for the worse off."[221] Health benefits matter equally both for the well off and the worst off. Contrary to Cooley's interpretation of distributive justice, the concept of equity as maximization does not demand contributions from individuals in order to be entitled to their improved health status. The idea of justice supported by this perspective requires that both the entire population in need and the research subjects who have contributed to development of a successful product are entitled to the benefits of research.[222]

Another perspective for a principle of justice or equity which applies some criteria of urgency to levels of health is known as equity as priority to the sickest.[223] This perspective focuses on the urgency of people's needs, "those who are threatened with the worst harms – who have the shortest life expectancy and most serious diseases and injuries – should count as the worst off."[224] Based on this account, we should give priority to those with the greatest urgent need, the sickest people. In clinical trials, this creates a problem because the research subjects who have received the benefits during the research of a successful product are no longer the sickest. Those who are the sickest are people who were not recruited in the research and as a result did not have access to any treatments because none of the treatments were available outside research.

The implication of this principle of priority to the sickest is that the rest of the populations who did not participate in the research and did not receive any beneficial product deserve priority for receiving the product. The principle did not fully account for the research subjects who may still need the product when the research is concluded. It points to the fact that no single principle of justice accounts for all situations. There is a consensus that research subjects should not be left worse off after a clinical trial is concluded than they were, while they were participating in the trial. In this context, another principle of justice is needed which accounts for a situation where the research subjects will continue to require a successful product that they received during their participation in the clinical trials.

The National Bioethics Advisory Commission (NBAC) report situates post-trial obligations to research subjects in the principle of justice as reciprocity which reechoes Cooley's line of thought earlier discussed. The NBAC report articulates that justice as reciprocity in the context of clinical trials could imply that something is owed to research participants at the conclusion of the trial, because their shouldering of research risks and burdens made it possible for the researchers to generate findings necessary to advance knowledge and develop new medical interventions.[225]

---

220 Marchand, Wikler, and Landesman, "Class, Health, and Justice," 456.

221 Marchand, Wikler, and Landesman, "Class, Health, and Justice," 456.

222 Macklin, *Double Standards in Medical Research in Developing Countries*, 79–80.

223 Marchand, Wikler, and Landesman, "Class, Health, and Justice," 461.

224 Marchand, Wikler, and Landesman, "Class, Health, and Justice," 461.

225 National Bioethics Advisory Commission, (NBAC), *Ethical and Policy Issues in International Research: Clinical Trials in Developing Countries* (Bethesda, MD: NBAC, 2001), 59.

The NBAC report indicates that the principle of reciprocity relate to research subjects who participated in a clinical trial that yielded successful products or not. In both instances, the subjects bore burdens and risks of the research.

Another notion of justice that may also apply to clinical trials in developing countries is compensatory justice. The concept of compensatory justice can be applied to the research context in two situations. The first situation refers to compensating research subjects for the injuries they sustained in course of the research. The second situation relates to research subjects who have lost time due to their participation in the research or have spent some money out of pocket for research related expenses like travel to the research site and childcare. Macklin contends that "in both situations, compensation is owed or provided to research subjects for something they did or that happened to them."[226]

Compensatory justice has also been applied to a broader context to encompass situations that have happened in the past. The idea of compensatory justice has been broadened to go beyond fairness in distribution and to include an attempt to remedy or redress past wrongs. An example usually cited is the monetary payments made to survivors of the Tuskegee syphilis study or to their relatives, to compensate them for the harm or wrong done by the study.[227] Paying the survivors of the Tuskegee syphilis study for the harm they incurred is in line with the concept of compensatory justice that compensates the subjects for something that happened to them, although in this case the events occurred in the past. On the other hand, there has been a debate whether compensating the relatives of Tuskegee syphilis study participants was a stretch in the idea of compensatory justice. Macklin argues that, "a case could be made for benefiting developing countries or the communities where research is conducted today as compensation for past research from which no benefits flowed to those communities or countries."[228] In this context, compensatory justice calls for providing benefits from current research because of injustices of the past. The inability of the host population to access successful products, e.g. anti-retroviral drugs at the conclusion of trials, has been acknowledged as a problem with the clinical trials in developing countries.

1.3.3 Research Responsiveness to Host Populations

1.3.3.1 Ethical Issues

Responsiveness of research to the needs and priorities of the host population has been hotly debated especially with regard to clinical trials in developing countries. Grady expresses concerns about exploitation and double standard in international

[226] Macklin, *Double Standards in Medical Research in Developing Countries*, 81.

[227] Mastroianni, Faden and Federman, ed., *Women and Health Research: Ethical and Legal Issues of Including Women in Clinical Studies*, 76.

[228] Macklin, *Double Standards in Medical Research in Developing Countries*, 81.

clinical research. She contends that "research participants and populations in developing countries may be particularly vulnerable to exploitation due to poverty; illiteracy; limited resources, education, and access to health care; and lack of familiarity or experience with research."[229] In the past, communities in developing countries who have participated in research have too often not enjoyed the benefits of the research. Instead, the benefits of research hosted in developing countries have been primarily enjoyed by developed countries.[230] London argues that international collaborative research "must be conducted in such a manner as to leave the host community better off than it was, or at least not worse off."[231]

The requirement for responsiveness of research to the health needs and priorities of the host population is supported by several groups who have grappled with how to minimize exploitation in international clinical research. For example, the National Bioethics Advisory Commission recommends clinical trials conducted in developing countries that are only responsive to the health needs of the host country.[232] The World Medical Association's 2000 version of the Declaration of Helsinki also stipulates that the justification for a medical research will be based on the reasonable likelihood that the populations involved in research will benefit from the results of the research.[233] The United Kingdom's Nuffield Council on Bioethics advises national priority-setting for health care research so that it will be, "easier for host countries to ensure that research proposed by external sponsors is appropriate and relevant to its national health care needs."[234] The Council of International Organizations of Medical Sciences (CIOMS) in their international guidelines recommends two essential requirements before engaging in a research in a population or community with limited resources. First, the sponsor and researcher must make every effort to guarantee that the research is responsive to the health needs and the priorities of the population or community in which it is conducted. Second, any intervention or product developed or knowledge generated will be made reasonably available for the benefit of that population or community.[235] There is a general consensus among scholars regarding the support for the requirement of responsiveness

[229] Grady, "Ethics of International Research: What Does Responsiveness Mean?," 235–236.

[230] Alex John London, "Responsiveness to Host Community Health Needs" in *The Oxford Textbook of Clinical Research Ethics*, ed. Ezekiel J. Emmanuel et al., (New York: Oxford University, 2008), 738.

[231] London, "Responsiveness to Host Community Health Needs" 738.

[232] National Bioethics Advisory Commission, (NBAC*), Ethical and Policy Issues in International Research: Clinical Trials in Developing Countries*, iii, 7–8.

[233] World Medical Association, *Declaration of Helsinki, Ethical Principles for Medical Research Involving Human Subjects*, (Edinburgh, Scotland: World Medical Association, 2000), 3, Principle 19, http:www.wma.net/e/ethicsunit/helsinki.htm (Accessed May 3, 2013).

[234] Nuffield Council of Bioethics, *The Ethics of Research Related to Healthcare in Developing Countries,* (London, UK: Nuffield Council on Bioethics, 2002), http://www.nuffieldbioethics.org/go/ (Accessed May 3, 2013).

[235] Council for International Organizations of Medical Sciences, *International Ethical Guidelines for Biomedical Research Involving Human Subjects*, (Geneva: CIOMS, 2002), http://www.cioms.ch/frame_guidelines_nov_2002.htm. (Accessed May 3, 2013).

in international clinical research, but the interpretation of what it entails in actual practice varies.

1.3.3.2 Different Perspectives

Different perspectives have emerged in the debate about the responsiveness requirement in international clinical research. One perspective stresses that research is responsive to the health needs of the population whenever it addresses health problems of the population.[236] Grady articulates, "The requirement to be responsive suggests that research should address a question that is relevant and important to a host country and that the answer should be of potential benefit to that country."[237] For example, a study of a less toxic malaria treatment or a strategy for preventing malaria is responsive to an important health need, in a country where malaria is prevalent and a major cause of mortality in children.[238] London describes health needs as, "concerns that are particularly important or urgent because of their close relationship to the ability of persons to be free from medical conditions that shorten their lives or prevent them from functioning in ways that are basic or fundamental to their pursuit of a reasonable life plan."[239] Developing countries have been ravaged by diseases such as malaria, tuberculosis and HIV, and finding new strategies for treating these diseases would constitute health needs that are also health priorities of such countries. Disease burden or prevalence is not the only criterion for defining the requirement of responsiveness in international clinical research.

Another perspective defines the criterion for responsive research as ensuring that successful products, e.g. anti-retroviral drugs are available at the conclusion of the research.[240] This perspective is in line with CIOMS guidelines that requires that the research project should not leave low-resource countries or communities worse off. It should be responsive to their health needs and priorities in such a way that any product developed is made reasonably available to them, and as much as possible leave the population in a better position to obtain effective health care and protect its own health.[241]

This statement attests to the fact that medical research can play a significant role in assisting communities protect their own health by finding new therapies. London argues that, "the requirement to ensure reasonable availability seems most appropriate when combined with the requirement that such research actually focus on health

---

236 Grady, "Ethics of International Research: What Does Responsiveness Mean?," 236.

237 Grady, "Ethics of International Research: What Does Responsiveness Mean?," 236.

238 Grady, "Ethics of International Research: What Does Responsiveness Mean?," 236–237.

239 London, "Responsiveness to Host Community Health Needs" 738.

240 Macklin, *Double Standards in Medical Research in Developing Countries*, 25.

241 Council for International Organizations of Medical Sciences, *International Ethical Guidelines for Biomedical Research Involving Human Subjects*, quoted in London, "Responsiveness to Host Community Health Needs" 739.

needs that are also health priorities of that community."[242] The results of research should be made reasonably available in such a way that maximizes their value and usefulness.[243] Collaboration with host country researchers, institutions, health policy makers, community groups, and others all through the phases of research will guarantee dissemination of results and assimilation of important new knowledge.[244]

In some situations, research may not yield successful products but may ensure that other benefits are provided for host communities. Some scholars argue that something is owed to the community or country at the end of the research, but it may not be necessarily successful products of the research.[245]

In line with this notion, another perspective links other benefits of the research with responsiveness. CIOMS guidelines indicates that the ethical requirement of responsiveness can only be realized if successful interventions or other kinds of health benefits are made available to the population.[246] Grady argues that if the goal is to minimize potential exploitation of research participants, benefits are certainly critical but the emphasis is on the level, not the type, of benefits that participants receive.[247] She further suggests several types of possible benefits associated with clinical research, therapeutic benefits to research participants, useful and generalizable knowledge for the community, infrastructure and capacity building, the inclusion of required public measures, training of research and clinical staff, ancillary medical benefits to participants or others, the post-trial benefits of new drugs and other products, economic benefits and increased business employment.[248] Similarly, Emmanuel et al. argue, that guaranteeing a type of benefit, the proven intervention instead of a fair level of benefits does not necessarily prevent exploitation.[249] They proposed what is usually referred to as "the fair benefits framework."[250]

Proponents of the fair benefits framework contend that it "Supplements the usual conditions for the ethical conduct of research trials, such as independent review by an institutional review board or research ethics committee and individual informed consent."[251] The fair benefits framework builds upon three background principles

---

[242] London, "Responsiveness to Host Community Health Needs" 739.

[243] Grady, "Ethics of International Research: What Does Responsiveness Mean?," 237.

[244] Grady, "Ethics of International Research: What Does Responsiveness Mean?," 237.

[245] Macklin, *Double Standards in Medical Research in Developing Countries,* 25.

[246] Council for International Organizations of Medical Sciences, *International Ethical Guidelines for Biomedical Research Involving Human Subjects*, http://www.cioms.ch/frame_guidelines_nov_2002.htm. (Accessed May 3, 2013).

[247] Grady, "Ethics of International Research: What Does Responsiveness Mean?," 237.

[248] Grady, "Ethics of International Research: What Does Responsiveness Mean?," 237.

[249] Ezekiel J. Emmanuel et al., "Addressing Exploitation: Reasonable Availability Versus Fair Benefits," in *Exploitation and Developing Countries: The Ethics of Clinical Research,* ed. Jennifer S. Hawkins and Ezekiel J. Emmanuel (Princeton, New Jersey: Princeton University, 2008), 295.

[250] Ezekiel J. Emmanuel et al., "Addressing Exploitation: Reasonable Availability Versus Fair Benefits, 299.

[251] Ezekiel J. Emmanuel et al., "Addressing Exploitation: Reasonable Availability Versus Fair Benefits," 299.

that are generally considered as requirements for ethical research. First, the research should have social value by focusing on a health problem of the developing country population. Second, fair subject selection guarantees that the scientific objectives of the research itself, not poverty or vulnerability, support a strong justification for carrying out the research in a specific population. Third, the research must have a favorable risk-benefit ratio, which entails that benefits to participants outweigh the risks, or the net risks are appropriately low.[252] The fair benefits framework emphasizes additional three principles such as fair benefits, collaborative partnership and transparency that are specified by multiple benchmarks.[253]

Principle 1: Fair Benefits

This requires a complete outline of tangible benefits that may accrue to the research participants and the population from the conduct and results of the research. These benefits comprise of three types: "(1) benefits to research participants during the research, (2) benefits to the population during the research, or (3) benefits to the population after completion of the research."[254] Since, exploitation is a major concern in international clinical research; the emphasis is on providing a fair level of benefits rather than types of benefits. Emmanuel et al. write, "… it would seem fair that as the burdens and risks of the research increase, the benefits should also increase. Similarly, as the benefits to the sponsors, researchers, and others outside the population increase, the benefits to the host population should increase."[255] The fair benefits framework tackles the issue of exploitation by allowing the host population that bears the burden of the research to receive benefits as well as determine their fairness. Nevertheless, every benefit of research may not directly flow to research participants but may also benefit the entire community. For example, capacity development realized through improving health care infrastructure, training of health and research personnel and training of personnel in research ethics could be provided to the community.[256]

Principle 2: Collaborative Partnership

It is important to note that only the host population can determine the adequacy and fairness of the level of benefits for itself. Emmanuel et al. articulate, "outsiders are likely to be poorly informed about the health, social and economic context in which the research is being conducted, and are unlikely to fully appreciate the importance

[252] Ezekiel J. Emmanuel et al., "Addressing Exploitation: Reasonable Availability Versus Fair Benefits," 299.

[253] Ezekiel J. Emmanuel et al., "Addressing Exploitation: Reasonable Availability Versus Fair Benefits," *299.*

[254] Ezekiel J. Emmanuel et al., "Addressing Exploitation: Reasonable Availability Versus Fair Benefits," 299.

[255] Ezekiel J. Emmanuel et al., "Addressing Exploitation: Reasonable Availability Versus Fair Benefits," 299.

[256] Ezekiel J. Emmanuel et al., "Addressing Exploitation: Reasonable Availability Versus Fair Benefits," 302.

of the proposed benefits to the population."[257] The choice of the host population to participate in research must be free and uncoerced, and refusing to participate must be a realistic alternative.[258]

Principle 3: Transparency
Transparency similar to the full information requirement for ideal market transactions allows the host population to compare benefits agreements in similar transactions. An independent body, for example, WHO should establish and operate a publicly accessible repository of all formal and informal benefits agreements. A central repository allows independent assessment of the fairness of benefits agreements by populations, researchers, governments and others, such as non-governmental bodies.[259] A series of community consultations are required in order to inform populations in developing countries about previous benefits agreements in other research projects.[260] Emmanuel et al., argue that the three background conditions and the three principles of the fair benefits framework guarantee the realization of essential requirements of research. The essential requirements of research include (1) the selection of the population based on good scientific reasons; (2) the research presenting limited net risks to the research participants; (3) presence of adequate and long lasting benefits to the population; (4) ensuring that the population is not subject to a coercive choice; (5) guaranteeing that the population freely determines whether to participate and whether the level of benefits is fair given the risks of the research; and (6) ensuring that the repository offers the opportunity for comparative assessments of the fairness of the benefit agreements.[261] They further argue that in comparison with reasonable availability requirement, the three principles – fair benefits, collaborative partnership, and transparency – are more inclined to guarantee that populations in developing countries are not exploited, benefit from clinical research, and retain decision-making responsibility.[262]

[257] Ezekiel J. Emmanuel et al., "Addressing Exploitation: Reasonable Availability Versus Fair Benefits," 302.

[258] Ezekiel J. Emmanuel et al., "Addressing Exploitation: Reasonable Availability Versus Fair Benefits," 302.

[259] Ezekiel J. Emmanuel et al., "Addressing Exploitation: Reasonable Availability Versus Fair Benefits," 303.

[260] Ezekiel J. Emmanuel et al., "Addressing Exploitation: Reasonable Availability Versus Fair Benefits," 303.

[261] Ezekiel J. Emmanuel et al., "Addressing Exploitation: Reasonable Availability Versus Fair Benefits," 303.

[262] Ezekiel J. Emmanuel et al., "Addressing Exploitation: Reasonable Availability Versus Fair Benefits," 307.

1.4 Summary of Book Chapters

The book presents an ethical argument for post-trial access to drugs for participants and host populations in developing countries based on the obligation of justice. Humanitarian assistance is not sufficient in itself because it has a limited term and lacks the capacity to regulate the relevant inequalities between societies. The analysis contrasts two dominant perspectives that address the issue of post-trial access to drugs: the statist approach of Rawls and the cosmopolitan approach of Pogge. In contrast to the approaches of Rawls and Pogge, the book argues for a paradigm involving a sliding scale of national and global responsibilities for the right to health. A summary of the book's chapters presents the argument in more detail, as follows.

Chapter 1 presents a general introduction to the analysis focusing on ethical issues in global health inequalities and global health justice. Chapters 2 and 3 present the context of the ethical problem under discussion: the process of international clinical research that tends to inadequately respect the health rights of local populations (Chap. 2); and the right of local populations that undergo clinical research to have post-trial access to resulting drugs, such as anti-retroviral drugs (Chap. 3). Chapters 4 and 5 present two contrasting but inadequate approaches to global justice that address post-trial access to drugs for host populations in developing nations. Chapter 6 which tackles global health responsibility presents a paradigm that combines national and international responsibilities for global justice to address the ethical problem of post-trial access. Chapter 7 concludes the book with a discussion on UNESCO Declaration on Bioethics and Human Rights. It highlights the significant role of the landmark intergovernmental document on biomedical research. The UNESCO Declaration aims at setting global minimum standards in biomedical research and clinical practice. As the first international legal, non-binding instrument, it grapples with linking human rights and bioethics. It resorts to international human rights law as a means of protecting responsible biomedical activities. The UNESCO Declaration emphasizes the principles of human dignity, human rights and fundamental freedom in its efforts to promote responsible biomedical research and clinical practice. It further stresses the priority of the individual over science or society. A more detailed explanation of the argument in the main chapters follows.

Chapter 2
Global Clinical Research

2.1 Introduction

The first highly publicized violation of the rights of research subjects was the atrocities committed by Nazi research physicians with non-consenting subjects under the pretense of medical experimentation. The startling revelations of such abuses of human rights at the Nuremberg war crime trials led to global outrage and the urgent need to craft a code of conduct for human research known as the Nuremberg code.[1] The Belmont report refers to this code as the "prototype of many later codes intended to assure that research involving human subjects would be carried out in an ethical manner."[2]

The World Medical Association adopted the Declaration of Helsinki in 1964, which further addressed the issue regarding rights of research subjects. The Declaration of Helsinki emphasized basic principles for the conduct of human biomedical research.[3] The issue of research in developing countries was finally addressed by the council for International Organization of Medical Sciences (CIOMS), in collaboration with the World Health Organization (WHO). In 1982, CIOMS and WHO proposed international guidelines for biomedical research involving human subjects. The guidelines were aimed at establishing ethical principles that should guide the conduct of biomedical research involving human subjects in the international arena. Furthermore, the wide publication of abuses in

---

[1] Zulfiquar Ahmed Bhutta, "Ethics in International Health Research: A Perspective from the Developing World," *Bulletin of the World Health Organization* 80, no. 2 (January 2002): 114–115.

[2] The National Commission for the Protection of Human Subjects of Biomedical and Behavioral Research, *The Belmont Report: Ethical Principles and Guidelines for the Protection of Human Subjects* (Washington, D.C. U.S. Government Printing Office, 1978), 1.

[3] World Medical Association, *Declaration of Helsinki, Ethical Principles for Medical Research Involving Human Subjects* , (Seoul: World Medical Association, 2008), http://www.wma.net/en/30publications/10policies/b3/index.html (Accessed July19, 2013).

E.C. Obi, *Post-Trial Access to Drugs in Developing Nations*, Advancing Global Bioethics 7, DOI 10.1007/978-3-319-60028-4_2

human research triggered establishment of a national commission in the United States of America to develop principles and guidelines for the protection of human subjects of biomedical and behavioral research.

The rights of research subjects are protected by the requirement of two safeguards, voluntary informed consent and review of research. Both safeguards are required by U.S. regulations and International declarations and guidelines for research.[4] Regarding protection of the rights of research subjects, current U.S. regulations require that Institutional Review Boards (IRBs) approve a research when, "risks to subjects are reasonable in relation to anticipated benefits, if any, to subjects, and the importance of the knowledge that may reasonably be expected to result."[5]

Similarly, international ethical guidelines for biomedical research involving human subjects prepared by the Council for International Organizations of Medical Sciences in collaboration with the World Health Organization emphasize review of all proposals to conduct research involving human subjects for scientific merit and ethical acceptability.[6] CIOMS highlights that, "the ethical review committee is responsible for safeguarding the rights, safety, and well-being of the research subjects."[7] Despite the general consensus regarding the need for these two safeguards of voluntary informed consent and review of research, enough evidence abounds that they are sometimes faulty, insufficient or even non-existent in the conduct of clinical trials in developing countries.[8]

The implementation of consent in developing countries which is one of the core responsibilities of ethics review committee must take into account their cultural values and cultural diversities. In developing countries, inadequate resources may create barriers to effective independent review of research protocols.

Post-trial access obligations to the participants and host populations in developing nations should be addressed in the informed consent process. Ethics Review Committee also should review plans for Post-trial access in advance as a critical element of the approval of the protocol. A more detailed discussion of these two safeguards follows.

---

[4] Macklin, *Double Standards in Medical Research in Developing Countries*, 131.

[5] U.S. Department of Health and Human Services, *Code of Federal regulations Title 45 Public Welfare, Part 46 Protection of Human Subjects*, (Washington, DC: U.S. Department of Health and Human Services, 2009), 7.

[6] Council for International Organizations of Medical Sciences, *International Ethical Guidelines for Biomedical Research Involving Human Subjects*, (Geneva: CIOMS, 2002), 24.

[7] Council for International Organizations of Medical Sciences, *International Ethical Guidelines for Biomedical Research Involving Human Subjects*, 25.

[8] Macklin, *Double Standards in Medical Research in Developing Countries*, 131.

2.2 Informed Consent

2.2.1 *Ethical Components of Informed Consent*

2.2.1.1 Different Senses of Informed Consent

Informed consent is an important aspect in discussing the ethical conduct of the clinical trials. The requirement to obtain voluntary informed consent from individuals before enrolling them in clinical trial is a fundamental principle of research ethics.[9] The requirement for informed consent in clinical trials reflects the ethical principles of respect for persons, human dignity, and autonomy.[10]

Two different senses of informed consent are presented by authors in literature and practices.[11] In the first sense, the informed consent is explained within the context of autonomous choice. An autonomous action refers to "normal choosers who act (1) intentionally, (2) with understanding, and (3) without controlling influences that determines their actions."[12] The first of these three conditions dealing with acting intentionally does not accommodate any degree, because acts are either intentional or unintentional. On the other hand, the two other conditions of understanding and absence of controlling influences can be met by acts to a greater or lesser degree.[13] Beauchamp and Childress articulate, "for an action to be autonomous in this account, it needs only a substantial degree of understanding and freedom from constraint, not a full understanding or a complete absence of influence."[14]

Informed consent is an autonomous authorization by an individual for participation in research or to receive medical care. In this sense, informed consent entails that a patient or a subject must authorize a professional to do something through an act of informed or voluntary consent.[15] The implication is that health professionals or researchers will explain the purpose, risks, benefits and alternatives of medical intervention or research to patients or subjects and ensure they understand before voluntarily consenting.

The doctrine of informed consent prescribes that research subjects should participate in research voluntarily and with adequate information about the research. Noteworthy, is that only a competent individual gives an informed consent, after

---

[9] National Bioethics Advisory Commission, (NBAC*), Ethical and Policy Issues in International Research: Clinical Trials in Developing Countries*, http://bioethics.georgetown.edu/nbac/clinical/chap3.html (Accessed July 19, 2013).

[10] National Bioethics Advisory Commission, (NBAC*), Ethical and Policy Issues in International Research: Clinical Trials in Developing Countries*, http://bioethics.georgetown.edu/nbac/clinical/chap3.html (Accessed July 19, 2013).

[11] Tom L. Beauchamp and James F. Childress, *Principles of Biomedical Ethics*, (New York: Oxford University Press, 2001), 78.

[12] Beauchamp and Childress, *Principles of Biomedical Ethics*, 59.

[13] Beauchamp and Childress, *Principles of Biomedical Ethics*, 59.

[14] Beauchamp and Childress, *Principles of Biomedical Ethics*, 59.

[15] Beauchamp and Childress, *Principles of Biomedical Ethics*, 78.

receiving necessary information, adequately understanding the information and considering the information in order to arrive at a decision without being subjected to coercion, undue influence, inducement or intimidation.[16] Informed consent recognizes that, "individuals have the right and the ability to make decisions in their own interest and to act upon them."[17]

The second sense of informed consent is explained within the context of the social rules of consent in institutions that require obtaining effective consent legally or institutionally from patients or subjects before initiating medical or research procedures.[18] In this sense, informed consent as an effective consent is a policy oriented approach whose conditions are not deducible from analyses of autonomy and authorization, or from extensive ideas of respect for autonomy.[19] Informed consent in this context deals with institutional requirement and practices of informed consent in health care or in research which mandates group of patients and subjects to be treated in accordance with rules, policies, and standard practices.[20]

Professionals are required according to the social and legal practice to obtain informed consent in institutional settings. Conforming to such policies and procedures satisfies the conditions of informed consent in the second sense. From this perspective, informed consents are not usually autonomous acts or meaningful authorizations. The second sense of informed consent focuses on regulating the behavior of the professional seeking the consent and on establishing procedures and rules for the context of consent. The requirements of such professional behavior and procedure are easily tracked and implemented by institutions.[21]

Informed consent in the second sense deals with effective authorization of either a patient or a subject as governed by institutional rules such as federal and state regulations and hospital policies. In this sense, "a patient or subject can autonomously authorize an intervention, and so give an informed consent in sense one, and yet not effectively authorize that intervention in sense two."[22] The code of regulations for medical and research interventions as well as case law develop models of consent that are outlined in a sense two informed consent. A typical example is disclosure criteria for informed consent which are integral to the history of informed consent. The disclosure requirement constitutes a necessary condition of effective informed consent. The legal doctrine of informed consent is clearly articulated in a law of disclosure, since the fulfillment of disclosure rules devour informed consent in law. The rules of informed consent in sense two concentrate on disclosure,

---

[16] Lema, Mbondo and Kamau, "Informed Consent for Clinical Trials: A Review," 135.

[17] Lema, Mbondo and Kamau, "Informed Consent for Clinical Trials: A Review," 135.

[18] Beauchamp and Childress, *Principles of Biomedical Ethics*, 78.

[19] Ruth R. Faden and Tom L. Beauchamp, "The Concept of Informed Consent." *In Contemporary Issues in Bioethics*, ed. Tom L. Beauchamp and LeRoy Walters, (Belmont, CA: Wadworth-Thomson Learning, 2003), 147.

[20] Faden and Beauchamp, "The Concept of Informed Consent," 145.

[21] Faden and Beauchamp, "The Concept of Informed Consent," 147.

[22] Faden and Beauchamp, "The Concept of Informed Consent," 147.

comprehension, the minimization of potentially controlling influences and competence.[23]

Faden and Beauchamp also discuss the relationship between informed consents in both sense one and sense two.[24] Informed consent in sense one may not meet the criteria to be an informed consent in sense two due to a lack of compliance with relevant rules and requirements. Likewise, an informed consent in sense two may not be an informed consent in sense one. The rules and requirements that regulate informed consents in sense two may not necessarily result in autonomous authorizations in any way, in order, to be considered as informed consents.[25] There is a general consensus among some scholars that the model of autonomous choice which reflects informed consent in sense one need to serve as the standard for the moral adequacy of institutional rules.[26] The rules and requirements of informed consent in sense two should be devised to comply with the standards of informed consent in sense one. This reinforces the fact that the objective of informed consent or the purpose for obtaining informed consent in both medical care and in research is to allow potential patients and subjects to make autonomous decisions relating to whether to participate or not in medical or research interventions.[27] Three conditions are essential to informed consent: disclosure of information, comprehension of information and voluntary participation.[28]

2.2.1.2 Essential Components of Informed Consent

Disclosure of Information

NBAC indicates that requirements for disclosing information in research settings generally surpass those for disclosing information in clinical contexts.[29] Most codes and regulations of research outline critical elements of disclosure in order to guarantee that potential subjects are given adequate information. NBAC highlights four principal types of disclosures that are crucial to the process of informed consent in the research setting: "(1) disclosure of diagnosis and risk; (2) disclosure of the use of placebos and randomization; (3) disclosure of alternative treatments and (4)

---

[23] Faden and Beauchamp, "The Concept of Informed Consent," 148.

[24] Faden and Beauchamp, "The Concept of Informed Consent," 148.

[25] Faden and Beauchamp, "The Concept of Informed Consent," 148.

[26] Beauchamp and Childress, *Principles of Biomedical Ethics*, 79.

[27] Faden and Beauchamp, "*The Concept of Informed Consent,* "148.

[28] Patricia A. Marshall, "Ethical Challenges in Study Design and Informed Consent for Health Research in Resource-Poor Settings," Social, Economic and Behavioral Research, Special Topics no. 5 (2007): 23.

[29] National Bioethics Advisory Commission, (NBAC), *Ethical and Policy Issues in International Research: Clinical Trials in Developing Countries,* http://bioethics.georgetown.edu/nbac/clinical/chap3.html (Accessed July 19, 2013).

disclosures about possible post-trial benefits."[30] Related to the issue about specific and detailed items for disclosure, the Belmont report gave a different list of important types of disclosures: the research procedure, purposes, risks and anticipated benefits, alternative procedures, and a statement that gives the research participant the opportunity to ask questions and to withdraw from the research at any time.[31]

The disclosure requirement also grapples with the standard of how much information and what type of information should be provided to potential subjects. Two competing standards of disclosure have emerged, the professional practice standard and the reasonable person standard.[32] The professional practice standard maintains that adequate disclosure is determined by a professional community's customary practices. The implication is that professional custom determines the amount and kinds of information to be disclosed.[33] In research, the professional practice standard that is, the information usually provided by professionals in the field is insufficient because research occurs when a common understanding does not exist.[34] The reasonable person standard requires the professional to divulge adequate information that reasonable persons would wish to be aware of in order to make informed decision regarding their medical care. The reasonable person standard is also considered inadequate in the research setting. This is supported by the reason that the research subject is typically a volunteer, who may wish to know a lot more about risks undertaken voluntarily than patients who are seeking needed care from a clinician. A third standard was also proposed as the reasonable volunteer. The reasonable volunteer standard requires that the nature and extent of information should entail that such persons aware that the procedure is neither necessary for their care nor fully comprehended, can decide whether they wish to participate in advancing of knowledge. More so, the subjects should understand clearly the range of risks and the voluntary nature of participation in research settings, when some direct benefit to them is anticipated.[35]

A related issue pertinent to the disclosure of information is that some types of research involve incomplete disclosure. A typical example of such type of research is the therapeutic use of placebos. The therapeutic use of placebos usually entails intentional deception or incomplete disclosure. A placebo is a substance or

---

[30] National Bioethics Advisory Commission, (NBAC*), Ethical and Policy Issues in International Research: Clinical Trials in Developing Countries,* http://bioethics.georgetown.edu/nbac/clinical/chap3.html (Accessed July 19, 2013).

[31] The National Commission for the Protection of Human Subjects of Biomedical and Behavioral Research, *The Belmont Report: Ethical Principles and Guidelines for the Protection of Human Subjects* (Washington, D.C.: U.S. Government Printing Office, 1978), 11.

[32] Beauchamp and Childress, *Principles of Biomedical Ethics*, 81.

[33] Beauchamp and Childress, *Principles of Biomedical Ethics,* 81–82.

[34] The National Commission for the Protection of Human Subjects of Biomedical and Behavioral Research, *The Belmont Report: Ethical Principles and Guidelines for the Protection of Human Subjects*, 11.

[35] The National Commission for the Protection of Human Subjects of Biomedical and Behavioral Research, *The Belmont Report: Ethical Principles and Guidelines for the Protection of Human Subjects*, 11.

intervention considered by the health care professional as biomedically or pharmacologically inactive for the condition being treated.[36]

Data also shows that an improvement in the patient after use of a placebo usually referred to as the placebo effect can occur in some situations without nondisclosure, incomplete disclosure, or deception. However, in many cases a placebo is less probably to be effective, if utilized with the knowledge of the patient.[37] In such cases informing subjects of some relevant aspects of the research is more likely to weaken the validity of the research. It is enough in such cases to notify subjects that they are being invited to participate in research of which some aspects will not be disclosed until the research is finished.[38] In all cases of research, incomplete disclosure should only be justified under the following conditions: (1) it is important to obtain vital information; (2) no significant risk is involved; (3) subjects are informed that deception or incomplete disclosure is part of the research, and (4) subjects consent to participate under these conditions.[39] Information regarding research risks should never be withheld for the purpose of obtaining the cooperation of subjects. Attention should be paid to differentiate cases in which disclosure would damage the research from cases in which disclosure would simply inconvenience the researcher.[40] A related issue in the discussion of disclosure requirements focuses on the impact of cultural differences in determining the scope of information to be disclosed to potential subjects of research in developing countries.

Cultural Barriers and Disclosure Requirement

Cultural barriers relating to disclosure requirement can constitute a challenge in obtaining informed consent in developing countries. Macklin acknowledges challenges encountered by researchers who conduct clinical trials in developing countries, in dealing with cultural practices that depart from the requirements of informed consent expressed in international and national research guidelines and regulations.[41] Ethical relativists argue for the need to withhold key information from potential research subjects. They contend that departures from substantive ethical standards of voluntary informed consent are justified by the cultural context in the country or community hosting the research. They further argue that cultural relativity justifies ethical relativism.[42] On the contrary, the type of relativism supported by

[36] Beauchamp and Childress, *Principles of Biomedical Ethics*, 84.

[37] Beauchamp and Childress, *Principles of Biomedical Ethics*, 85.

[38] The National Commission for the Protection of Human Subjects of Biomedical and Behavioral Research, *The Belmont Report: Ethical Principles and Guidelines for the Protection of Human Subjects*, 12.

[39] Beauchamp and Childress, *Principles of Biomedical Ethics*, 88.

[40] The National Commission for the Protection of Human Subjects of Biomedical and Behavioral Research, *The Belmont Report: Ethical Principles and Guidelines for the Protection of Human Subjects*, 12.

[41] Macklin, *Double Standards in Medical Research in Developing Countries*, 139–140.

[42] Macklin, *Double Standards in Medical Research in Developing Countries*, 140.

many other scholars is the increasing need to adapt the form and content of procedures for obtaining informed consent to the educational level and understanding of the potential research subjects. The justification of this perspective stems from the fact that the method and type of informed consent must be relative to the literacy level of the subjects, their ease with signing documents and other cultural conditions.[43]

Deviating from the accepted standard of disclosure of information in the voluntary informed consent process has sometimes been justified by cultural considerations especially in developing countries. It is still accepted in some developing countries for physicians to withhold certain information such as diagnoses and prognoses of cancer and other serious medical conditions from patients. Professionals prefer to provide such information to family members.[44] Sugarman et al. articulate, "in one country, complete information about medical diagnoses and prognoses are withheld routinely from patients with certain diseases, such as cancer. Consequently, valid informed consent for either treatment or research participation can be difficult or impossible."[45] Such cultural practices of withholding information from patients in clinical care are pertinent to research subjects who are involved in similar circumstances.

NBAC report recommends that "research should not deviate from the substantive ethical standard of voluntary informed consent."[46] The commission refutes the assertion that cultural standards about the inappropriateness of providing diagnoses and prognoses to patients or research subjects justify deviation from the substantive ethical standard of informed consent in research. The commission further argued that lack of information regarding diagnoses and prognoses by potential subjects impedes understanding of the purpose of research, any potential benefits, the risks of not participating or the alternatives to participation. The potential subjects cannot make an informed decision to enroll in the research without an understanding that they may not receive a proven therapy. It is a departure from the substantive ethical standard of disclosure required for adequate informed consent to enroll individuals in research without giving them the opportunity to understand essential features of the information regarding the research. The practice of disclosing information in different ways in the clinical context does not change the requirements for such disclosure in the research context.[47]

---

[43] Macklin, *Double Standards in Medical Research in Developing Countries,* 140.

[44] National Bioethics Advisory Commission, (NBAC*), Ethical and Policy Issues in International Research: Clinical Trials in Developing Countries,* http://bioethics.georgetown.edu/nbac/clinical/chap3.html (Accessed July 19, 2013).

[45] Jeremy Sugarman, et al., "International Perspectives on Protecting Human Research Subjects," in *Ethical and Policy Issues in International Research: Clinical Trials in Developing Countries*, Vol. II, Commissioned Papers and Staff Analysis, (Bethesda, MD: National Bioethics Advisory Commission, 2001), 6.

[46] National Bioethics Advisory Commission, (NBAC), *Ethical and Policy Issues in International Research: Clinical Trials in Developing Countries*, http://bioethics.georgetown.edu/nbac/clinical/chap3.html (Accessed July 19, 2013).

[47] National Bioethics Advisory Commission, (NBAC), *Ethical and Policy Issues in International Research: Clinical Trials in Developing Countries*, http://bioethics.georgetown.edu/nbac/clinical/chap3.html (Accessed July 19, 2013).

Cultural Barriers and Disclosure of Risks and Research Study Design

Another issue presented by cultural differences is the disclosure of potential risks and harms associated with treatment or research. International guidelines for informed consent require that all potential risks including the possibility of death must be disclosed to potential subjects in the informed consent process. Marshall acknowledges that disclosing all potential risks associated with research in a direct way may be alarming and frightening for many individuals. She further describes frustrations of Nigerian researchers regarding the lengthy and complex disclosure requirements of informed consent which included information that were considered irrelevant and culturally inappropriate for potential subjects. One of the researchers indicated that in Nigerian cultural norms, disclosing all possible risks would unnecessarily scare potential research subjects associated with the research.[48]

Furthermore, the language utilized to communicate risks in informed consent documents may be difficult to understand because of different views that researchers and the general public have regarding the idea of risks. Researchers, unlike the general public usually understand risks in terms of statistical probabilities. More so, it may be hard to communicate potential risks that cannot be easily measured or that may be difficult for individuals to comprehend or realistically anticipate. For instance, the risks of side-effects from taking medications utilized in a protocol may raise some concerns among some people, but there is a clear connection between the procedure, that is, taking the medicine and the possibility of risk, that is, getting sick from the drug.[49] On the other hand, it may be difficult to communicate the potential for group risks that might occur in the future in genetic epidemiological research as genetic research findings are reported.[50]

A related issue in the discussion of the communication of risks is cultural and social factors that impact beliefs about what really constitutes a risk or potential harm. For example, in developed countries drawing a sample of blood is depicted usually in consent documents as only posing a minimal risk for individuals. But, in developing countries blood and other bodily fluids or tissues are considered to have great symbolic power. There are sometimes concerns that they may be used in sorcery practices or in other means to harm people. This belief heightens the perception of risks among potential subjects that it affects their understanding and signing of informed consent in HIV/AIDS clinical studies.[51] Researchers from Nigeria that conducted community based studies on diabetes and hypertension reported the

---

[48] Marshall, "Ethical Challenges in Study Design and Informed Consent for Health Research in Resource-Poor Settings," 25.

[49] Marshall, "Ethical Challenges in Study Design and Informed Consent for Health Research in Resource-Poor Settings," 26.

[50] Marshall, "Ethical Challenges in Study Design and Informed Consent for Health Research in Resource-Poor Settings," 26.

[51] Mystakidou et al., "Ethical and Practical Challenges in Implementing Informed Consent in HIV/AIDS Clinical Trials in Developing or Resource- Limited Countries." 49.

concerns of patients regarding the amount of blood drawn and the possibility that blood samples could be utilized in sorcery practices.[52] Concerns among Kenyan parents regarding the amount of blood drawn from their children for research were reported. Furthermore, there were indications that some parents were perplexed about the blood samples, due to their belief that blood drawn from their children might be combined and given to other patients.[53]

Cultural barriers are also experienced in clinical trials in developing countries regarding the requirement of disclosure of information about the use of placebo, the randomization of subjects and uncertainty about the efficacy of an experimental intervention. Sugarman et al. describe ideas of local population regarding cultural barriers to randomization and use of placebos. For example, investigators deliberately omitted the use of randomization in their research in one of the cases because they felt it would have posed a major obstacle in obtaining valid informed consent for a randomized trial. Investigators in another case utilized placebos, despite their conception that research subjects did not understand the implication of doing so.[54] NBAC articulates, "despite these barriers, cultural differences do not provide adequate justification for foregoing the requirement to disclose key elements of the nature of the clinical trial, such as the use of a placebo or the randomization of participants into different trial arms."[55] The commission further indicated that the challenges of cultural differences in clinical trials in developing countries for obtaining informed consent do not adequately justify foregoing the requirement for disclosure of alternative therapies available to potential subjects.[56] Cultural differences make it more imperative for investigators to explore innovative strategies for presenting information to participants in order to enhance understanding of disclosed information to research participants.

[52] Patricia A. Marshall, " *The Relevance of Culture for Informed Consent in US-Funded International Health Research*," in *Ethical and Policy Issues in International Research: Clinical Trials in Developing Countries*, Vol. II, Commissioned Papers and Staff Analysis, (Bethesda, MD: National Bioethics Advisory Commission, 2001), 28.

[53] C.S. Molyneux, N. Peshu and K. Marsh, "Understanding of Informed Consent in a Low-Income Setting: Three Case Studies from the Kenyan Coast," *Social Science and Medicine* 59, no. 12 (December 2004): 2547–2559.

[54] Sugarman et al., "International Perspectives on Protecting Human Research Subjects," E 8–9.

[55] National Bioethics Advisory Commission, (NBAC*), Ethical and Policy Issues in International Research: Clinical Trials in Developing Countries*, http://bioethics.georgetown.edu/nbac/clinical/chap3.html (Accessed July 19, 2013).

[56] National Bioethics Advisory Commission, (NBAC*), Ethical and Policy Issues in International Research: Clinical Trials in Developing Countries,* http://bioethics.georgetown.edu/nbac/clinical/chap3.html (Accessed July 19, 2013).

Comprehension of Information

Comprehension is a key element of informed consent process. The notion of informed consent in clinical trials stems from the fact that research subjects giving the consent understand the purpose and nature of the study, what is expected of them and the potential benefits and risks resulting from the study.[57] There are concerns regarding research subjects consenting to participate in clinical trials without adequate understanding about the nature and purpose of the research. The comprehension issue is further exacerbated in international clinical trials where subjects have different language and culture from researchers. The situation is even worse in developing countries where high poverty and low literacy levels and poor access to quality health care services make them vulnerable to exploitation in biomedical research.[58]

Language Barriers and Lower Level of Literacy

Communication between researchers and potential subjects may be difficult to attain when they are from different cultures. Misunderstandings and miscommunication about biomedical research are more likely to happen when researchers and potential subjects speak different languages.[59] Language barriers create significant concerns regarding adequate understanding of the nature and purpose of the research by potential subjects. In most developing countries, people speak and live for the most part of their lives in a different language from the language of the researchers and practitioners, and with an educational level far below desired standard.[60] Interpretation of study purposes involves not only the translation of language, but also cultural.[61] Dawson and Kass observe that potential subjects may lack education or exposure to western scientific concepts in biomedical research and their language might lack terminology for these concepts.[62]

In international clinical trials, informed consent documents are generally translated into the host country's national language and in some instances to the local community language or dialect. Translating sophisticated scientific or medical concepts presents a serious challenge in the comprehension of informed consent. There are some concerns that some local dialects do not have written form. For example,

---

[57] Lema Mbondo and Kamau, "Informed Consent for Clinical Trials: A Review," 137.

[58] Lema, Mbondo and Kamau, "Informed Consent for Clinical Trials: A Review," 137.

[59] Marshall, " The Relevance of Culture for Informed Consent in US-Funded International Health Research," C7.

[60] Mystakidou et al., "Ethical and Practical Challenges in Implementing Informed Consent in HIV/AIDS Clinical Trials in Developing or Resource- Limited Countries." 51.

[61] J.M. Kaufert, R.W. Putsch and M. Lavallee, "Experience of Aboriginal Health Interpreters in Mediation of Conflicting Values in End of Life Decision Making," *International Journal of Circumpolar Health* 57, no 1(1998): 43–48.

[62] L. Dawson and N.E. Kass, "Views of U.S. Researchers about Informed Consent in International Collaborative Research," *Social Science and Medicine* 61, no 6 (September 2005): 1211–1222.

the Bambara in West Africa do not have written form. Furthermore, some scientific or medical concepts do not have direct translation in local languages or dialects. Some examples are randomization, placebo and clinical trial. Attempts by researchers in some cases to explain what one word in a foreign language means in a local language may need a lengthy paragraph, causing the consent document to be too long and not user friendly. The prevalence of low literacy levels among potential subjects in developing countries may make them not to be able to read and understand what they are consenting to.[63]

Language barriers may be significantly reduced by the use of native language interpreters. Elementary language should be utilized in communicating with research subjects, rather than the technical or high level language used in informed consent forms of developed countries. Dialects that adequately accommodate scientific or medical concepts can as well be used.[64] The training of translators in research methods may assist in eradicating or decreasing the introduction of personal interpretations and attitudes.[65]

However, potential problems continue to exist in the use of an interpreter. Marshall identifies a dual problem for health researchers created by a clinical trial requiring a translator. First, the researcher relies on the translator for communicating the research objectives correctly and effectively. Second, the researcher as well relies on the translator to follow through with the consent, which entails presenting the information and consenting to participate in the research. Consent can only be assumed if the respondent agrees to participate, especially in cases where a translator is used.[66]

Generally, there is an assumption that translators are straightforward interpreters of information exchanged between health researchers and subjects. The implication is that this perspective underestimates the complexities of the process of interpretation, which requires the translator to negotiate not only language, but also cultural and contextual factors.[67] Putsch observes that there may be inclination for family members or friends to conceal, overstate, or minimize information, if they are used as translators.[68] Research shows that interpreters have a significant influence on medical interactions and their outcomes. It shows further that native interpreters, apart from mediating the explanatory models of illness held by clinicians and

[63] Lema, Mbondo and Kamau, "Informed Consent for Clinical Trials: A Review," 139.

[64] Mystakidou et al., "Ethical and Practical Challenges in Implementing Informed Consent in HIV/AIDS Clinical Trials in Developing or Resource- Limited Countries." 51.

[65] N.J. Crigger, L. Holcomb and J. Weiss, "Fundamentalism, Multiculturalism and Problems of Conducting Research with Populations in Developing Nations," *Nursing Ethics* 8, no.5 (September 2001): 459–468.

[66] Marshall, "The Relevance of Culture for Informed Consent in US-Funded International Health Research," C7.

[67] Marshall, "The Relevance of Culture for Informed Consent in US-Funded International Health Research," C7.

[68] R.W. Putsch, "Cross-Cultural Communication: The Special Case of Interpreters in Health Care," *Journal of the American Medical Association* 254, no. 23 (December 1985): 3344–3348.

patients, they also usually introduce their own beliefs and personal agendas into the interaction.[69]

It is evident that interpreters wield some degree of influence over the communication between researchers and potential research subjects. Their influence on communication between researchers and potential subjects is clearly shown through the function of gatekeeping, where interpreters make critical decisions regarding the selection of information to communicate the terminology to express concerns, and the clarification of information to fit particular interactions.[70] There are also indications that interpreters impact the communication process by acting as cultural brokers, patient advocates and counselors.[71]

Cultural Beliefs and Health, Disease and Biomedical Procedures

Another major barrier to comprehension of information for potential research subjects is their belief system about health, disease and biomedical procedures. In some cultures, the belief system of potential research subjects does not explain health, and disease utilizing the concepts and terms of modern science and technology.[72] Kass and Hyder acknowledge the overwhelming challenge of obtaining voluntary informed consent when people do not comprehend or accept scientific and western explanations of health and disease.[73] Marshall also indicates that in some circumstances cultural beliefs regarding the cause and treatment of disease may differ completely from western views about underlying disease etiology which are consistent with medical and scientific explanations.[74] She further cites a physician's perspective on the potential subjects' understanding of the nature of the research thus, "... Indeed, what I worry about is whether we are really informing them. We are talking to a society that does not believe in the germ theory of disease so it's difficult to explain."[75] There is a pervasive belief in most cultures of developing countries

---

[69] J.M. Kaufert and J.D. O'Neil, "Biomedical Rituals and Informed Consent: Native Canadians and the Negotiation of Clinical Trust," In *Social Science Perspectives on Medical Ethics*, ed. G. Weisz, (Philadelphia: University of Pennsylvania Press, 1990), 41–64.

[70] D.M. Barnes et al., "Informed Consent in a Multicultural cancer Patient Population: Implications for Nursing Practice," *Nursing Ethics* 5, no. 5 (1998): 412–423.

[71] Marshall, "The Relevance of Culture for Informed Consent in US-Funded International Health Research," C8.

[72] National Bioethics Advisory Commission, (NBAC*), Ethical and Policy Issues in International Research: Clinical Trials in Developing Countries*, http://biorthics.georgetown.edu/nbac/clinical/chap3.html (Accessed July 19, 2013).

[73] Nancy Kass and Adnan A. Hyder, "Attitudes and Experiences of U.S. and Developing Country Investigators Regarding U.S. Human Subjects Regulations," In *Ethical and Policy Issues in International Research: Clinical Trials in Developing Countries,* Vol. II, Commissioned Papers and Staff Analysis, (Bethesda, MD: National Bioethics Advisory Commission, 2001), B4–5.

[74] Marshall, "The Relevance of Culture for Informed Consent in US-Funded International Health Research," C28.

[75] Marshall, "The Relevance of Culture for Informed Consent in US-Funded International Health Research," C28.

that a person's death is usually as result of sorcery, rather than an underlying disease. Potential subjects also believe that illness and disease can be inflicted on individuals by the use of sorcery.[76] Marshall quotes a physician reflecting on patient's belief about the cause of hypertension as follows: "some people ask us what causes hypertension… whether it's inherited or whether it's caused by someone thinking something as in sorcery."[77]

Researchers from developed countries should respect the culture and the belief system of the local population involved in research in developing countries. They need to refrain from attacking their belief system, and focus on explaining the safety and efficacy of the interventions being tested. Sommer articulates a similar view thus, "we do not want to fight a belief system. We simply say we have this pill. We believe it is safe. We think it may reduce the recurrence of the following thing. We would like you to take it."[78]

A related issue in the discussion of barrier to comprehension is the cultural belief about the biomedical procedures, especially with regard to blood drawn for laboratory tests. There are grave concerns that some potential subjects believe that blood drawn for tests could be used for sorcery practices against them. Marshall cites a research assistant's view regarding this issue, "there are concerns about drawing too much blood. People are worried about the effect on their health, and also what you are going to do with it, some might think it could be used for sorcery."[79] Beliefs about the potential harm linked with the misuse of blood specimens play a major role in informed consent process. It affects potential subjects' comprehension of research goals and their decision to participate. Potential subjects should be educated and reassured about the purpose of drawing blood and how it might be used.[80] Their fears should be allayed about the potential harm that may result from their involvement in the research. The comprehension of research goals and procedures can also be impacted by problems, linked with the misunderstanding of potential subjects that participating in clinical trials is the same as receiving routine medical care or treatment.

---

[76] Marshall, "The Relevance of Culture for Informed Consent in US-Funded International Health Research," C28.

[77] Marshall, "The Relevance of Culture for Informed Consent in US-Funded International Health Research," C28.

[78] A. Sommer, *Testimony before NBAC*, September 16, 1999, Arlington, Virginia, Meeting transcript, 214.

[79] Marshall, "The Relevance of Culture for Informed Consent in US-Funded International Health Research," C29.

[80] Marshall, "The Relevance of Culture for Informed Consent in US-Funded International Health Research," C30.

Therapeutic Misconception

A key feature of informed consent to participate in biomedical research is the understanding of the difference between clinical research and ordinary treatment. In some cases, research subjects fail to appreciate the difference between research and treatment, and this condition is dubbed therapeutic misconception.[81] Confusion regarding the purpose of research is critical in any definition of therapeutic misconception. Therapeutic misconception is prevalent when a subject is primarily motivated by a desire for personal benefit to enroll in a research, even in studies with minimum chance of benefit.[82] Applelbaum et al., define therapeutic misconception as "when a research subject fails to appreciate the distinction between the imperatives of clinical research and ordinary treatment, and therefore inaccurately attributes therapeutic intent to research procedures."[83] In the same vein, the NBAC defines therapeutic misconception as "the belief that the purpose of a clinical trial is to benefit the individual patient rather than to gather data for the purpose of contributing to scientific knowledge."[84]

Applebaum and Lidz documented a study in which the patients interviewed were enrolled in clinical trials that involved randomization, placebo, non-treatment control groups and double blind procedures. The study showed that 69% of the research subjects did not know that their allocation to control and experimental groups would be randomized. Furthermore, 40% of the research subjects thought that treatment assignment's decision would be made based on their therapeutic needs. Finally, 44% of the research subjects did not appreciate that the use of placebos and non-treatment control groups implied that some subjects who wanted experimental intervention would not receive it.[85] Literature highlights some essential features which indicate the prevalence of therapeutic misconception in clinical research; people often overestimate the benefit of enrolling in a study, [86] underestimate the risks,[87] are muddled regarding how treatments will be allocated either to control or

---

[81] Paul S. Appelbaum, Loren H. Roth, and Charles Lidz, "The Therapeutic Misconception: Informed Consent in Psychiatric Research," *International Journal of Law and Psychiatry* 5, nos. 3–4 (1982): 319–329.

[82] B.R. Cassileth et al., "Attitudes toward Clinical Trials among Patients and the Public," *JAMA* 248, no. 8 (August 1982): 968–970.

[83] Charles W. Lidz and Paul S. Appelbaum, "The Therapeutic Misconception: Problems and Solutions," *Medical Care* 40, no. 9 (September 2002): V.55-V63.

[84] National Bioethics Advisory Commission, (NBAC), *Ethical and Policy Issues in International Research: Clinical Trials in Developing Countries*, http://bioethics.georgetown.edu/nbac/clinical/chap3.html (Accessed July 19, 2013).

[85] Paul S. Appelbaum and Charles W. Lidz, "Therapeutic Misconception,"" in *The Oxford Textbook of Clinical Research Ethics*, ed. Ezekiel J. Emmanuel et al., (New York: Oxford University, 2008), 637.

[86] Monica H. Schaeffer et al., "The Impact of Disease Severity on the Informed Consent Process in Clinical Research," *The American Journal of Medicine* 100, no. 3 (March 1996): 261–268.

[87] S. Joffe et al., "Quality of Informed Consent in Cancer Clinical Trials: A Cross-Sectional Survey," *Lancet* 358, no. 9295 (November 2001): 1772–1777.

experimental group,[88] and usually inclined to confuse research with ordinary treatment.[89]

However, some authors do not agree that overestimation of direct benefit or underestimation of risk or both together constitutes an essential feature of therapeutic misconception. Horng and Grady contend that there is a clear distinction between therapeutic misconception and therapeutic misestimation. They argue that therapeutic misconception focuses on the nature or intent of clinical research, while therapeutic misestimation deals with misunderstanding the probability of direct benefit or risk that may result from participating in research. They further contend that the heterogeneity of clinical trial design makes inferences about realistic expectation of direct benefits very challenging.[90] Some authors have identified five draft dimensions of research that trial participants should understand before enrolling in clinical research including scientific purpose, study procedures, uncertainty, adherence to protocol and clinician as investigator. They indicate that specific questions to assess therapeutic misconception should be developed within the scope of five draft dimensions of research already outlined.[91] Potential subjects who lack adequate understanding of the purpose and methods of research are not equipped to evaluate risks and benefits of research participation. They also may not be able to appreciate how personal care may be compromised by research procedures such as randomization and use of placebos.[92]

The prevalence of therapeutic misconception in clinical trials conducted in developing countries is exacerbated by the severity of the disease. Study conducted by Schaeffer et al., shows that disease severity affects comprehension of information, and that the most severely sick research subjects are likely to attribute therapeutic intent to research that has remote chance of benefit as in phase 1 trials. They postulate that a subject with an immediate life-threatening disease and no therapeutic alternative might retain less information from a consent document, than a subject with less severe disease and more therapeutic options. In the same vein, research subjects with life-threatening conditions are less autonomous than healthy volunteers in their decision making. Their motivation for enrolling in the trials is care of their health condition and may consent without considering the potential risks involved. The study shows that healthy volunteers retained the most information about risks and side effects, while severely ill subjects retained the least. Furthermore,

[88] C. Snowdon, J. Garcia and D. Elbourne, "Making Sense of Randomization: Responses of Parents of Critically Ill Babies to Random Allocation of Treatment in a Clinical Trial," *Social Science and Medicine* 45, no. 9 (November 1997): 1337–1355.

[89] K. Cox, "Informed Consent and Decision-making: Patients' Experiences of the Process of Recruitment to Phase I and II Anti-Cancer Drug Trials," *Patient Education and Counseling* 46, no. 1 (January 2002): 31–38.

[90] Sam Horng and Christine Grady, "Misunderstanding in Clinical Research: Distinguishing Therapeutic Misconception, Therapeutic Misestimation, and Therapeutic Optimism," *IRB: Ethics and Human Research* 25, no. 1 (January–February 2003): 11–16.

[91] Gail E. Henderson et al., "Clinical Trials and Medical Care: Defining the Therapeutic Misconception," *PLoS Medicine* 4, no. 11 (November 2007): 1737.

[92] Lidz and Appelbaum, "The Therapeutic Misconception: Problems and Solutions," V.55–V63.

more sick subjects than healthy volunteers reported that the informed consent document had no effect on their decision to participate in the trial. Subjects with severe health conditions like cancer and HIV/AIDS may enroll in clinical trials with goals that are different from the goals of the research. Research subjects with advanced disease conditions rate the consent document as less relevant in their decision to participate in a trial which was primarily motivated by expectations and hopes of recovery. They have poor retention of disclosed risk information probably due to denial of unpleasant realities, or avoidance of disturbing thoughts related to risk information.[93]

The onslaught of HIV/AIDS and lack of adequate access to anti-retroviral treatments in developing countries for most of the population result in desperation for potential subjects. In most developing countries, potential subjects are usually not able to access adequate and quality treatment outside the research context. The implication is that potential subjects in resource limited countries count solely on being enrolled in clinical trials in order to access better health care. NBAC affirms, "it is not a misconception to believe that participants probably will receive good clinical care during research. But, it is a misconception to believe that the purpose of clinical trials is to administer treatment, rather than to conduct research."[94] Despite potential barriers to adequate comprehension of information for research subjects in clinical trials conducted in developing countries, there is an indication that those barriers can be surmounted by innovative ways of presenting information to potential subjects.

Innovative Strategies

Comprehension of information by research subjects can be enhanced when researchers engage the community in which research is conducted in active discussions of project goals and procedures through meetings with community leaders or public forums, and when information is provided to potential subjects prior to obtaining informed consent.[95] Woodsong and Karim indicate that community involvement is a prerequisite to achieving high quality informed consent especially in circumstances in which the cultural norms of the researchers and participants differ significantly. They proposed a model designed to enhance informed consent process that occurs during three phases of the research period, pre-enrollment, enrollment and post-enrollment, and at two levels, individual and community. Individual participants are familiarized with a study before enrollment, and the larger community in which participants are drawn are also involved in order to be cognizant of and

[93] Schaeffer et al., "The Impact of Disease Severity on the Informed Consent Process in Clinical Research," 261–268.

[94] National Bioethics Advisory Commission, (NBAC), *Ethical and Policy Issues in International Research: Clinical Trials in Developing Countries*, http://bioethics.georgetown.edu/nbac/clinical/chap3.html (Accessed July 19, 2013).

[95] Marshall, "Ethical Challenges in Study Design and Informed Consent for Health Research in Resource-Poor Settings," 25.

support the research effort.[96] Respecting the community and its values is essential. More so, the research protocol should start and end with the community.[97] In another study of HIV-1 transmission in Haiti, Fitzgerald et al. report that the understanding of the content of the consent forms by research participants increased considerably after meetings with a counselor in which information was provided concerning the study. In this study, 80% of the 30 individual participants passed an oral examination before enrollment in the research project.[98]

Comprehension of information by research subjects can also be enhanced through consultation with cultural experts and local representatives concerning the most effective ways of communicating with potential research participants regarding the purpose of the study and the importance of obtaining consent.[99] Researchers can conduct focus groups with representatives of potential subjects for comprehending issues and concerns related to preparing the consent form and developing approaches to obtaining consent.[100]

Comprehension may be increased using a continuous consent process. Vallely et al. write, "providing information to trial participants in a focused, locally appropriate manner, using methods developed in consultation with the community, and within a continuous informed consent framework resulted in high levels of comprehension and message retention in this setting."[101] They describe a study aimed at investigating the effectiveness of a continuous informed consent process adopted during microbicide trial in Mwanza, Tanzania. In this study, participatory community research methods were used to develop a locally-appropriate pictorial flipchart in order to communicate key messages regarding the trial to potential participants. Pre-recorded audio tapes were also used to promote understanding and compliance with instructions pertinent to the trial. A comprehension checklist was also administered to all participants at different stages of the trial. In depth interviews were used to measure how well participants internalize and retain key messages provided in a continuous informed consent process.[102]

---

[96] Cynthia Woodsong and Quarraisha Abdool Karim, "A Model Designed to Enhance Informed Consent: Experiences from the HIV Prevention Trials Network," *American Journal of Public Health* 95, no. 3 (March 2005): 413.

[97] Crigger, Holcomb and Weiss, "Fundamentalism, Multiculturalism and Problems of Conducting Research with Population in Developing Nations," 459–468.

[98] D.W. Fitzgerald et al., "Comprehension during Consent in a less developed Country," *Lancet* 360, no. 9342 (October 2002): 1301–1302.

[99] P.A. Marshall and C. Rotimi, "Ethical Challenges in Community Based Research," *American Journal of Medical Sciences* 322, no. 5 (November 2001): 259–263.

[100] Marshall, "Ethical Challenges in Study Design and Informed Consent for Health Research in Resource-Poor Settings," 25.

[101] Andrew Vallely et al., "How Informed is Consent in Vulnerable Populations? Experience Using a Continuous Consent Process during the MDP301 Vaginal Microbicide Trial in Mwanza, Tanzania," *BMC Medical Ethics* 11, no. 10 (June 2010): 1–15.

[102] Vallely et al., "How Informed is Consent in Vulnerable Populations? Experience Using a Continuous Consent Process during the MDP301 Vaginal Microbicide Trial in Mwanza, Tanzania," 1–15.

Researchers also facilitate comprehension by using concepts and terms understandable to the community hosting the research, to explain complex issues in biomedical research. For example, the principle of randomization and the possibility that one of the vaccines might fail were explained with a familiar agricultural example, the analogy of testing fertilizers or new seeds on randomized plots, a procedure familiar to farmers in the area.[103] The concepts of immunology and immune cells were explained with familiar analogy, people who guard houses, but it's a particular kind of watchman in your body.[104] Pertinent to note also is that adequate comprehension of information by research participants facilitates either voluntary informed consent or refusal.

Voluntary Participation

The requirement of voluntary participation in clinical research is a critical component of informed consent. Clinical research in developing countries is confronted with several challenges especially with regard to voluntary consent. Marshall highlights four major challenges to voluntary consent, (1) the ability of a person to understand study objectives and the risks involved; (2) the vulnerability of potential subjects to incentives including money, drugs or medical treatment; (3) the power and authority of researchers to impact a potential subject's decision to participate due to their professional background and social status; (4) the influence of community pressure to participate in a study especially when community elders give permission.[105] In developing countries, diminished autonomy for research participants is prevalent. There are concerns that due to high levels of poverty in developing countries that payment provided to research participants may unduly induce them to enroll in HIV/AIDS clinical trial.[106] Payment is possibly viewed as coercion when those recruited are very poor or if the benefits are considerable.[107]

Furthermore, voluntary participation in research is more challenging for potential participants when no other treatment options are available. In developing countries, people may feel unduly induced to enroll in HIV clinical research because of limited affordable access to anti-retroviral drugs. They consider enrolling in clinical research as the only option to receive anti-retroviral treatment which makes voluntary informed consent or refusal more difficult. However, we need to resist the line

---

[103] M.P. Preziosi et al., "Practical Experiences in Obtaining Informed Consent for a Vaccine Trial in Rural Africa," *New England Journal of Medicine* 336, no. 5 (January 1997): 370–373.

[104] Kass and Hyder, "Attitudes and Experiences of U.S. and Developing Country Investigators Regarding U.S. Human Subjects Regulations," B–26.

[105] Marshall, "The Relevance of Culture for Informed Consent in US-Funded International Health Research," C–32.

[106] Mystakidou et al., "Ethical and Practical Challenges in Implementing Informed Consent in HIV/AIDS Clinical Trials in Developing or Resource- Limited Countries." 50.

[107] Crigger, Holcomb and Weiss, "Fundamentalism, Multiculturalism and Problems of Conducting Research with Population in Developing Nations," 459–468.

of argument that one's consent is not voluntary just in case one has no acceptable alternative.[108] Voluntary participation in research requires conditions free of coercion and undue inducement or influence. A detailed analysis of the impact of coercion and undue inducement in voluntary choice to participate in clinical research follows.

Coercion and Undue Inducement

Coercion occurs when one person intentionally uses a credible and severe threat of harm or force to control another or to compel him or her to do something.[109] Coercion is also defined as the presence of a threat of harm or force that could make the coerced person worse off in some way.[110] On the other hand, undue influence deals with an offer that is considered excessive, unwarranted or inappropriate or improper reward or other overture for obtaining compliance.[111] Beauchamp and Childress identify three forms of influence, coercion, persuasion and manipulation that can void autonomous decision to participate in clinical research.[112]

Most decisions that individuals make, including the decision whether to participate in a research study, are affected by multiple influences. Generally, people choose and act in consonance with their wants and needs, which are usually influenced by their physical, psychological, social, economic, and cultural experiences and circumstances.[113] In a similar vein, Faden and Beauchamp acknowledge that influences on peoples' decisions can come in many forms, and from many sources. They can differ significantly in degree of influence actually exercised.[114] Some influences may be adequately strong to constitute inducements, motivations, or stimuli for action. It is also pertinent to note that inducements do not necessarily invalidate or preclude voluntary choice. We encounter and respond to inducements all of the time in various areas of life, including selecting employment, making purchases, participating in research, and other choices. There is usually no single reason for doing something, since human motivation is complex and most times entails multiple considerations.[115]

---

[108] Alan Wertheimer, "Exploitation in Clinical Research," in *Exploitation and Developing Countries: The Ethics of Clinical Research*, ed. Jennifer S. Hawkins and Ezekiel J. Emmanuel (Princeton, New Jersey: Princeton University, 2008), 91.

[109] Beauchamp and Childress, *Principles of Biomedical Ethics*, 94.

[110] A. Wertheimer, *Coercion* (Princeton, New Jersey: Princeton University, 1987), 7.

[111] The National Commission for the Protection of Human Subjects of Biomedical and Behavioral Research, *The Belmont Report: Ethical Principles and Guidelines for the Protection of Human Subjects*, 14.

[112] Beauchamp and Childress, *Principles of Biomedical Ethics*, 94.

[113] Christine Grady, "Money for Research Participation: Does it Jeopardize Informed Consent?," *The American Journal of Bioethics* 1, no. 2 (Spring 2001): 41.

[114] Ruth Faden and Tom Beauchamp, *A History and Theory of Informed Consent* (New York: Oxford University Press, 1986), 256.

[115] Grady, "Money for Research Participation: Does it Jeopardize Informed Consent?," 41.

Coercion in the sense of researchers threatening to make anyone worse off for refusing to participate in a research study is not a common problem. Coercion may be an issue in relation to research conducted with prisoners or other captive populations, in which refusal to participate in research could result in punishment or retaliation. However, most institutional review boards (IRBs) ban threats of harm for refusal, whenever a power differential exists between researcher and participant. Even in circumstances when there is no threat, people may sometimes fear they will be treated worse. Both perceived and real coercion do not have any link to payment because payment should never be a threat itself. There may be possibility of a third party coercion in some cases, if the spouse of someone refusing to participate in a paid research study threatens the refusing spouse. But, in such cases researchers should not be held responsible for coercion. Payment may decrease perceived coercion in doctor-patient relation by completely changing the exchange into one that is less personal and unrelated to medical care.[116]

There is an ongoing debate as to what makes certain offers undue. Generally, offers are considered unduly influential if they are so enticing that they lead individuals to participate in clinical research they would otherwise preferred not to participate.[117] Certain conditions may raise concerns for the possibility of undue inducements to participate in clinical research, even if they pose significant risk of harm. Such conditions may include offers of medical care not otherwise available or offers of money. CIOMS guidelines document recognizes that, "it may be difficult to distinguish between suitable recompense and undue influence to participate in research… someone without access to medical care may or may not be unduly influenced to participate in research simply to receive such care."[118] This situation is prevalent in developing countries, in which most people living with HIV/AIDS have limited or no access to anti-retroviral treatments. In general terms, the provision of medical care or treatment that would not otherwise be available to research participants should not be interpreted as an undue influence to participate.[119] Researchers from developing countries surveyed by Kass and Hyder supported this conclusion. In the survey, 64 percent of the researchers indicated that participants joined research projects in order to obtain compensation, medical care or other benefits.[120] Most researchers interviewed in focus groups for this same study were of the opinion that it was satisfactory, given the general risk/benefit ratio of the research. Some focus group respondents mentioned that providing significant benefits basically gave

[116] Neal Dickert and Christine Grady, "Incentives for Research Participants," in *The Oxford Textbook of Clinical Research Ethics,* ed. Ezekiel J. Emmanuel et al., (New York: Oxford University, 2008), 388.

[117] Dickert and Grady, "Incentives for Research Participants," 388.

[118] Council for International Organizations of Medical Sciences, *International Ethical Guidelines for Biomedical Research Involving Human Subjects*, 46.

[119] National Bioethics Advisory Commission, (NBAC), *Ethical and Policy Issues in International Research: Clinical Trials in Developing Countries*, http://bioethics.georgetown.edu/nbac/clinical/chap3.html (Accessed July 19, 2013).

[120] Kass and Hyder, "Attitudes and Experiences of U.S. and Developing Country Investigators Regarding U.S. Human Subjects Regulations," B–37.

potential participants no reasonable choice except to participate, but they did not construe this as undue inducement.[121] NBAC indicates that even though the potential benefits of participation in research for those in developing countries who lack access to medical care may be an inducement to participate in research, this does not adequately diminish the voluntariness of their decision in a way that would make their consent ethically invalid.[122]

On the other hand, being attracted to the money offered for participation in a research does not necessarily exclude the possibility of other influential motivations and considerations. Research subjects may participate in research for numerous reasons other than money.[123] Grady argues, "if inducements can be compatible with voluntary choice, then money, as an inducement, does not inherently obviate or compromise voluntariness."[124] Apparently, most subjects who are attracted to participate in research partly because of money, still have the freedom to refuse. Potential subjects are usually advised of their right to exercise this freedom in the process of obtaining voluntary informed consent. They are reminded about their voluntary choice to participate and that they have the right to refuse or withdraw at any time without punishment.[125]

Furthermore, many people who are attracted to research because of money usually have other options for acquiring money, generally from other full or part-time unskilled jobs. Potential subjects may choose research participation because of other considerations such as flexible hours, limited time, or that it seems more interesting and easier. More so, subjects who volunteered to participate in research can exercise their freedom to refuse when they decide that participating in the particular research study is not in their advantage.[126] Concerns about potential for money serving as an undue inducement in the sense of making research an irresistible offer have been acknowledged for persons who are poor and have no other means of obtaining comparable amounts of money. Eliminating the option of obtaining money through research participation which has been propounded by some people does not resolve the issue. In the process of obtaining informed consent, the emphasis is better focused on more appropriate and effective strategies such as, a subject's reason for participating, his or her understanding and expectations of research, and his or her impression of freedom to choose whether to participate or not.[127]

The ethical concern about money being an inappropriate motivating factor for research participation has been identified. Money is known for getting people to do

---

[121] Kass and Hyder, "Attitudes and Experiences of U.S. and Developing Country Investigators Regarding U.S. Human Subjects Regulations," B–38.

[122] National Bioethics Advisory Commission, (NBAC), *Ethical and Policy Issues in International Research: Clinical Trials in Developing Countries*, http://bioethics.georgetown.edu/nbac/clinical/chap3.html (Accessed July 19, 2013).

[123] Grady, "Money for Research Participation: Does it Jeopardize Informed Consent?," 41.

[124] Grady, "Money for Research Participation: Does it Jeopardize Informed Consent?," 41.

[125] Grady, "Money for Research Participation: Does it Jeopardize Informed Consent?," 41.

[126] Grady, "Money for Research Participation: Does it Jeopardize Informed Consent?," 41–42.

[127] Grady, "Money for Research Participation: Does it Jeopardize Informed Consent?," 42.

things they would prefer not to do, and in some instances for getting people to do something they know is wrong. Money is as well considered capable of inappropriately distorting people's judgments and motivations.[128] The U.S. Office of Protection from Research Risks (OPRR), currently known as the Office of Human Research Protections (OHRP) points out the problematic nature of money as a possible undue inducement for research participation. In this context, money can diminish or weaken an individual's judgment about what is at stake in the research or blind him or her to the potential risks of research participation. OPRR also indicates that offers of money could influence potential participants to distort something about themselves in order to acquire or maintain enrollment in a research study and receive the money. Distortion may not only endanger the informed consent of participants, but perhaps also their well-being as well as the integrity of the study.[129] The vulnerability of potential subjects to distorted judgment because of money is relative not only to their particular circumstances but more importantly to their values. The implication is that even in very desperate situations, some people cannot be bought. Nevertheless, the bigger the sum of money involved, the greater the tendency for altering judgment or prompting potential participants to lie or ignore risks.[130] CIOMS notes, "the payments should not be so large … or the medical services so extensive as to induce prospective subjects to consent to participate in the research against their better judgment."[131] There is a consensus among some scholars that limiting the amount of money paid for research participation decreases the possibility that money will alter judgment and induce people to engage in deception.[132] Payment to research participants as acknowledgement for their contribution which may be calculated according to locally acceptable standard is probably more modest. The implication is that modest payment is less likely to alter judgment than amounts designed exclusively to attract subjects and exceed the competition in relation to recruitment.[133] Random or huge amounts of money intended clearly to attract, to overpay other studies, or to compensate for risk should not be allowed. Modest payment considered as compensation for the participant's contribution decreases the likelihood of undue inducement, because the offer of money is neither excessive nor inappropriate.[134] Voluntary participation in clinical research can also be enhanced

---

[128] Grady, "Money for Research Participation: Does it Jeopardize Informed Consent?," 42.

[129] U.S. Department of Health and Human Services, *Office of Protection from Research Risks, Protecting Human Research Subjects: IRB Guidebook*, (Washington: Government Printing Office, 1993), 3–44.

[130] Grady, "Money for Research Participation: Does it Jeopardize Informed Consent?," 43.

[131] Council for International Organizations of Medical Sciences, *International Ethical Guidelines for Biomedical Research Involving Human Subjects*, 45.

[132] Ruth Macklin, "Due" and "Undue" Inducements: On Paying Money to Subjects," *IRB: A Review of Human Subjects* 3, no. 5 (1981): 1–6.

[133] Neal Dickert and Christine Grady, "What's the price of a research subject? Approaches to Payment for Research Participation," *The New England Journal of Medicine* 341, no. 3 (July 1999): 198–203.

[134] Grady, "Money for Research Participation: Does it Jeopardize Informed Consent?," 43.

or impaired by influences related to decisional authority for consent to research in developing countries.

Decisional Authority

Freedom of choice and personal decision-making are usually emphasized in the discussion of voluntary informed consent. Marshall articulates that, "beliefs about individual autonomy and decisional capacity are embedded within the social and cultural patterns of community obligations and family ties."[135] Personal autonomy is highlighted in western industrialized countries. The implication is that individuals are anticipated to make decisions about research participation for themselves or through chosen surrogates. On the other hand, in numerous non-western countries, family members, or community leaders may play a major part in decisions regarding medical care and medical research.[136] There are two points to be considered, first, influences from community leaders; and second, influences from family members in the decision to participate in clinical research.

In most cultures in developing countries researchers must seek permission from a community leader or village council before any interactions with potential research subjects. A clear distinction should be made between obtaining permission to enter a community for conducting research and for acquiring individual voluntary informed consent.[137] CIOMS highlights, the importance of meeting with community leaders, councils of appointed or elected elders or other designated authorities to seek formal or informal approval of a study in some contexts, but also emphasizes that it does not replace obtaining individual voluntary consent.[138] In meeting with community and village leaders, researchers explain and discuss the details of the proposed study with them. The leaders are given opportunity to discuss and ask questions regarding the proposed study, before reaching a consensus whether to approve or reject permission for the research to be conducted in their community. When the permission has been granted by community and village leaders, the researcher could easily approach individuals for their participation. Individuals have the choice to decline participation in the research despite their village or community leader's approval of the study.[139]

While researchers are encouraged to obtain permission from community leaders or designated authorities before engaging in research, an ethical problem is

[135] Marshall, "*Ethical Challenges in Study Design and Informed Consent for Health Research in Resource-Poor Settings,*" 27.

[136] Marshall, "Ethical Challenges in Study Design and Informed Consent for Health Research in Resource-Poor Settings," 27.

[137] National Bioethics Advisory Commission, (NBAC), *Ethical and Policy Issues in International Research: Clinical Trials in Developing Countries*, http://bioethics.georgetown.edu/nbac/clinical/chap3.html (Accessed July 19, 2013).

[138] Council for International Organizations of Medical Sciences, *International Ethical Guidelines for Biomedical Research Involving Human Subjects*, 35.

[139] Kass and Hyder, "Attitudes and Experiences of U.S. and Developing Country Investigators Regarding U.S. Human Subjects Regulations," B–79.

encountered when the community leader wields undue influence on the community in a way that impedes the voluntariness of individual consent.[140] In some situations, the head of the village or a group of elders makes a joint decision for the village. The implication is that almost everyone will participate if they make decision to approve participating in the research. The people in the community are very unwilling to withdraw from the research because of the shared nature of community activities.[141] Marshall acknowledges that in some settings, authorization by the community leader is consistent with individuals' right to decline and authorization in a context in which the chief has the final say.[142] One researcher articulated two levels of consent or permission: "One is community and the other is individual…. When you leave the chief, the chief is expected to open households, so there is really another level of consent in between…the chief and council, the household head, then the individual."[143] The impact of the approval of the study by a chief on the community response was explored. One researcher indicated that community members for the most part consent to participate when researchers obtain prior approval from the chief or household heads. There were also some doubts regarding the degree to which individual consent to participate is voluntary.[144]

There were also discussions on differences between the rural and urban settings concerning the significance of obtaining permission from local community leaders. The strict requirement of community consent is stronger in rural setting than urban setting.[145] One physician commenting on the difference between the process of obtaining consent in urban and rural settings indicated: "In the rural area, community consent is stronger than the urban area. In Ibadan, some neighborhoods have traditional leadership, some modern, some have a traditional chief and the community structure still holds."[146] In some cultures, the processes for recruiting participants include community leaders who use their authority to prevent individuals from refusing to participate in research for which permission has been granted. Furthermore, authoritarian governments in some countries may restrict autonomous decision-making by their citizens, which may influence their participation in

[140] National Bioethics Advisory Commission, (NBAC), *Ethical and Policy Issues in International Research: Clinical Trials in Developing Countries*, http://bioethics.georgetown.edu/nbac/clinical/chap3.html (Accessed July 19, 2013).

[141] National Bioethics Advisory Commission, (NBAC*), Ethical and Policy Issues in International Research: Clinical Trials in Developing Countries,* http://bioethics.georgetown.edu/nbac/clinical/chap3.html (Accessed July 19, 2013).

[142] Marshall, "The Relevance of Culture for Informed Consent in US-Funded International Health Research," C–17.

[143] Marshall, "The Relevance of Culture for Informed Consent in US-Funded International Health Research," C–17.

[144] Marshall, "The Relevance of Culture for Informed Consent in US-Funded International Health Research," C–17–18.

[145] Marshall, "The Relevance of Culture for Informed Consent in US-Funded International Health Research," C–19.

[146] Marshall, "The Relevance of Culture for Informed Consent in US-Funded International Health Research," C–19.

research.[147] A related issue with regard to decisional capacity for the consent to participate in research in developing countries is the influences from family members in personal decision-making.

In developing countries, family-centered decision-making is customary than individual decision-making.[148] Family members of potential research participants are usually involved in the informed consent process. Potential research participants generally seek permission from a family member in order to be enrolled in clinical trials.[149] In most cases, the need to include the family is not intended as a replacement for individual consent, but rather as an additional step in the process. For example, researchers in one country pointed out that research participants frequently become doubtful of researchers who attempt to enroll participants in a biomedical study without involving the family in the decision-making process.[150] Sugarman et al. also described a research project in another country that involves a multistep consent process which starts with community consent and followed by individual parental consent for research involving children. Lastly, village elders were involved in sessions with children, due to community worries that children might be kidnapped and used as servants or be subjected to harvesting of their organs.[151]

Loue, Okello and Kawuna acknowledge the importance of involving family members in the informed consent process in Uganda. Even though Ugandan civil law requires an 18 year old male residing at home to make his decisions, it is normal for the son to seek his father's consent before engaging in any obligation or contract, including research participation. Furthermore, some Ugandan women seek the approval of their husband prior to making a decision concerning their participation in research. In numerous traditional societies in developing countries, obtaining approval from one's husband may be routine.[152] In another study of anti-malarial drugs conducted in Kenya, the research assistants indicated that the women were hesitant to discuss with the researchers before obtaining approval from their husbands.[153]

Researchers in developing countries emphasize the need to involve the family members in the informed consent process, without compromising the requirement of individual voluntary informed consent. For example, Marshall indicated that obtaining the approval of a woman's husband before enrolling her in a research

[147] Kass and Hyder, "Attitudes and Experiences of U.S. and Developing Country Investigators Regarding U.S. Human Subjects Regulations," B–28–29.

[148] Mystakidou et al., "Ethical and Practical Challenges in Implementing Informed Consent in HIV/AIDS Clinical Trials in Developing or Resource- Limited Countries." 50.

[149] Mystakidou et al., "Ethical and Practical Challenges in Implementing Informed Consent in HIV/AIDS Clinical Trials in Developing or Resource- Limited Countries." 50.

[150] Sugarman et al., "International Perspectives on Protecting Human Research Subjects," E–6.

[151] Sugarman et al, "International Perspectives on Protecting Human Research Subjects," E–6.

[152] Sana Loue, David Okello and Medi Kawuma, "Research Bioethics in the Ugandan Context: A Program Summary," *Journal of Law, Medicine and Ethics* 24, no. 1 (March 1996): 47–53.

[153] Marshall, "Ethical Challenges in Study Design and Informed Consent for Health Research in Resource-Poor Settings," 41.

might be a requirement in Nigeria, where traditional cultural norms are strong. A Nigerian physician engaged in a breast cancer study described that cancer patients frequently require the permission of their husbands to participate in clinical research. Nevertheless, the physician also stressed that in such cases, the individual consent of the woman is still crucial. It is also pertinent to note that most researchers have devised strategies that accommodate and encourage discussion about study participation with family members. In the hypertension study, for instance, the study is explained to patients and they are handed over the information to take home. The patient is given an appointment for a later date in order to obtain his or her consent.[154]

Marshall also explained the difficulties involved in the negotiation of permission. For example, in one case, a woman described different strategies she can use in order to convince her hesitant husband to grant her an approval to participate in a clinical research. She listed strategies such as cooking him a good meal before asking him again, giving him time to think about it, and bringing it up again and seeking the assistance of individuals he respects.[155] In this context, persuasion can be an effective instrument. It implies also that obtaining approval does not essentially indicate a loss of personal autonomy. In some cases, outcomes of interviews conducted with women in Nigeria indicate that they differentiate between their experience of self-determination and their need, to persuade their husband of the relevance and significance of the research to them personally.[156] It is also pertinent to acknowledge the changeability that occurs not just across cultures, but within specific social settings. A study focusing on informed consent in genetic research conducted in Nigeria, reported that all the women interviewed were not required to obtain approval from their husbands to participate in the study.[157]

Various international guidelines and recommendations such as CIOMS, 2002; Nuffield Council on Bioethics, 2002, 2005; National Bioethics Advisory Commission, 2001, emphasized the significance of individual consent to research.[158] For example, CIOMS guidelines indicated that the woman's informed consent is the only thing required for her participation in research. The permission of either a spouse or a partner will not on any occasion substitute for the requirement of individual informed consent. Women were highly encouraged on their own to consult with their husbands or partners before making a decision to enroll in research. It was also made categorically clear that a strict requirement of authorization of spouse or

[154] Marshall, "The Relevance of Culture for Informed Consent in US-Funded International Health Research," C–19.

[155] Patricia A. Marshall, "The Individual and the Community in International Genetic Research," *Journal of Clinical Ethics* 15, no. 1 (Spring 2004): 79.

[156] Marshall, "Ethical Challenges in Study Design and Informed Consent for Health Research in Resource-Poor Settings," 27.

[157] Marshall, "The Individual and the Community in International Genetic Research," *Journal of Clinical Ethics* 15, no. 1 (Spring 2004): 76–86.

[158] Marshall, "Ethical Challenges in Study Design and Informed Consent for Health Research in Resource-Poor Settings," 28.

partner, infringes on the substantive principle of respect for persons.[159] The implication is that CIOMS guidelines give a stronger defense of women's autonomy than NBAC recommendations, which allow a research ethics committee to make determination on the need for spousal authorization with appropriate documentation.[160] Macklin expresses her uncertainty regarding allowing exceptions that would involve approaching a woman's husband or father before speaking to her. She strongly supports not permitting any exceptions, so as not to perpetuate or reinforce the practice prevalent in developing countries where women are considered subservient or inferior, which violates the principles of respect for autonomy and equal respect for women. Macklin also acknowledged that spousal authorization could be justified from the utilitarian perspective, if the consequences of not conducting the research would be serious.[161] Related to the issue of informed consent is the protection of the rights of potential human subjects with an adequate and thorough independent review of research protocol by a well constituted ethics review committee.

2.3 Ethics Review Committees

Preamble

U.S. federal regulations generally referred to as the "Common Rule" require that research funded by U.S. government or conducted by a government agency or institutions that comply with common rule or intended for submission to the Food and Drug Administration must be reviewed by a U.S. Institutional Review Board (IRB) and also by a local, regional, or national committee in the country where the research is conducted.[162] However, the Common Rule only offers guidance about researchers informing participants in advance of any intervention that will be provided in the course of the trial if a participant is injured,[163] but it does not address the issue of what would happen after a clinical trial. International ethical guidelines and declarations also emphasized the requirement of review of research involving human subjects. For example, CIOMS stipulates that, "all proposals to conduct research involving human subjects must be submitted for review of their scientific merit and ethical acceptability to one or more scientific review and ethical review committees."[164] In the same vein, the 1993 CIOMS did not address the issue of post-

[159] Council for International Organizations of Medical Sciences, *International Ethical Guidelines for Biomedical Research Involving Human Subjects*, 73.

[160] Macklin, *Double Standards in Medical Research in Developing Countries*, 145.

[161] Macklin, *Double Standards in Medical Research in Developing Countries*, 144–145.

[162] U.S. Department of Health and Human Services, *Code of Federal regulations Title 45 Public Welfare, Part 46 Protection of Human Subjects*, 3–7.

[163] U.S. Department of Health and Human Services, *Code of Federal regulations Title 45 Public Welfare, Part 46 Protection of Human Subjects*, 8.

[164] Council for International Organizations of Medical Sciences, *International Ethical Guidelines for Biomedical Research Involving Human Subjects*, 24.

trial access to participants. The CIOMS guidelines also addressed the issue of compensation for injury sustained in the course of the research and discussed the idea that the sponsoring agency of externally sponsored research "should agree in advance of the research that any product developed through such research will be made reasonably available to the inhabitants of the host community or country at the completion of successful testing."[165]

Debate in the late 1990s regarding the ethics of international HIV clinical research attracted more attention to the critical issue of post-trial benefits.[166] Many experts indicate that reducing the possibility of exploiting participants in developing nations would require a plan for how the benefits of clinical research would be made reasonably available to the developing nation or community.[167] The discussion concentrated on the requirement that interventions proven effective through clinical research be made available to the host community from which participants were drawn.

Allegations of violations regarding ethical review of research protocols for trials conducted in developing countries abound. Macklin reported two cases that involved U.S. researchers from U.S. institutions that violated the requirement of review of research in conducting international collaborative research. In one case, a researcher from the Harvard School of Public Health conducted a series of epidemiologic genetic studies in a rural province in China. The studies recruited vulnerable subjects who were poor and illiterate. The studies entailed taking subjects' blood samples and carrying out lung function tests and x-rays. One of the violations reported about this research was that the researcher failed to submit some studies to the IRB at Harvard School of Public Health and to obtain approval from the committee before conducting the study. The researcher also made changes in the research in some of the studies after the initial approval, but failed to obtain approval for the changes he effected. Informed consent violations were also reported. The informed consent documents were considered insufficient, both because they had no information about the right of subjects to refuse to participate, and because they were too complicated for rural Chinese farmers to comprehend. There was also another concern about the subjects risking being discriminated against in the area of job, if their employers discovered health problems diagnosed in the studies. No information was furnished regarding the level to which the subjects' confidentiality could be sufficiently protected in China.[168]

The second case of violation involved a professor of microbiology at the University of California at Los Angeles (UCLA), who failed to obtain IRB approval

---

[165] Council for International Organizations of Medical Sciences, *International Ethical Guidelines for Biomedical Research Involving Human Subjects*, 43.

[166] Marcia Angell, "The Ethics of Clinical Research in the Third World," The New England Journal of Medicine 337, no.12 (September 1997): 847.

[167] National Bioethics Advisory Commission, (NBAC), *Ethical and Policy Issues in International Research: Clinical Trials in Developing Countries*, http://bioethics.georgetown.edu/nbac/clinical/chap4.html (Accessed November 12, 2016).

[168] Macklin, *Double Standards in Medical Research in Developing Countries,* 150–151.

for a research he conducted in collaboration with another researcher in China. The study involved analyzing data and blood samples of research subjects. The study also involved injecting malaria-infected blood into Chinese AIDS patients which has been confirmed by many scientists as fraudulent medical practice.[169] These violations more prevalent in research conducted in developing countries, where subjects are more vulnerable calls for safeguarding their rights. Protecting the rights of research subjects requires a properly constituted ethics review committee, so that it will be adequately equipped to discharge its responsibilities.

2.3.1 Composition and Responsibilities of Ethics Review Committees

2.3.1.1 Composition of Ethics Review Committees

The composition of Ethics Review Committee or IRB has direct relationship with its function, because function which is usually regarded as what an entity does, can as well in this context be defined in terms of structure. The implication is that proper composition of ethics review committee directly influences the quality of the ethical review of research protocols. For example, it will be hard for an ethics review committee to adequately deliberate on the needs and viewpoints of adults with diminished capacity for decision making, if none of the members has expertise or experience dealing with this particular population.[170] Lo articulates, "as a group, IRB members must have the expertise, experience, and diversity of backgrounds needed to review the typical types of research conducted at an institution. Diversity of backgrounds should include not only areas of scientific or professional expertise but also culture, race and gender."[171]

U.S. federal regulations stipulate the composition of an IRB to be a minimum of five members from diversified backgrounds, whose primary responsibility is to promote complete and adequate review of research carried out by the institution.[172] The ethics review committee must have at least one member who focuses mainly on scientific themes. The ethics review committee is required to comprehend the science fundamental to the studies it usually reviews, in order to conclude whether the study will generate valid, generalizable knowledge, and whether minimized risky methods could be utilized without compromising the science. The ethics review

[169] Macklin, *Double Standards in Medical Research in Developing Countries*, 152.

[170] Marjorie A. Speers, "Evaluating the Effectiveness of Institutional Review Boards," " in *The Oxford Textbook of Clinical Research Ethics*, ed. Ezekiel J. Emmanuel et al., (New York: Oxford University, 2008), 561.

[171] Bernard Lo, *Ethical Issues in Clinical Research: A Practical Guide* (Philadelphia: Lippincott Williams & Wilkins, 2010), 28.

[172] U.S. Department of Health and Human Services, *Code of Federal regulations Title 45 Public Welfare, Part 46 Protection of Human Subjects*, 6.

committee must also have at least one member whose emphasis is on issues related to nonscientific themes. At least one committee member must not be affiliated with the institution conducting the research.[173]

In general, ethics review committee members may include physicians, scientists, and other professionals like nurses, lawyers, ethicists and clergy, also lay persons qualified to stand for the cultural and moral values of the community and to guarantee that the rights of the research subjects will be esteemed. The committee members should consist of men and women. When a study focuses on illiterate persons, they should be recruited as committee members or given opportunity for their perspectives to be represented.[174]

A national or local ethics review committee charged with reviewing and approving protocols for research sponsored by developed countries should recruit members who are very acquainted with the customs and traditions of the population or community involved. Ethics review committees that usually review research protocols focusing on particular diseases or diminished mental or physical abilities, such as HIV/AIDS or paraplegia, should recruit members who are knowledgeable and experienced working with such populations or pay attention to the perspectives of individuals or organizations advocating for patients with such diseases or impairments. On the other hand, committees should call or listen to the perspectives of people who represent or advocate for vulnerable subjects such as children, students, elderly persons or employees involved in research.[175] Members of ethics review committee should not participate in any review of protocols in which they have conflict of interest and they are required to offer background information to the committee.[176]

2.3.1.2 Responsibilities of Ethics Review Committees

There is a consensus among most scholars that the primary responsibility of an ethics review committee is to protect the rights of research participants.[177] Marshall articulates that, "the primary aim of ethical review committees (ERCs) is to ensure the protection of human participants by safeguarding their rights and determining that the risks associated with participation in the study do not endanger the safety of individuals and are reasonable in relation to the anticipated benefits."[178] Guillemin

---

173 Lo, *Ethical Issues in Clinical Research: A Practical Guide*, 28.

174 Council for International Organizations of Medical Sciences, *International Ethical Guidelines for Biomedical Research Involving Human Subjects*, 27.

175 Council for International Organizations of Medical Sciences, *International Ethical Guidelines for Biomedical Research Involving Human Subjects*, 27.

176 Lo, *Ethical Issues in Clinical Research: A Practical Guide*, 29.

177 Gerry Kent, "The Views of Members of Local Research Ethics Committees, Researchers, and Members of the Public towards the Roles and Functions of LRECs," *Journal of Medical Ethics* 23, no. 3 (June 1997): 186.

178 Marshall, "Ethical Challenges in Study Design and Informed Consent for Health Research in Resource-Poor Settings," 15.

et al. cautioned regarding ethics review committees' sometimes being overprotective and paternalistic toward research participants.[179] Ethics review committee accomplishes its task of protecting the rights of research participants through providing oversight, review and approval of research protocols. Before the approval of any research study by ethics review committee, the following criteria must be fulfilled, " (1) the risks of the study are minimized; (2) risks are reasonable in relation to the anticipated benefits; (3) selection of participants is equitable; (4) informed consent is obtained; (5) confidentiality is maintained; (6) data are adequately monitored."[180] Ethics review committee must also determine that research participation is voluntary and that withdrawing from the research will not lead to any adverse consequences for the participants.[181] The ethics review committee in the host country must also determine whether the goals of the research are responsive to the health needs and priorities of the identified host country.[182]

Ethics review committees especially in developing countries encounter many significant challenges in their oversight and review of research. For instance, there may be structural obstacles that are essentially inherent in their institutions that make it hard to satisfy international ethical regulatory requirements. Some of these obstacles include lack of resources, inadequate training among ethics review committee members, and insufficient infrastructure. Accomplishing adequate ethical review of protocols requires well equipped and trained ethics review members, which comprehend the necessity of ethical review of protocols and the responsibilities linked with it. Furthermore, it requires that institutions possess the technological resources that will enable them to carry out effective reviews including funds for photocopying materials and staff for managing and tracking protocols. Nevertheless, it may be predominant in many developing countries that there are no ethics review committees or that they are not well equipped to implement adequate reviews due to inadequate resources or the absence of trained professionals.[183] Hyder et al. reported their findings from a survey that involved 670 health researchers in developing countries, which focused on the role of IRBs in guaranteeing the adequacy of ethical standards in research carried out in those countries. Forty four percent of the researchers surveyed indicated that their studies were neither reviewed by an IRB from a developing country nor by ministry of health and one third of these studies received their funding from organizations in the U.S. Their findings also revealed that IRBs in the U.S. were significantly more likely to raise issues related to the

[179] Marilys Guillemin et al., "Human Research Ethics Committees: Examining Their Roles and Practices," *Journal of Empirical Research on Human Research Ethics* 7, no. 3 (July 2012): 41.

[180] Lo, *Ethical Issues in Clinical Research: A Practical Guide*, 19.

[181] Council for International Organizations of Medical Sciences, *International Ethical Guidelines for Biomedical Research Involving Human Subjects*, 37.

[182] Council for International Organizations of Medical Sciences, *International Ethical Guidelines for Biomedical Research Involving Human Subjects*, 31.

[183] Marshall, "Ethical Challenges in Study Design and Informed Consent for Health Research in Resource-Poor Settings," 16.

need for consent forms in the local language and approval letters from developing country representatives, and the protection of confidentiality than by IRBs in the host country.[184]

There were also significant concerns regarding conflicts of interests for IRBs in developing countries. Kass and Hyder reported that some respondents indicated that local review committees in some countries stress scientific, political, or funding issues rather than ethical considerations.[185] The political nature of decisions made by local IRB was highlighted by some respondents. One respondent commented, "it is a political approval. It is not an approval that is about ethics. It was more about whether we would be spies or we would be real researchers that would benefit Asian Country."[186] In developing countries with limited resources, corruption and bribes for government officials constitute major concerns in the establishment and implementation of standards for research ethics at the national level. Some respondents felt that external organizations exploited resource-poor countries, and imposed their wishes on them through controlling some government officials.[187] Some respondents also mentioned power differences between United States and developing countries, which established a paternalistic situation for negotiating the terms of research. The disparities in power unfairly impacted decision making about research. One respondent remarked that, "the biggest problem in developing countries is that our poverty puts us in a situation where the beggar has no choice."[188]

Ethics review committees in developing countries also encounter difficulties regarding essential features required for their proper establishment and functioning. Macpherson discusses the uncertainties that confronted the establishment of a research ethics committee in Grenada. Some of the unanticipated issues focused on specific guidelines and procedures to adopt and the appointment and training of members.[189] In another work, Macpherson indicates that international guidelines do not deal with issues such as the connection between the IRBs and the governments that do not mandate them, and what kind of procedures or documentation will function in a developing country. She discusses doubts related to whether ways of guaranteeing confidentiality and obtaining informed consent will be effective due to socio-cultural impacts in local circumstances, and whether departures from western standards are justifiable. International guidelines are considered beneficial in dealing with these issues, but they are prone to diverse interpretations. She reports that

[184] A. A. Hyder et al., "Ethical Review of Health Research: A Perspective from Developing Country Researchers," *Journal of Medical Ethics* 30, no. 1 (February 2004): 68–72.

[185] Kass and Hyder, "Attitudes and Experiences of U.S. and Developing Country Investigators Regarding U.S. Human Subjects Regulations," B–107.

[186] Kass and Hyder, "*Attitudes and Experiences of U.S. and Developing Country Investigators Regarding U.S. Human Subjects Regulations*," B–107.

[187] Kass and Hyder, "Attitudes and Experiences of U.S. and Developing Country Investigators Regarding U.S. Human Subjects Regulations," B–108.

[188] Kass and Hyder, "Attitudes and Experiences of U.S. and Developing Country Investigators Regarding U.S. Human Subjects Regulations," B–108.

[189] Cheryl Cox Macpherson, "Research Ethics Committee: Getting Started," *West Indian Medical Journal* 50, no. 3 (September 2001): 186–188.

the Grenada experience shows that it is possible for IRBs in developing countries to adopt international standards. She further contends that there is a significant need for educational programs not only to improve the capacity of IRBs, but as well to guarantee that leaders of developing countries are knowledgeable regarding the importance of international research guidelines for their nations.[190]

Most national and international guidelines recommend dual independent ethical review of research protocols both in the host country and sponsor country, for externally sponsored research conducted in developing countries.[191] Review of externally sponsored research conducted in developing countries with poor resources poses several ethical challenges. One problem identified in this area deals with the responsibilities for research oversight when multiple ethics review boards are involved.[192] Adhering to strict U.S regulations by host country IRBs may present a significant challenge. Researchers from Nigeria explained administrative concerns about the process of securing approval from ethics review committees. Some researchers described the problems related to responding to the requirements of U S funding agencies and at local institutions in Nigeria. A physician researcher from Lagos, expressed frustrations regarding fighting with Washington to change the consent form and adapting the form to be useful and suitable for his Nigerian patients. He expressed frustrations also regarding the administrative aspects of the process, comprising of paperwork and committee negotiation.[193] Discrepancies between ethics review committees in the developed and developing countries should be resolved. Mechanisms for effectively dealing with such conflicts between multiple ethics review committees are currently nonexistent. In situations where ethics review committees cannot resolve differences between themselves, a committee may decide not to approve the research. The implication is that when a committee from sponsor's country decides not to approve the research, the sponsor cannot fund it. Conversely, when an ethics review committee from a developing county chooses not to approve the research, at that point the research cannot be carried out within that country.[194]

Apart from ethics review committees' primary responsibility to protect the rights of research participants, they also have responsibilities to society and to researchers. Ethics review committees have a responsibility to society, since it provides the resources for carrying out the research and it can be significantly influenced by research findings.[195] This entails that ethics review committees have a responsibility

[190]Cheryl Cox Macpherson, "Ethics Committees. Research Ethics: Beyond the Guidelines," *Developing World Bioethics* 1, no. 1 (November 2001): 57–68.

[191]Nuffield Council on Bioethics, *The Ethics of Research Related to Healthcare in Developing Countries*, 107.

[192]Marshall, "Ethical Challenges in Study Design and Informed Consent for Health Research in Resource-Poor Settings," 16.

[193]Marshall, "The Relevance of Culture for Informed Consent in US-Funded International Health Research," C–25.

[194]Nuffield Council on Bioethics, *The Ethics of Research Related to Healthcare in Developing Countries*, 107–108.

[195]Leslie Gelling, "Role of the Research Ethics Committee," *Nurse Education Today* 19, no. 7 (October 1999): 566.

to evaluate the scientific merit of research protocols. Any research that does not possess potential benefit to society should not be approved.[196] Kent reports divergent views among local research committee members, researchers and the public regarding the role of the ethics review committees in assessing and maintaining the scientific merit of research.[197] Some scholars argue that there is a clear distinction between assessing the ethics of a proposal and evaluating the scientific merit of a proposal.[198] Researchers contend that the scientific value of a research is usually recognized before requesting for ethical review and approval, and that it is not within the scope of the ethics review committee's responsibility to evaluate scientific merit. Ethics review committees may consider scientific aspects of research, but it has not been determined to what extent this constitutes part of their function.[199]

CIOMS international guidelines stipulate that ethics review committees in both sponsor country and host country have obligation to carry out both scientific and ethical review of research protocols. They are also vested with the authority to refuse approval of research protocols that did not satisfy criteria for their scientific and ethical standards.[200] Gelling articulates that ethics committee is discharging its obligation to society, if it refuses to grant approval to research proposals that did not meet criteria for scientific standards. The implication is that ethics committee is protecting society from research that is not beneficial, so that available limited resources would not be squandered on it.[201] Benatar describes this function of ethics review committees to the society as monitoring and auditing research, as well as providing accountability to the public.[202]

Ethics review committees as well have responsibility to researchers. Kent acknowledges researchers' right to have their protocols treated with respect and due consideration.[203] Researchers expect ethics review committees to avoid unnecessary delays in the review and approval of their protocols, since they are compelled to produce outcomes.[204] Benatar identified this role of ethics review committees to researchers as educating and assisting researchers and the community in comprehending and appreciating the ethics of research.[205] Ethics review committee should

---

196 Gelling, "Role of the Research Ethics Committee," 566.

197 Kent, "The Views of Members of Local Research Ethics Committees, Researchers, and Members of the Public towards the Roles and Functions of LRECs," 188.

198 M. E. Redshaw, A. Harris and J. D. Baum, "Research Ethics Committee Audit: Differences between Committees," *Journal of Medical Ethics* 22, no. 2 (April 1996): 78–82.

199 J.B. Cookson, "Auditing a Research Ethics Committee," *Journal of the Royal College of Physicians of London* 26, no. 2 (April 1992): 181–183.

200 Council for International Organizations of Medical Sciences, *International Ethical Guidelines for Biomedical Research Involving Human Subjects,* 31.

201 Gelling, "Role of the Research Ethics Committee," 567.

202 S. R. Benatar, "Reflections and Recommendations on Research Ethics in Developing Countries," *Social Science & Medicine* 54, no. 7 (April 2002): 1137–1138.

203 Kent, "The Views of Members of Local Research Ethics Committees, Researchers, and Members of the Public towards the Roles and Functions of LRECs," 186.

204 Gelling, "Role of the Research Ethics Committee," 567.

205 Benatar, "Reflections and Recommendations on Research Ethics in Developing Countries," 1137.

also address the issue of post-trial access to participants and host populations both in the informed consent process and in the review of the protocol.

2.3.1.3 Ethics Review Committee and Post-trial Access

The increasing concerns about the exploitation of the participants and host populations of developing nations at the end of the clinical research that yields proven effective intervention points to the importance of post-trial access to safeguard the rights of such communities. In recent years, some research ethics guidance documents,[206] reports,[207] and national guidelines[208] have started to address the issue of post-trial access of participants to beneficial treatment.

The Declaration of Helsinki of World Medical Association (WMA), 2000 states that "at the conclusion of the study, every patient entered into the study should be assured of access to the best proven prophylactic, diagnostic, and therapeutic methods identified by the study."[209] The primary objective of the paragraph is to prevent research sponsors from conducting clinical trials in populations that would not usually have access to the beneficial treatment, only to stop access to the intervention at the end of the trial.[210] This happened after some of the HIV/AIDS clinical trials conducted in developing African nations in the 1990s. The implication of the Declaration of Helsinki provision is that "a person who participates in a trial should have a chance to benefit from what is learned from the trial – a principle that is particularly important for participants in the developing world."[211] Clinical trial participants in developed nations will normally receive the best available intervention at the end of the clinical trial. On the other hand, in developing nations, participants will be left with nothing when researchers pack up and go home.[212] A clarification of paragraph 30 was issued by the WMA in October 2004. The WMA reaffirmed its position that "it is necessary during the study planning process to identify post-trial

[206] World Medical Association, *Declaration of Helsinki, Ethical Principles for Medical Research Involving Human Subjects*, (Edinburgh, Scotland: World Medical Association, 2000): http:www.wma.net/e/ethicsunit/helsinki.htm (Accessed November 12, 2016).

[207] National Bioethics Advisory Commission, (NBAC), *Ethical and Policy Issues in International Research: Clinical Trials in Developing Countries*, http://bioethics.georgetown.edu/nbac/clinical/chap4.html (Accessed November 12, 2016).

[208] Medical Research Council of South Africa (MRC-SA), Guidelines on Ethics for Medical Research, (South Africa: MRC-SA, 1993).

[209] World Medical Association, *Declaration of Helsinki, Ethical Principles for Medical Research Involving Human Subjects*, (Edinburgh, Scotland: World Medical Association, 2000): Paragraph 30, http:www.wma.net/e/ethicsunit/helsinki.htm (Accessed November 12, 2016).

[210] Jeff Blackmer and Henry Haddad, "The Declaration of Helsinki: An Update on Paragraph 30," Canadian Medical Association Journal (CMAJ) 179, no. 9 (October 2005): 1052–1053.

[211] Christine Grady, "The Challenge of Assuring Continued Post-Trial Access to Beneficial Treatment." Yale Journal of Health Policy, Law, and Ethics 5, no.1 (2005): 429.

[212] The Lancet, "One Standard, Not Two," Lancet 362, (November 2003): 1005.

access by study participants to prophylactic, diagnostic and therapeutic procedure identified as beneficial in the study or access to other appropriate care. Post-trial access arrangements or other care must be described in the study protocol so the ethical review committee may consider such arrangements during its review."[213] Some authors argued for the ethical considerations for post-trial access to treatment. They argue that arrangements for the continuation of intervention at the end of the clinical trial must be described in the research protocol.[214] The information about post-trial access arrangements must also be disclosed to participants during the informed consent process.[215]

Guidelines from developing nations have taken it a step further not just to emphasize that the issue of post-trial access be addressed, but they also impose positive obligations to provide effective interventions to participants and in certain cases to the host population.[216] For example, the Ugandan document Guidelines for the Conduct of Health Research Involving Human Subjects in Uganda obliges researchers to "make every effort to ensure its (an effective intervention, if available) provision, without charge, to participants in the trial following the conclusion of the trial."[217]

Several international and national guidelines also acknowledge post-trial obligations to host communities and countries. The commentary on guideline 15 of CIOMS 1993 regarding "Obligations of Sponsoring and Host Countries" recognize the obligation of sponsors to make the products of research available.[218] Another important international document issued by the WHO puts the consideration of the availability of successful interventions in the host community as a critical part of the ethical review process. The document indicates that "a description of the availability and affordability of any successful study product to the concerned communities following the research"[219] should be considered as part of the ethical review process.

---

213 World Medical Association, *Declaration of Helsinki, Ethical Principles for Medical Research Involving Human Subjects*, Clarification on Declaration of Helsinki, http://www.wma.net/en/40news/20archives/2004/2004_24/ (Accessed November 12,2016).

214 Dinesh Chandra Doval, Rashmi Shirali, and Rupal Sinha, "Post-trial Access to Treatment for Patients Participating in Clinical Trials," Perspectives in Clinical Research 6, no. 2 (April-June, 2015): 82–85.

215 World Medical Association, *Declaration of Helsinki, Ethical Principles for Medical Research Involving Human Subjects*, Clarification on Declaration of Helsinki, http://www.wma.net/en/30publications/10policies/b3/ (Accessed November 12, 2016).

216 National Bioethics Advisory Commission, (NBAC), *Ethical and Policy Issues in International Research: Clinical Trials in Developing Countries*, http://bioethics.georgetown.edu/nbac/clinical/chap4.html (Accessed November 12, 2016).

217 National Consensus Conference on Bioethics and Health Research in Uganda, Guidelines for the Conduct of Health Research Involving Human Subjects in Uganda, (Kampala, Uganda: National Consensus Conference, 1997), Section V. Part D. 4.

218 Council for International Organizations of Medical Sciences, *International Ethical Guidelines for Biomedical Research Involving Human Subjects*, (Geneva: CIOMS, 1993), Commentary on Guideline 15.

219 World Health Organization (WHO), Operational Guidelines for Ethics Committees That Review Biomedical Research, (Geneva: WHO, 2000), para. 6.2.6.6.

The recent WMA clarification indicates that the adequacy of arrangements for post-trial access to beneficial interventions should be decided by ethics review committees.[220] Ethics review committees need direction to determine the adequacy of such arrangements for post –trial intervention. The decision should consider the need for monitoring and administration of delivered interventions.[221]

In recent years, some concerted efforts have been made to delineate the arrangements for making proven treatments available at the end of a successful clinical trial in the form of prior agreements. Prior agreements can be utilized in several ways to provide the benefits of the proposed research to the population from which the participants are drawn. For example, prior agreements can be designed so that the experimental treatment that is being tested would be made available to participants and their communities at a cost the developing nation can afford.[222]

Some have argued that for the research sponsored by developed nations and conducted in developing nations to be ethically acceptable, it must "offer the potential of actual benefits to the inhabitants,"[223] of that nation by providing affordable access to the treatment to those communities where the treatment has been tested. The developing nation receives little benefit, if the treatment is provided only to the participants without any guarantee of affordable access to the host population. If the knowledge obtained from the research is utilized mainly for the benefit of the developed nation, the research may be properly described as exploitative and therefore unethical.[224] Some authors have taken the argument a step further that it is ethically not adequate to make a proven treatment available to a developing nation by eliminating the financial barrier to access if there is no means of providing the treatment to the host population.[225] A feasible plan for distribution of the proven treatment must be offered as a critical part of the study review process to determine that there will be enough potential benefit to justify carrying out the research. Glantz et al., articulate "where the infrastructure is so undeveloped that it would be impossible to deliver the intervention even if it were free, research would be unjustified in the absence of a plan to improve the country's health care delivery capabilities."[226] Plans that are designed for the funding, distribution, and use of successful treatments

[220] World Medical Association, Workgroup Report on the Revision of Paragraph 30 of the Declaration of Helsinki, (2004), http://www.wma.net/e/ethicsunit/pdf/wgdohjan2004.pdf (Accessed November 12, 2016).

[221] Grady, "The Challenge of Assuring Continued Post-Trial Access to Beneficial Treatment." 432.

[222] National Bioethics Advisory Commission, (NBAC), *Ethical and Policy Issues in International Research: Clinical Trials in Developing Countries*, http://bioethics.georgetown.edu/nbac/clinical/chap4.html (Accessed November 12, 2016).

[223] L.H. Glantz, et al., "Research in Developing Countries: Taking Benefit Seriously," Hastings Center Report 28, no. 6 (November-December, 1998): 39.

[224] C. Del Rio, "Is Ethical Research Feasible in Developed and Developing Countries?" Bioethics 12, no. 4 (October, 1998): 328–330.

[225] L.H. Glantz, et al., "Research in Developing Countries: Taking Benefit Seriously," 41.

[226] L.H. Glantz, et al., "Research in Developing Countries: Taking Benefit Seriously," 41.

before the research starts can help to surmount some of the main obstacles to making treatments widely available in the nations in which they are tested.[227]

2.4 Conclusion

Concluding reflections on the global clinical research emphasize the priority of safeguarding the rights of research participants and host populations in the design and implementation of research protocols. Obtaining voluntary informed consent from research participants and thorough review of research protocols by well constituted and competent ethics review committee were considered desiderata in conducting clinical trials in developing countries.

Cultural and language barriers were highlighted as challenges in conducting clinical research in developing countries. However, researchers and sponsors from developed countries were encouraged to respect the local culture and values of research participants and host populations and to adapt standards of informed consent to the cultural norms and practices of developing countries.

The prominent roles of the family and the community in personal decision-making for consenting to research participation in most cultures of developing countries were acknowledged. Creative strategies for presenting information to research participants in developing countries were strongly recommended as effective means of improving comprehension of essential information regarding research study goals and procedures.

Post-trial access to proven interventions to participants and host populations was also considered as safeguarding their rights at the end of the clinical trial. The informed consent must also thoroughly address the issue of post-trial access to successful interventions. The ethics review committee must include in its review process concrete and feasible plans for the provision of post-trial access to successful interventions to participants and host populations after the end of the trial. However, clinical trials in developing countries are still plagued by the inability of participants and host populations to access successful drugs at the completion of the trials. A related issue in the discussion of international clinical research is the significant impact of the intellectual property law and international trade agreements on post-trial access to drugs in poor developing countries as discussed in the next chapter.

---

[227] National Bioethics Advisory Commission, (NBAC), *Ethical and Policy Issues in International Research: Clinical Trials in Developing Countries*, http://bioethics.georgetown.edu/nbac/clinical/chap4.html (Accessed November 12, 2016).

Chapter 3
Post-Trial Access to Drugs in Developing Nations

3.1 Introduction

The pandemic nature of HIV/AIDS which decimates millions of people in developing countries creates an urgent need for post-trial access to anti-retroviral drugs. Conducting clinical trials has been defended as a major means of providing medical benefits to poor populations in developing countries through development of cheaper and affordable drugs.[1] Lavery articulates that, "it has become increasingly well-recognized in recent years that an equitable distribution of the benefits of research is an important component of international research ethics.".[2]

International guidelines such as CIOMS, Declaration of Helsinki and the United Nations Joint Programme on AIDS (UNAIDS) Ethical Considerations in HIV Preventive Vaccine Research allocate to researchers and sponsors the job of guaranteeing and accomplishing benefits of research for participants and the host communities.[3]

The history of international clinical research is marred by poor record with regard to the transfer of benefits to the communities in LMIC which have helped in producing interventions, especially novel drugs and vaccines.[4]

The use of placebo controls in clinical research has been identified as a viable way of developing cheaper drugs in developing countries, where majority of the populations get nothing as the standard of care. There was disagreement among

---

[1] Macklin, *Double Standards in Medical Research in Developing Countries,* 163.

[2] James V. Lavery, "Putting International Research Ethics Guidelines to Work for the Benefits of Developing Countries," *Yale Journal of Health Policy, Law and Ethics* 4, no. 2 (Summer 2004): 319–336.

[3] Lavery, "Putting International Research Ethics Guidelines to Work for the Benefits of Developing Countries," 319.

[4] Reidar K. Lie, Justice and International Research, In *Biomedical Research Ethics: Updating International Guidelines*, eds. Robert J. Levine, Samuel Gorowitz, and J. Gallagher (Geneva: CIOMS, 2000), 27–40.

E.C. Obi, *Post-Trial Access to Drugs in Developing Nations*, Advancing Global Bioethics 7, DOI 10.1007/978-3-319-60028-4_3

scholars regarding the use of placebo-controlled research design in HIV clinical trials in developing countries.

The lack of access to essential drugs has three components. First, pharmaceutical research neglects drugs for diseases that have high prevalence among the poor.[5] The second component of the access problem of the poor highlights that existing drugs are priced out of reach for buyers in developing countries during the patent years on the market. The third component of lack of access to essential drugs by poor people in developing countries is lack of adequate local health infrastructure. The post-trial access to drugs must be discussed within the broader context of the impacts of intellectual property law and international trade agreements. The non-patent factors must also be explored in our discussion of affordable access to drugs in developing countries.

3.2 Intellectual Property Law

3.2.1 Origin and Meaning of Intellectual Property Rights

The publication of John Locke's Second Treatise on Government[6] emphasizes the priority of individual rights to property, and specifically private property, as one of the tenets of philosophy of rights theory and one of the bases for justifying Western-style free enterprise. The Western thinking developed the notion of intellectual property rights, proprietary rights to what one invents, writes, paints, composes, or creates, from the idea of property in Locke.[7] Intellectual property is defined as "creations of the mind, that is, intellectual creations, such as literary and artistic works, inventions and more.".[8]

Property rights regardless of how they are socially defined by a particular society, establish obligations both for others and the state to protect property interests.[9] The advent of industrial revolution ushered in an era that emphasized the protection of

[5] Thomas Pogge, Matthew Rimmer and Kim Rubenstein, "Access to Essential Medicines: Public Health and International Law," In *Incentives for Global Public Health: Patent Law and Access to Essential Medicines* ed. Thomas Pogge, Matthew Rimmer and Kim Rubenstein (New York: Cambridge University Press, 2010), 4.

[6] John Locke, *The Second Treatise of Government, 1764*, ed. P. Laslett, (Cambridge: Cambridge University Press, 1983).

[7] Patricia H. Werhane and Michael E. Gorman, "Intellectual Property Rights, Access to Life-Enhancing Drugs, and Corporate Moral Responsibilities," in *Ethics and the Pharmaceutical Industry*, ed. Michael A. Santoro and Thomas M. Gorrie, (Cambridge: Cambridge University Press, 2005), 260.

[8] Zainatul A. Zainol et al., "Pharmaceutical Patents and Access to Essential Medicines in sub-Saharan Africa," *African Journal of Biotechnology* 10, no. x (September 2011): 12,377.

[9] Werhane and Gorman, "Intellectual Property Rights, Access to Life-Enhancing Drugs, and Corporate Moral Responsibilities," 261.

ideas as well as material property.[10] There is recognition that patent protection promotes invention and creativity by safeguarding ownership of new ideas, as well as authorizing the inventor or creator to obtain benefits from that idea, in the same way the farmer benefits from good agricultural practices on his or her land.[11] However, distinguishing farm land from ideas, Jefferson writes, "… ideas should freely spread from one to another over the globe, for the moral and mutual instruction of man, and improvement of his condition, seems to have been peculiarly and benevolently designed by nature …. Inventions, then, cannot, in nature, be a subject of property."[12] In contrast to the farmer, the inventor is encouraged to publicize her or his innovation while at the same time safeguarding the right to copy or reproduce the invention.[13]

Intellectual property has been defended from two different foundations, a standard rights-based defense and utilitarian justification. The rights-based perspective is derived from Locke's theory of rights, which stresses that inventors have rights to what they create. The utilitarian justification associates rights with utility, which implies that inventors may not be likely creative without intellectual protection, since they will not essentially reap honor or the benefits of their inventions.[14] A more detailed discussion on two different grounds for intellectual property follows.

3.2.1.1 Rights–Based Defenses

The perspective that intellectual property is a type of ownership which entitles one to exclusive rights to use, copying, or distribution is usually the way intellectual property is considered, especially in countries and companies that sponsor the development of new processes or products. The implication is that from this point of view, if a person or company creates a patentable process or product, due to the creativity and work involved, the person or organization has exclusive rights to that creativity.[15] Governments grant specific rights to the creators of intellectual property, in order to motivate the continuous and useful enhancement of society with

[10] David B. Resnik, "A Pluralistic Account of Intellectual property," *Journal of Business Ethics* 46, no. 4 (September 2003): 319–335.

[11] Werhane and Gorman, "Intellectual Property Rights, Access to Life-Enhancing Drugs, and Corporate Moral Responsibilities," 261.

[12] Thomas Jefferson, "Letter to Isaac McPherson," 1813, reprinted in A. Knock and W. Peden, *The Life and Selected Writings of Thomas Jefferson* (New York: Modern Library, 1972), cited in Werhane and Gorman, "Intellectual Property Rights, Access to Life-Enhancing Drugs, and Corporate Moral Responsibilities," 262.

[13] Werhane and Gorman, "Intellectual Property Rights, Access to Life-Enhancing Drugs, and Corporate Moral Responsibilities," 262.

[14] Werhane and Gorman, "Intellectual Property Rights, Access to Life-Enhancing Drugs, and Corporate Moral Responsibilities," 262.

[15] Werhane and Gorman, "*Intellectual Property Rights, Access to Life-Enhancing Drugs, and Corporate Moral Responsibilities*," 262.

such creations. These rights are termed intellectual property rights (IPRs), and entitle the holders to avert misuse of their creation for a specified period of time by others.[16]

Although intellectual property (IP) rights are considered as time-limited protected claims, they are in some situations especially Western countries conceived as perfect rights in such a way that violations of copyrights, trademarks, or patents are always wrong without exception.[17] Rand captures this view when she writes, "patents and copyrights are the legal implementation of the base of all property rights: man's right to the product of his mind… patents are the heart and core of property rights, and once they are destroyed, the destruction of all other rights will follow automatically, as a brief postscript."[18] Rand argues that intellectual property rights are the most fundamental rights, and the implication is that without them all other rights are endangered. This implies that intellectual property rights might forestall other significant rights such as right to life and right to liberty. This perspective supports that one's liberty is given up when one gives up control or some control over products of one's mind.

There is a contention that some fundamental liberty rights are given up when our intellectual agreements are violated. A pertinent distinction is relevant here between liberty and creativity usually considered as acts of the mind, and the productivity or products of the mind.[19] Werhane and Gorman articulate, "I can sell, give away or sacrifice my property or my creation, but I cannot, without being enslaved, give up my entitlement to liberty and free choice."[20] There is a consensus that without intellectual property rights we may not acknowledge our creativity and the fruits or our labor, but it is rather more debatable whether we are forfeiting all our fundamental liberties. Locke's perspective indicates that we have rights to our bodies and to liberty and consequently we can claim ownership of our own labor and its productivity, and are in position to use and entitled to property rights. Locke contends that life, labor, and liberty are the grounds for property rights and not the opposite.[21] Werhane and Gorman argue that, "without rights to liberty, I can be enslaved, and slavery erodes the justification for the natural or human right to private ownership and thus for ownership of products of the mind.".[22]

---

[16] Zainol et al., "Pharmaceutical Patents and Access to Essential Medicines in sub-Saharan Africa," 12,377.

[17] Werhane and Gorman, "Intellectual Property Rights, Access to Life-Enhancing Drugs, and Corporate Moral Responsibilities," 262.

[18] Ayn Rand, *Capitalism: The Unknown Ideal* (New York: New American Library, 1966), 125, 128.

[19] Werhane and Gorman, "Intellectual Property Rights, Access to Life-Enhancing Drugs, and Corporate Moral Responsibilities," 263.

[20] Werhane and Gorman, "Intellectual Property Rights, Access to Life-Enhancing Drugs, and Corporate Moral Responsibilities," 263.

[21] Werhane and Gorman, "Intellectual Property Rights, Access to Life-Enhancing Drugs, and Corporate Moral Responsibilities," 263.

[22] Werhane and Gorman, "Intellectual Property Rights, Access to Life-Enhancing Drugs, and Corporate Moral Responsibilities," 263.

It may be questionable to argue as Rand implies that intellectual property rights are the foundation for liberty, because that will entail that those without property are less free. It may also be considered a stretch to argue that intellectual property rights can override rights to life in some circumstances. In the past, it has been argued that liberty rights can override rights to life, but it is not plausible to argue that property rights, including the right to the product of our minds, override rights to life or liberty. It is rather more defensible to argue that property rights grow out of, but are not the foundation for, rights to life and liberty.[23]

3.2.1.2 Utilitarian Defenses

Protection of intellectual property rights from utilitarian perspective has been defended with a number of strong arguments. It is usually argued that protection of intellectual property is crucial for the ongoing innovation, creation, and development of novel ideas. Inventors and companies contend that they have rights to protect the patent of their process and product and invariably control the access, because without such protection there will be very limited incentives for new product or development of innovative ideas.[24] Werhane and Gorman argue that, "few people will write new material, create new art, or invent new products without such protections, because there is little in the way of honor, recognition, or profit in such activities.".[25]

A second argument in defense of intellectual property from utilitarian point of view focuses on the importance of patent protection. Patent protection is built on the notion that patents are private property rights that grant unconditional rights over inventions and discoveries.[26] Patent protection is critical to the survival and innovation of pharmaceutical companies that usually require big amounts of money for research and development. It helps pharmaceutical companies and other patent holders to recoup profit on their investments in research and development. The provision gives them the incentive to continue investing in research and development for new drugs.[27]

A third argument regarding the utilitarian defense of intellectual property emphasizes consumer benefits. There is a contention that consumers would benefit more in

---

[23] Werhane and Gorman, "Intellectual Property Rights, Access to Life-Enhancing Drugs, and Corporate Moral Responsibilities," 263.

[24] Werhane and Gorman, "Intellectual Property Rights, Access to Life-Enhancing Drugs, and Corporate Moral Responsibilities," 264.

[25] Werhane and Gorman, "Intellectual Property Rights, Access to Life-Enhancing Drugs, and Corporate Moral Responsibilities," 264.

[26] James Thuo Gathii, "Third World Perspectives on Global Pharmaceutical Access," *in Ethics and the Pharmaceutical Industry*, ed. Michael A. Santoro and Thomas M. Gorrie, (Cambridge: Cambridge University Press, 2005), 337.

[27] James Thuo Gathii, "The Legal Status of the Doha Declaration on TRIPS and Public Health Under the Vienna Convention on the Law of Treaties," *Harvard Journal of Law and Technology* 15, no. 2 (Spring 2002): 294.

the short term, if patents on drugs are removed, which will result in increase in the competition with generic products. Implementation of this strategy will lead to lowering of costs of all drugs. Nevertheless, in the long run consumers would be worse off. This stems from the fact that pharmaceutical companies would not be able to recoup adequate revenues in order to continue to invest in research and development which is crucial for the development of new drugs. The implication is that the development of new drugs would decline slowly, and new life-saving and life-enhancing interventions would not be accessible to future generations.[28]

The fourth argument for the utilitarian defense of intellectual property argues that in the absence of intellectual property protection, pharmaceutical companies such as Pfizer, which relies on patent protection to earn profits and develop products, will not enter countries such as India where patents are not strictly enforced. In countries such as India, reverse engineering of the product development is allowed and it is not illegal under Indian patent law for companies to copy other companies' products, resulting in decreased market share and hindrance from recouping company research and development (R&D) investments.[29] Data from World Bank's early 1990s survey of international executives shows that tax rates and intellectual property protection were the major factors in making decisions regarding global corporate investment.[30] The World Bank survey further indicates that lack of IP protection harms investment in less developed countries, since companies such as Pfizer will not want to invest in countries where patent protections are not enforced.[31] Bale articulates that, "without strong and effective global intellectual property rules, the gap between developed and developing countries will only grow in the future."[32] The same line of argument was supported by the World Intellectual Property Organization (WIPO). In a new book sponsored by WIPO, Idris contends that altering natural resources and products of indigenous populations into intellectual property as well as safeguarding those ideas and others with a patent law can significantly contribute to the affluence of any nation.[33] There are two major classifications of IP namely industry property and artistic and literary property. These properties were formerly governed by the Paris Convention (1883) and

[28] J.W. Hughes, M.J. Moore, and E.A. Snyder, "Napsterizing Pharmaceuticals Access, Innovation, and Consumer Welfare," National Bureau of Economic Research, Working Paper 9229 (2002).

[29] C. L. Clemente, "Intellectual Property: The Patent on Prosperity," *Pfizer Forum* (2001), http://www.pfizerforum.com (Accessed January 17, 2014).

[30] World Bank, (1999), http://www.worldbank.org (Accessed May 25, 2005), cited in Werhane and Gorman, "Intellectual Property Rights, Access to Life-Enhancing Drugs, and Corporate Moral Responsibilities," 265.

[31] World Bank, (1999), http://www.worldbank.org (Accessed May 25, 2005), cited in Werhane and Gorman, "Intellectual Property Rights, Access to Life-Enhancing Drugs, and Corporate Moral Responsibilities," 265.

[32] H.E. Bale, "Patents and Public Health: A Good or Bad Mix?" Pfizer Forum (2001), http://www.pfizerforum.com (Accessed January 17, 2014).

[33] Kamul Idris, *Intellectual Property: A Tool for Economic Growth* (Geneva, Switzerland: World Intellectual Property Organization, 2003), cited in Werhane and Gorman, "Intellectual Property Rights, Access to Life-Enhancing Drugs, and Corporate Moral Responsibilities," 265.

the Berne Convention (1896) respectively. Both conventions have undergone many revisions and are currently administered by the World Intellectual Property Organization which is based in Geneva.[34] The emphasis in our discussion of affordable access to drugs is on the industry property. Rewarding pharmaceutical innovation through strict enforcement of patent protection and ensuring affordable access to essential drugs for the poor especially in developing countries stand in some tension with each other. The World Trade Organization (WTO) has been a major actor in the debate concerning patent law and access to essential medicines.[35] The Trade-Related Aspects of Intellectual Property Rights Agreement (TRIPS) established by WTO was a significant attempt to deal with the issue of global enforcement of patent protection for pharmaceutical products, which adversely impacted affordable access to essential drugs in poor developing countries.

3.2.2 Trade-Related Aspects of Intellectual Property Rights Agreement (TRIPS)

3.2.2.1 Background

The end of the Uruguay Round of Multilateral Trade Negotiations resulted in the establishment of the WTO, with the signing of the Marrakesh Agreement on April 15, 1994. TRIPS Agreement was also included in the new international trading regime, governed by the WTO.[36] All the members of WTO are required to be signatories and are also obliged by TRIPS Agreement, which is managed by the TRIPS Council located in Geneva. Members are required to conform to the TRIPS provisions, but with certain exceptions and emphasis on the way they are implemented. Currently, TRIPS represents a global indication that stresses the protection and enforcement of Intellectual Property Rights (IPRs) at national levels for WTO members.[37]

Before the establishment of TRIPS Agreement, the protection of IPRs in various countries differed significantly. In developed countries such as U.S. there was effective protection,[38] while in many developing countries, protection was either

[34] Zainol et al., "Pharmaceutical Patents and Access to Essential Medicines in sub-Saharan Africa," 12,377.

[35] Thomas Pogge, Matthew Rimmer and Kim Rubenstein, "Access to Essential Medicines: Public Health and International Law," in *Incentives for Global Public Health: Patent Law and Access to Essential Medicines*, ed. Thomas Pogge, Matthew Rimmer and Kim Rubenstein (Cambridge: Cambridge University Press, 2010), 8.

[36] Zainol et al., "Pharmaceutical Patents and Access to Essential Medicines in sub-Saharan Africa," 12,377.

[37] Zainol et al., "Pharmaceutical Patents and Access to Essential Medicines in sub-Saharan Africa," 12,377.

[38] Christopher Lea Lockwwod, "Biotechnology Industry Organization v. District of Columba: A Preemptive Strike Against State Price Restrictions on Prescription Pharmaceuticals," *Albany Law Journal of Science and Technology* 19, no. 1 (April 2009): 148–149.

nonexistent or enforcement was tepid.[39] There was a consideration by developed countries such as U.S., that this situation was militating against their interests and consequently with U.S. at the forefront they fought hard and succeeded in incorporating TRIPS in the Marrakesh Agreement.[40]

The TRIPS Agreement includes a variety of intellectual property issues outside patents, such as trademarks, industrial designs, and copyright applicable to any sector.[41] It offers minimum standards for intellectual property law, procedures and solutions and grants rights' holders exclusive rights to effectively enforce their rights. The chief rule of TRIPS for patents encompasses their availability for any invention, either product or process, in all fields of technology with exception. Inventions included under the patent law have to fulfil the standards for novelty, inventive step and industrial applicability.[42] Article 7 of TRIPS makes provision for the global effective protection and enforcement of IPRs, with emphasis also on easing technological innovation and diffusion of technology.[43]

Some scholars contend that the primary objective of establishing TRIPS was the promotion of the global protection and enforcement of IPRs.[44] However, other relevant provisions of TRIPS, such as Articles 8, 30 and 73, as well as the Doha Ministerial Declaration of November 2001, attest to the fact that the primary objective of TRIPS goes beyond the protection of IPRs.[45] Article 8 of the TRIPS Agreement makes provision for members to adopt measures essential to promote public health and nutrition, and to promote the public interest in important areas relevant to their socio-economic and technological development, as far as such measures conform to the provisions of this Agreement.[46] More so, at Doha representatives of various countries agreed that the least developed country members classified by United Nations based on several indicators comprising income, nutrition, health,

---

[39] Baris Karapinar and Michelangelo Temmerman, "Benefiting from Biotechnology: Pro-Poor Intellectual Property Rights and Public-Private Partnerships," *Biotechnology Law Report* 27, no. 3 (June 2008): 189–202.

[40] Marla L. Mellino, "The TRIPS Agreement: Helping or Hurting Least Developed Countries' Access to Essential Pharmaceuticals? *Fordham Intellectual Property, Media and Entertainment Law Journal* 20, no. 4 (Summer 2010): 1349–1388.

[41] Jillian Clare Cohen and Patricia Illingworth, "The Dilemma of Intellectual Property Rights for Pharmaceuticals: The Tension Between Ensuring Access of the Poor to Medicines and Committing to International Agreements," *Developing World Bioethics* 3, no. 1(May 2003): 28.

[42] Cohen and Illingworth, "The Dilemma of Intellectual Property Rights for Pharmaceuticals: The Tension Between Ensuring Access of the Poor to Medicines and Committing to International Agreements," 29.

[43] Zainol et al., "Pharmaceutical Patents and Access to Essential Medicines in sub-Saharan Africa," 12,377.

[44] Mellino, "The TRIPS Agreement: Helping or Hurting Least Developed Countries' Access to Essential Pharmaceuticals? 1349–1388.

[45] Zainol et al., "Pharmaceutical Patents and Access to Essential Medicines in sub-Saharan Africa," 12,377.

[46] Pogge, Rimmer and Rubenstein, "Access to Essential Medicines: Public Health and International Law," 9.

education, literacy, and economic vulnerability, were not required to implement patent law for pharmaceuticals until January 1, 2016. Among the 50 least developed countries, 32 of them are WTO members.[47]

Article 27 (1) of TRIPS Agreement obliges all members to broaden patent protection for a minimum period of 20 years to any invention in all fields of technology. It encompasses pharmaceutical patents, which grant the holders exclusive rights recognized globally to manufacture, use, sell and import patented medicines.[48] These developments are unprecedented. Before TRIPS, the Paris Convention did not mandate the broadening of patents to any area of technology, neither did it require the transfer of exclusive patents, nor stipulate a minimum duration for such rights.[49] For example, more than forty countries did not patent pharmaceuticals, several others such as India patented only pharmaceutical processes, and others offered shorter patent periods.[50] Countries such as Thailand granted pharmaceutical patents only for 3 years, while South Africa limited the duration of patents to only 16 years. Some countries in sub-Saharan Africa, Angola, Ghana and Malawi did not patent pharmaceutical products before the introduction of TRIPS.[51] Forman articulates that "introducing patents where there were previously none drives up drug prices by enabling monopoly pricing and excluding cheaper generic alternatives. Given how price sensitive drug access is in poor countries, higher prices can significantly limit access for the poor.".[52]

However, the TRIPS Agreement makes provisions for protecting public health and for governments to effectively respond to national health emergencies. Article 31 (f) authorizes governments to issue a compulsory license for a patented drug without the permission of the patent holder whenever it can be justified on the grounds of public interest.[53] The implication here is that in the event of a national health emergency, a government who is a member of WTO is allowed to break a patent by authorizing a third party to produce a generic version of patented drug without the permission of the patent holder.[54] For example, the threat of avian bird flu pandemic resulted in the pressure for Roche to relax patent restrictions on a drug

---

[47] Michael Westerhaus and Arachu Castro, "How Do Intellectual Property Law and International Trade Agreements Affect Access to Antiretroviral Therapy? *PLoS Medicine* 3, no. 8 (August 2006): e332.

[48] Zainol et al., "Pharmaceutical Patents and Access to Essential Medicines in sub-Saharan Africa," 12,377.

[49] Zainol et al., "Pharmaceutical Patents and Access to Essential Medicines in sub-Saharan Africa," 12,377.

[50] Forman, "Trade Rules, Intellectual Property, and the Right to Health," 337–357.

[51] Zainol et al., "Pharmaceutical Patents and Access to Essential Medicines in sub-Saharan Africa," 12,377.

[52] Forman, "Trade Rules, Intellectual Property, and the Right to Health," 337–357.

[53] Cohen and Illingworth, "The Dilemma of Intellectual Property Rights for Pharmaceuticals: The Tension Between Ensuring Access of the Poor to Medicines and Committing to International Agreements," 29–30.

[54] Zainol et al., "*Pharmaceutical Patents and Access to Essential Medicines in sub-Saharan Africa*," 12,378.

effective against bird influenza known as oseltamivir.[55] In another instance, during the fall of 2001 anthrax attacks, the US government under pressure decisively stepped in to break Bayer's patent on ciprofloxacin for the benefit of increasing availability of the drug.[56] TRIPS also makes provision for parallel imports in the interest of public health and social welfare. Parallel importing authorizes countries to import cheaper priced patented drugs without any restrictions.[57]

Conforming to the TRIPS provisions is mandatory for all WTO members, and is a requirement for the membership of the Organization. Members are sanctioned if they do not comply with the provisions. A member who violates trade agreements could receive summons from another aggrieved member for dispute settlement enforced by WTO dispute panel known as Dispute Settlement Understanding (DSU). DSU is a method devised by WTO for resolving IP and other trade disputes among members. WTO authorizes a member to use retaliatory trade measures against another erring member, which usually have severe adverse effects on the domestic economy of the latter. The erring member could also be impacted by other consequences such as negative international publicity and poor perception as an untrustworthy trade partner, inappropriate for foreign investment.[58] For example, South Africa passed an amendment to its Medicines and Related Substances Act, which permitted the utilization of compulsory licensing and parallel importing in order to provide cheap priced medications to South Africans in need.[59] The amendment was not fully implemented due to the pressure from the U.S. government and the multinational pharmaceutical companies opposing the overruling of patents in order to enhance affordable access to essential drugs in the world capital of the HIV/AIDS crisis.[60] The United States utilized trade sanctions against South Africa as a retaliatory measure by including it in the infamous section 301 watch list. Section 301 which permits the United States to utilize unilateral trade sanctions was a retaliatory measure against any trade partner that violates patents established under TRIPS Agreement.[61] It has been argued that extremely strong legal protection of patents realized through TRIPS Agreement under the auspices of WTO plays a critical role in limiting access to essential drugs, including anti-retroviral drugs for the developing countries.[62]

[55] Westerhaus and Castro, "How Do Intellectual Property Law and International Trade Agreements Affect Access to Antiretroviral Therapy? 1230.

[56] Westerhaus and Castro, "How Do Intellectual Property Law and International Trade Agreements Affect Access to Antiretroviral Therapy? 1230.

[57] Forman, "Trade Rules, Intellectual Property, and the Right to Health," 337–357.

[58] Zainol et al., "Pharmaceutical Patents and Access to Essential Medicines in sub-Saharan Africa," 12,378.

[59] Gathii, "Third World Perspectives on Global Pharmaceutical Access," 339.

[60] Rosalyn S. Park, "The International Drug Industry: What the Future Holds for South Africa's HIV/AIDS Patients," *Minnesota Journal of Global Trade* 11, no.1 (Winter 2002) 125, 136–139.

[61] Gathii, "Third World Perspectives on Global Pharmaceutical Access," 339.

[62] Gathii, "Third World Perspectives on Global Pharmaceutical Access," 337.

3.2.2.2 TRIPS Agreement and Access to Drugs in Developing Countries

There has been an ongoing vigorous debate on the impact of strict patent protection on affordable access to drugs in developing countries. Two major perspectives emerged. The first perspective championed by Pharmaceutical companies and developed countries is articulated by supporters of strong patent protection. The second perspective defended by human rights activists and developing countries is articulated by critics of strict patent protection. Ferreira writes, "The United Nations (U.N.) and non-governmental human rights organizations claim that patents are a major factor in the lack of access to HIV/AIDS drugs, a point hotly disputed by the drug industry and its proponents."[63] Pharmaceutical companies and developed countries promote strong patent protection, disapprove of compulsory licensing and parallel importing, and blame developing countries such as South Africa of violating their legal obligations under TRIPS Agreement for adopting the stalled Medicines Act Amendment.[64] Supporters of strict patent protection contend that patents are not responsible for lack of affordable access to HIV/AIDS drugs in developing countries, rather, they attribute it to non-patent factors such as poverty, poor or inadequate health infrastructure, the lack of political will and commitment on the part of government to fighting HIV/AIDS, and cultural barriers.[65]

On the other hand, developing countries and human rights activists strongly advocate for the use of compulsory licensing and parallel importing in order to enhance affordable access to HIV/AIDS drugs in developing countries. They also contend that laws created to increase affordable access to drugs are legal under various public and social welfare exceptions of TRIPS Agreement.[66] They blamed lack of affordable access to anti-retroviral drugs in developing countries on extremely strong legal protection of patents. Donald argues that despite the impact of non-patent factors, the broadening and strengthening of IP protection under TRIPS, especially patents, would further significantly impede lack of access to essential drugs such as anti-retroviral drugs in poor developing countries.[67] A detailed analysis of the role of patent and non-patent factors in impeding access to affordable HIV/AIDS drugs in developing countries follows.

---

[63] Lissett Ferreira, "Access to Affordable HIV/AIDS Drugs: The Human Rights Obligations of Multinational Pharmaceutical Corporations," *Fordham Law Review* 71, no. 3 (December 2002): 1137–1138.

[64] Ferreira, "Access to Affordable HIV/AIDS Drugs: The Human Rights Obligations of Multinational Pharmaceutical Corporations," 1138.

[65] Ferreira, "Access to Affordable HIV/AIDS Drugs: The Human Rights Obligations of Multinational Pharmaceutical Corporations," 1138–1139.

[66] Ferreira, "Access to Affordable HIV/AIDS Drugs: The Human Rights Obligations of Multinational Pharmaceutical Corporations," 1138.

[67] Anna Donald, "The Political Economy of Technology Transfer," *BMJ* 319, no. 7220 (November 1999): 1298.

Patents

Research and Development costs in pharmaceutical industry are very exorbitant and high, as well as the risk of failure. Pharmaceutical companies would not be able to recover their costs and make profit without patents.[68] The primary objective of patents is to offer a temporary monopoly to rights holders as a motivation to innovations and their commercialization.[69] The implication is that the monopoly rights enshrined in patents enable pharmaceutical companies to recoup research and development costs and generate profit. Therefore, patenting pharmaceutical drugs would motivate them to engage in more research in order to manufacture new drugs for the benefit of the society.[70]

However, it is debatable whether pharmaceutical patents offer incentives also for research on drugs for the treatment of diseases predominant in developing countries, considering their small market and weak purchasing power. It has been estimated that averagely, pharmaceutical companies require about a minimum profit of $1 billion in order to embark on any risk of researching a particular disease.[71] To this extent, pharmaceutical patents offer very limited commercial incentive to pharmaceutical companies to engage in research relevant to the diseases affecting majority of the poor people in developing countries. For example, the amount of money invested globally on pharmaceutical R&D for diseases prevalent in developing countries is estimated to be less than 5%.[72]

Furthermore, out of the 1393 drugs approved from 1975 to 1999, only 13 were related to diseases prevalent in developing countries.[73] There is very limited research on malaria, tuberculosis and sleeping sickness.[74] Conversely, the story is completely different for diseases such as HIV/AIDS affecting both developed and developing countries. For example, in the U.S. as at 2002, there were 64 approved drugs for the treatment of HIV/AIDS, while 103 are still in the process of development.[75]

---

[68] Angell Marcia, "The Truth About the Drug Companies," *New York Review of Books* (July 15 2004), http://www.nybook.com/articles/archives/2004/jul/15/the-truth-about-the-drug-companies/?pagination=false (Accessed January 29, 2014).

[69] Commission on Intellectual Property Rights, *Integrating Intellectual Property Rights and Development Policy*, (London: Commission on Intellectual Property Rights, 2002), 34.

[70] Commission on Intellectual Property Rights, *Integrating Intellectual Property Rights and Development Policy*, 34.

[71] Commission on Intellectual Property Rights, *Integrating Intellectual Property Rights and Development Policy*, 32.

[72] Commission on Intellectual Property Rights, *Integrating Intellectual Property Rights and Development Policy*, 32.

[73] Commission on Intellectual Property Rights, *Integrating Intellectual Property Rights and Development Policy*, 32.

[74] Commission on Intellectual Property Rights, *Integrating Intellectual Property Rights and Development Policy*, 32.

[75] Commission on Intellectual Property Rights, *Integrating Intellectual Property Rights and Development Policy*, 32.

Critics of patent protection have charged that monopoly pricing which has been made possible by patents impedes affordable access to drugs to those who need them most especially in developing countries.[76] Proponents of pharmaceutical patents countered that "poverty rather than patents is the main problem, and activists should focus their energy on poverty alleviation rather than IPR protection."[77] They reframed the debate on two grounds. First, that very few drugs are patented in developing countries and they cannot constitute a significant impediment in accessing drugs. Second, that even if many more drugs are patented that they do not become a determining factor in pricing, but rather that there are other superseding factors that impede access to drugs by the poor.[78]

The prevalence of patenting pharmaceutical products was explored as driving the debate. There was an indication that pharmaceutical companies do not usually seek to patent their products in developing countries because they have small markets and limited technological capacity. Pharmaceutical companies do not consider it lucrative to patent their drugs and enforce the patent when the potential market is small and the risk of infringement low.[79]

The International Intellectual Property Institute (IIPI), a pro-intellectual property think tank established in 1999, and currently located in Washington, D.C. was among the first to argue for lack of correlation between strong patent protection and impeding access to essential drugs especially by the poor in developing countries. In 2000, IIPI published a report that explores the prevalence of the HIV/AIDS epidemic in Africa from three different perspectives. First, it examined the international community's response, with particular reference to the levels of foreign aid offered by Western countries such as the United States.[80] Second, it analyzed patent systems in various countries in Africa. Third, it explored the number of HIV/AIDS drugs patented in these countries.[81] The report's conclusion was that access to essential drugs comprises "numerous and complex issues, including healthcare infrastructure, international pricing mechanisms, financing, debt, tariffs and patents.".[82]

---

[76] Michael J. Selgelid and Eline M. Sepers, "Patents, Profits, and the Price of Pills: Implications for Access and Availability," in *The Power of Pills: Social, Ethical and Legal Issues in Drug Development, Marketing and Pricing* ed. Jillian Clare Cohen, Patricia Illingworth and Udo Schuklenk (London: Pluto Press, 2006), 156.

[77] Selgelid and Sepers, "Patents, Profits, and the Price of Pills: Implications for Access and Availability," 156.

[78] Commission on Intellectual Property Rights, *Integrating Intellectual Property Rights and Development Policy*, 35.

[79] Commission on Intellectual Property Rights, *Integrating Intellectual Property Rights and Development Policy*, 35.

[80] International Intellectual Property Institute, *Patent Protection and Access to HIV/AIDS Pharmaceuticals in Sub-Saharan Africa,* (Washington, DC: IIPI, 2000), 6–8.

[81] International Intellectual Property Institute, *Patent Protection and Access to HIV/AIDS Pharmaceuticals in Sub-Saharan Africa,* 36–38.

[82] International Intellectual Property Institute, *Patent Protection and Access to HIV/AIDS Pharmaceuticals in Sub-Saharan Africa,* 2.

Furthermore, the report came to a conclusion that the TRIPS Agreement does not constitute a barrier to the distribution of HIV/AIDS drugs based on three reasons. First, the TRIPS Agreement was not implemented in most countries in sub-Saharan Africa. Second, the TRIPS Agreement allows adequate flexibility for countries to circumvent negative effects. Third, most pharmaceutical companies have not sought patents for their products extensively in Africa.[83] The report did not attribute the primary barrier to access of HIV/AIDS drugs to poverty, but rather stressed that the core issue stems from sufficient financing of the overall health system as well as the development of healthcare infrastructure. The report gave an indication for the need to do more research in order to conclusively establish whether or not patents and TRIPS Agreement played any critical role regarding access to affordable drugs.[84]

Two recent studies by Attaran and Gillespie-White have been cited to support proponents' argument. The first study examined the extent of patenting of 15 vital anti-retroviral drugs in 53 African countries. They found that most of the anti-retroviral drugs for treatment of HIV/AIDS were patented in only a few African countries estimated at about 21.6%. There were no patents at all on these drugs in 13 of the 53 countries. They concluded that because the patenting rate was very small, patents generally do not constitute a significant barrier to treatment access in Africa. However, it acknowledged that there would be an issue when TRIPS becomes fully implemented by all WTO members.[85]

The second study authored by Attaran explores the extent of patenting for essential medicines in low-income and middle-income countries. The study reveals that patenting for 319 drugs classified by WHO as essential drugs is infrequent in 65 low and middle income countries, with estimated population of four billion people. Patents exist for only 17 essential drugs, but most of them were not actually patented. The estimated rate of patent is as low as 1.4% and it was focused on countries with larger markets and adequate technological capacity. There was a conclusion in the study that patents for essential drugs are usually not common in poor countries and consequently cannot easily explain why access to those drugs is frequently lacking, indicating that poverty, not patents, accounts for more limitation on access.[86]

Both studies have been criticized on various grounds. The first study authored by Attaran and Gillespie-White fails to recognize that not all existing antiretroviral drugs are important to the same degree in treating the disease. Critics contend that the quantitative method employed by Attaran and Gillespie-White is deceptive because the most effective combinations of anti-retroviral drugs are usually

[83] International Intellectual Property Institute, *Patent Protection and Access to HIV/AIDS Pharmaceuticals in Sub-Saharan Africa,* 3.

[84] International Intellectual Property Institute, *Patent Protection and Access to HIV/AIDS Pharmaceuticals in Sub-Saharan Africa*, 3.

[85] Amir Attaran and Lee Gillespie-White, "Do Patents for Antiretroviral Drugs Constrain Access to AIDS Treatment in Africa?," *JAMA* 286, no. 15 (October 2001): 1886–1892.

[86] Amir Attaran, "How Do Patents and Economic Policies Affect Access to Essential Medicines in Developing Countries?," *Health Affairs* 23, no. 3 (May 2004): 155–166.

obstructed in many of the African countries.[87] Another criticism is that the study did not recognize the significant impact patents in one country can have on other countries. For example, South Africa has about 13 out of 15 antiretroviral drugs patented. Therefore, South Africa as the most affluent country in Africa would have been in the best situation with its strong technological capacity to manufacture and distribute generic drugs to its neighbors.[88]

Critics indicated that the first study conducted by Attaran and Gillespie-White has fundamental scientific limitations. The study did not consider extraneous variables such as levels of wealth and size of market and consequently it weakens any deduction that was made regarding correlation of geographic patent coverage with anti-retroviral treatment access in Africa. To examine whether patents cause lack of access to drugs in developing countries, it will be important to compare countries with the same characteristics such as equally wealthy or equally large markets.[89]

The second study was also found to have fundamental limitations. Attaran utilized WHO's essential medicines lists for examining the patency prevalence of drugs. The WHO considers the cost when preparing this list, which implies that cheap drugs are favored. Consequently, it does not come as a surprise that so few essential drugs are patented.[90] Selgelid and Sepers argue that "if patents increase prices and thus make medicines less likely to appear on the list, then it should be no surprise that few drugs on the list turn out to be patented."[91] This view undermines the significance of Attaran's finding that so few essential drugs are patented.[92] Critics also contend that Attaran and Gillespie-White did not recognize the apparent failure of patents to provide incentives that usually result in the global development of drugs. Patents provide slight incentive to develop medical technologies precisely needed by the poor. Patents facilitate price increases, but this does not result in profits if those in need are not able to pay high prices for the products.[93]

There is extensive evidence from developed countries that prices fall fairly sharply once drugs patents expire, and if there are generic competitors. The price fall appears to be larger with entry of more generic competitors into the market.[94] Governments can promote price reductions by easing generic producers' early entry

---

[87] M. Boelaert et al., "Letter to the Editor: Do Patents Prevent Access to Drugs for HIV in Developing Countries?," *JAMA* 287, no. 7 (February 2002): 840–841.

[88] Michael J. Selgelid and Udo Schuklenk, "Letter to the Editor: Do Patents Prevent Access to Drugs for HIV in Developing Countries?," *JAMA* 287, no. 7 (February 2002): 842–843.

[89] Selgelid and Sepers, "Patents, Profits, and the Price of Pills: Implications for Access and Availability," 158.

[90] E. Goemaere et al., "Patent Status Matters," *Health Affairs* 23, no. 3 (May 2004): 279–280.

[91] Selgelid and Sepers, "Patents, Profits, and the Price of Pills: Implications for Access and Availability," 158.

[92] Selgelid and Sepers, "Patents, Profits, and the Price of Pills: Implications for Access and Availability," 159.

[93] Selgelid and Sepers, "Patents, Profits, and the Price of Pills: Implications for Access and Availability," 159.

[94] Commission on Intellectual Property Rights, *Integrating Intellectual Property Rights and Development Policy*, 36.

into the market. For example, the 1984 Drug Competition and Patent Term Restoration Act in the United States popularly known as the Hatch-Waxman Act accomplished this objective, which lead to a significant increase in the delivery of generic versions of prescription drugs from 19% in 1984 to 47% in 2000. The size of market for generic drugs is even larger in other developed countries such as the United Kingdom.[95] Pharmaceutical companies have instituted law suits in order to delay or block generic entry of producers and to defend or extend a monopoly on a successful drug.[96] A recent study conducted in the US revealed that prices sharply fall when there is intense generic competition in the market, but a minimum of about five generic competitors are required to drive down prices to an extent.[97]

Developing countries can also mitigate the impact of patent protection for their population by easing generic entry and generic competition. However, in most instances their choices are strictly limited by the small size of their markets and lack of local technological, productive and regulatory capacity.[98] Commission on Intellectual Property Rights articulates, "It is this lack of capacity to create a competitive environment for both patented and generic products that makes the existence of patents more contentious than in developed markets with greater capacity to enforce a strongly pro-competitive regulatory environment.".[99]

Evidence abounds in the international arena that drugs patented in countries with strong patent protection are much cheaper in markets which do not provide patent protection. For example, the Indian market which does not offer product protection at a time has the cheapest drugs in the world.[100] Nevertheless, with the introduction of drug product patents in India in 2005, there were expectations of significant increase in drug costs. For example, a case study of the influence of introducing patents on four domestic antibiotics projected that the total annual welfare losses to the economy of India stemming from increases in price and limits in access would be about $305 million or around 50% of the sales of the whole systemic antibacterial section in 2000.[101]

---

[95] Commission on Intellectual Property Rights, *Integrating Intellectual Property Rights and Development Policy*, 36.

[96] Commission on Intellectual Property Rights, *Integrating Intellectual Property Rights and Development Policy*, 36.

[97] Commission on Intellectual Property Rights, *Integrating Intellectual Property Rights and Development Policy*, 36.

[98] Commission on Intellectual Property Rights, *Integrating Intellectual Property Rights and Development Policy*, 36.

[99] Commission on Intellectual Property Rights, *Integrating Intellectual Property Rights and Development Policy*, 36.

[100] Commission on Intellectual Property Rights, *Integrating Intellectual Property Rights and Development Policy*, 36.

[101] S. Chaudhuri, P.K. Goldberg, and P.J. Jia, "The Effects of Extending Intellectual Property Rights Protection to Developing Countries: A Case Study of the Indian Pharmaceutical Market," National Bureau of Economic Research Working Paper no. 10159 (December 2003), http://www.nber.org/papers/w10159 (Accessed February 1, 2014).

Furthermore, introducing global drug patents result in a systemic influence on the production and export of generic versions of new drugs. The implementation of TRIPS will ultimately results in phasing out the generic production of patented drugs entirely, unless this is completed through compulsory licensing. This will restrict domestic production of generic drugs, especially in India, which has been a principal source of generic antiretroviral drugs for other developing countries.[102] The full implementation of TRIPS by 2016 will specifically influence developing countries that depend on importation of generic copies of drugs currently patented.[103] There was an urgent need to develop measures that will continue to guarantee that the patent system supports the right of every country to protect human health and to enhance access to essential drugs in accordance with the Doha Declaration on TRIPS and Public Health and WTO General Council Decision of 30 August 2003.[104]

Doha Declaration and TRIPS and Public Health

The TRIPS Agreement made provisions for parallel importation in Article 6 and compulsory licensing in Article 31 as tools for protecting public health and increasing access to essential drugs especially anti-retroviral drugs.[105] However, some members of WTO did not interpret and implement TRIPS in a way that promotes public health. Two divergent interpretations of the TRIPS Agreement emerged at the special section of the TRIPS council held in June 2001. The purpose of this special section was to clarify the relationship between intellectual property rights and access to essential drugs under TRIPS Agreement.[106]

The objective of the African group and other developing countries was to elucidate the degree to which TRIPS Agreement permits members to promote and safeguard public health and other all-embracing public policy goals.[107] Furthermore, developing countries stressed that restrictive interpretation of TRIPS as advanced by the United States and other developed countries would excessively restrict their

[102] Medecins Sans Frontieres, "Will the Lifeline of Affordable Medicines for Poor Countries be Cut?: Consequences of Medicines Patenting in India," External Briefing Document (Geneva, Switzerland: MSF, 2005), 3, http://www.who.int/hiv/amds/MSFopinion.pdf (Accessed February 2, 2014).

[103] F.M. Scherer and J. Watal, "Post-TRIPS Options for Access to Patented Medicines in Developing Countries," Commission on Macroeconomics and Health Working Paper no. WG4, no.1 (June 2001): 13.

[104] Commission on Intellectual Property Rights, *Integrating Intellectual Property Rights and Development Policy,* 38.

[105] Ferreira, "Access to Affordable HIV/AIDS Drugs: The Human Rights Obligations of Multinational Pharmaceutical Corporations," 1140.

[106] Gathii, "The Legal Status of the Doha Declaration on TRIPS and Public Health Under the Vienna Convention on the Law of Treaties," 296.

[107] Gathii, "The Legal Status of the Doha Declaration on TRIPS and Public Health Under the Vienna Convention on the Law of Treaties," 296.

ability to tackle public health emergencies such as AIDS.[108] Developing countries emphasized that the TRIPS Agreement does not prevent members from taking measures to protect public health.[109] The implication is that "TRIPS does not remove a member's sovereign power to address public health emergencies within its own borders.".[110]

The United States and other developed countries argue that strict patent protection would offer benefits to all countries, but at the same time recognizing the interests of developing countries in access to essential drugs. They further argue that the TRIPS Agreement strikes a balance between incentives for innovation and affordable access to essential drugs.[111] Developed countries contend that the most effective strategy for tackling public health emergencies involves economic, social and health policies. These policies need strong patent protection to support the development of new drugs.[112]

Despite these divergent perspectives in the interpretation and implementation of the TRIPS Agreement, a Declaration on TRIPS and Public Health was released by a broad unanimity of all WTO members at the Doha Ministerial meeting in Qatar in November 2001.[113] Sell and Odell articulated that the Doha Declaration was enabled by a shared and united efforts from a coalition of civil society organizations, developing countries and mid-tier countries, such as Thailand, India and Brazil.[114] Sridhar indicates that although the Declaration may be imperfect and far from being an ideal document from moral view point, compromise was arguably essential to realize agreement on it, which was extremely more desirable to no Declaration at all.[115]

The Doha Declaration affirming support for public health clearly articulates, "We agree that the TRIPS Agreement does not and should not prevent members from taking measures to protect public health. Accordingly, while reiterating our commitment to the TRIPS Agreement, we affirm that the Agreement can and should be interpreted and implemented in a manner supportive of WTO members' right to

[108] Gathii, "The Legal Status of the Doha Declaration on TRIPS and Public Health Under the Vienna Convention on the Law of Treaties," 296.

[109] Gathii, "The Legal Status of the Doha Declaration on TRIPS and Public Health Under the Vienna Convention on the Law of Treaties," 298.

[110] Gathii, "The Legal Status of the Doha Declaration on TRIPS and Public Health Under the Vienna Convention on the Law of Treaties," 298.

[111] Gathii, "The Legal Status of the Doha Declaration on TRIPS and Public Health Under the Vienna Convention on the Law of Treaties," 296.

[112] Gathii, "The Legal Status of the Doha Declaration on TRIPS and Public Health Under the Vienna Convention on the Law of Treaties," 298.

[113] Gathii, "The Legal Status of the Doha Declaration on TRIPS and Public Health Under the Vienna Convention on the Law of Treaties," 298.

[114] John Odell and Susan Sell, "Reframing the Issue: The WTO Coalition on Intellectual Property and Public Health," in *Negotiating Trade: Developing Countries in the WTO and NAFTA* ed. John Odell (Cambridge: Cambridge University Press, 2006) cited in *Incentives for Global Public Health: Patent Law and Access to Essential Medicines* ed. Pogge, Rimmer and Rubenstein, 11.

[115] Devi Sridhar, "Improving Access to Essential Medicines: How Health Concerns can be Prioritised in the Global Governance System," *Public Health Ethics* 1, no. 2 (July 2008): 83–88.

protect public health and in particular to promote access to medicines for all."[116] The Declaration acknowledges that HIV/AIDS, tuberculosis, malaria, and other epidemics are grave public health problems afflicting developing and least developed countries. It also reasserts, "the right of the WTO Members to use, to the full, the provisions in the TRIPS Agreement, which provide flexibility for this purpose."[117] Paragraph 5(b) of the Declaration confirms Members' right to grant compulsory licenses, as well as the right to determine the grounds for granting such licenses. Paragraph 5(c) stresses members' right to determine what constitutes a national emergency, or other circumstances of extreme urgency, such as, but not limited to HIV/AIDS, malaria and TB. The Declaration also emphasized that members were free to establish their own regimes for parallel importation without challenge.[118] The implication is that the Doha Declaration made provisions for TRIPS Agreement to be responsive to the healthcare needs of developing countries and to underscore how members could utilize its flexibilities to achieve that purpose.[119]

The Doha Declaration also acknowledged the problem presented by the TRIPS requirement which specifies that compulsory licensing shall be "predominantly for the supply of the domestic market."[120] This implies that developing countries without domestic manufacturing capacity will not be able to access generic drugs for their population. Consequently, in paragraph 6, the Doha Declaration acknowledged the need for an expeditious solution to the problem encountered specifically by developing countries without local manufacturing capacity.[121]

Post Doha – The WTO Decision of August 2003

Arriving at a consensus on what was dubbed the paragraph 6 problem resulted in protracted debate among TRIPS council members. The US headed a relentless effort to limit the provisions of paragraph 6 of the Doha Declaration to particular

---

[116] World Trade Organization, *Doha Declaration on the TRIPS Agreement and Public Health*, (November 20, 2001), http://www.wto.org/english/thewto_e/minist_e/min01_e/mindecl_trips_e.htm. (Accessed February 5, 2014).

[117] World Trade Organization, *Doha Declaration on the TRIPS Agreement and Public Health*, (November 20, 2001), http://www.wto.org/english/thewto_e/minist_e/min01_e/mindecl_trips_e.htm. (Accessed February 5, 2014).

[118] World Trade Organization, *Doha Declaration on the TRIPS Agreement and Public Health*, (November 20, 2001), http://www.wto.org/english/thewto_e/minist_e/min01_e/mindecl_trips_e.htm. (Accessed February 5, 2014).

[119] Haochen Sun, "The Road to Doha and Beyond: Some Reflections on the TRIPS Agreement and Public," *European Journal of International Law* 15, no. 1 (2004): 123–125.

[120] World Trade Organization, *TRIPS Agreement on Trade –Related Aspects of Intellectual Property Rights, Part II- Standards Concerning the Availability, Scope and Use of Intellectual Property Rights*, (1994), http://www.wto.org/english/tratop_e/trips_e/t_agm3c_e.htm . (Accessed February 5, 2014).

[121] World Trade Organization, *Doha Declaration on the TRIPS Agreement and Public Health*, (November 20, 2001), http://www.wto.org/english/thewto_e/minist_e/min01_e/mindecl_trips_e.htm. (Accessed February 5, 2014).

diseases such as HIV/AIDS, malaria, tuberculosis, and other infectious diseases resulting in epidemics.[122] Apart from restricting the use of compulsory licenses, the US intensified efforts to restrict the number of countries that could benefit from the importation of generic drugs.[123]

In August 2003, the WTO General Council issued the decision on the implementation of Paragraph 6 of the Doha Declaration, which specifies that countries without local manufacturing capacity could issue compulsory licenses and on that foundation legally import generic drugs.[124] The export solution is aimed at authorizing developing countries without local production capacity to import generic drugs made under compulsory licensing in accordance with strict conditions.[125] For example, some conditions outlined include: both importing and exporting countries must declare compulsory licenses; eligible importing members other than least-developed countries must establish insufficient or no manufacturing capacities in the pharmaceutical sector for the products in question; such drugs are restricted to the amount required to fulfill the needs of the importing country and must be entirely imported to the member; the drugs must be clearly identified as manufactured under this system through labeling, marked by packaging and/or shaping and coloring; and importing countries must utilize reasonable measures to prevent re-exportation of products.[126]

However, the WTO August 2003 decision was only a temporary waiver, pending a consensus on a permanent amendment.[127] Efforts to reach a consensus on a permanent amendment to TRIPS were accompanied by further disagreement among WTO members. The US and other developed countries forcefully argued for formal approval of the temporary waiver as a permanent amendment. Conversely, developing countries, led by the African Group, contended that the temporary waiver contained too many procedural problems that would still impede access to essential drugs including anti-retroviral drugs for countries without local manufacturing capacity.[128] Furthermore, Medecins Sans Frontieres objected to making the temporary waiver permanent arguing that it would be imprudent because no country had

---

[122] Westerhaus and Castro, "How Do Intellectual Property Law and International Trade Agreements Affect Access to Antiretroviral Therapy? 1231.

[123] B. Loff, "No Agreement Reached in Talks on Access to Cheap Drugs," *Lancet* 360, no. 9349 (December 2002): 1951.

[124] Westerhaus and Castro, "How Do Intellectual Property Law and International Trade Agreements Affect Access to Antiretroviral Therapy? 1231.

[125] Forman, "Trade Rules, Intellectual Property, and the Right to Health," 337–357.

[126] WTO General Council, "Implementation of Paragraph 6 of the Doha Declaration on the TRIPS Agreement and Public Health: Decision of the General Council of 30 August 2003," WT/L/540, (September 1, 2003), paras 2(a) (ii), (b) (i) and (ii), and paras 4and 5, http://www.wto.org/English/tratop_etrips_e/implem_para6_e.htm (Accessed February 5, 2014).

[127] Westerhaus and Castro, "How Do Intellectual Property Law and International Trade Agreements Affect Access to Antiretroviral Therapy? 1231.

[128] Martin, Khor, Impasse on Talks on TRIPS and Health "Permanent Solution." Third World Network(October26,2005),http://www.health-now.org/site/article.php?articleId=499&menuId=1. (Accessed February 5, 2014).

really used it.[129] So far, the export solution has been scarcely used. As of 2009, Rwanda is the first and only country to utilize the WTO General Council Decision of 30 August 2003 with its application for the importation of inexpensive generic drugs from Canada.[130] Several factors contributed to failure of the WTO members especially developing countries to utilize the export solution, including constant threats of legal or economic sanctions from pharmaceutical companies and developed countries especially US, and the difficulty, cost and limited duration and scope of the rules.[131]

Regardless of protracted debate on the status of the temporary waiver for the export solution, WTO members reached a consensus in early December 2005, just before the WTO Ministerial Conference in Hong Kong, to make it permanent if at least two-thirds of the 148 WTO members formally approved the amendment by December 1, 2007.[132] Moreover, there have been concerns that the United States Trade Representative has negotiated TRIPS-Plus bilateral and regional trade agreements that undercut the goal and effect of the Doha Declaration and the WTO General Council Decision of the 30 August 2003.[133]

3.2.2.3 US Trade Policy

In January 2003, President Bush announced in the State of the Union address to Congress his five-year initiative for the United States to support the global effort to fight the HIV/AIDS epidemic. The proposed allocation fund for the initiative now known as the President's Emergency Plan for AIDS Relief (PEPFAR) was $15 billion in order to provide AIDS drugs for two million people living with HIV/AIDS, to provide education to prevent seven million new infections, and to support care for ten million AIDS patients and orphans.[134]

Eight months after the implementation of the US HIV/AIDS plan, significant progress report was given by PEPFAR in accomplishing its goals. By March 2005, 155, 000 people were receiving anti-retroviral drugs, 1.2 million women and babies had benefited from measures preventing mother-to-child transmission of HIV, and 1.7 people infected or affected by HIV/AIDS were receiving supportive care with

[129] Medecins Sans Frontieres, MSF to WTO: Rethink Access to Life-Saving Drugs Now, (October 24, 2005), http://www.msf.org/msfinternational/invoke.cfm?objectid=224B1730-E018-OC72-091E8829E29F80E6&ccomponent=toolkit.article&method=full_html. (Accessed February 5, 2014).

[130] Matthew Rimmer, "Race Against Time: The Export of Essential Medicines to Rwanda," *Public Health Ethics* 1, no. 2 (July 2008): 89–103.

[131] Forman, "Trade Rules, Intellectual Property, and the Right to Health," 337–357.

[132] S. Cage, *WTO Oks Measures to Improve Drug Access*, Associated Press, (December 6, 2005).

[133] Westerhaus and Castro, "How Do Intellectual Property Law and International Trade Agreements Affect Access to Antiretroviral Therapy? 1233.

[134] Rachel L. Swarns, "African Nations Applaud Bush Plan to Fight AIDS Epidemic," New York Times (January 29, 2003): A19.

the help of PEPFAR.[135] Furthermore, at the 2005 Summit of the Group of Eight Nations (G8), the heads of state of the eight affluent countries pledged extra aid for combating HIV/AIDS in Africa, and projected the possibility of universal access to HIV treatment by 2010.[136]

In spite of the aforementioned initiatives to promote universal access to anti-retroviral drugs especially for developing countries, recent US negotiated trade agreements threaten to undermine these gains in improving access to anti-retroviral drugs.[137] MSF writes "One by one, countries are trading away their people's health in free trade agreements with the United States. These countries are being pushed to accept extremely restrictive intellectual property provisions that could put an end to competition from generic medicine producers and to countries' ability to make use of existing safeguards against patent abuse."[138] The United States' failure to promote free trade in the hemisphere and the globe resulted in its engaging in a forceful campaign to liberalize trade through bilateral, regional, and multilateral trade agreements. These recent negotiated trade agreements by the United Sates have been based on the extension of Intellectual Property (IP) law for multinational pharmaceutical companies that hold patents for anti-retroviral drugs.[139] Forman articulates that the intellectual property rules in TRIPS are significantly less strict than the rules developing countries are more and more accepting in free-trade agreements with the United States and other developed countries. These TRIPS-plus measures require greater limitations on the use of TRIPS flexibilities.[140] The implication is that these Free Trade Agreements (FTAs) considerably restrict generic competition and consequently affordable access to essential drugs including anti-retroviral drugs.[141]

The United States plays a leading role in establishing bilateral and regional trade and IP agreements. It signed bilateral trade agreements with 42 countries between

[135] Office of the United States Global AIDS Coordinator, *Engendering bold Leadership, First Annual Report to Congress on the President's Emergency Plan for AIDS Relief*, (Washington D.C.: United States Department of State, 2005), http://www.sate.gov/s/gac/rl/c14961.htm . (Accessed February 5, 2014).

[136] The Group of Eight, *Gleneagles Summit Documents, Chair's Summary*, (July 8, 2005), http://www.g8.gov.uk/serlet/Front?pagename=OpenMarket/Xcelerate/ShowPage&cc=Page&cid=1119518698846. (Accessed February 5, 2014).

[137] Westerhaus and Castro, "How Do Intellectual Property Law and International Trade Agreements Affect Access to Antiretroviral Therapy? 1233.

[138] Medecins Sans Frontieres, *Access to Medicines at Risk Across the Globe: What to Watch Out for in Free Trade Agreements with the United States*, (Geneva, Switzerland: MSF, 2004), http://www.accessmed-msf.org/documents/ftabriefingenglish.pdf. (Accessed February 5, 2014)

[139] Westerhaus and Castro, "How Do Intellectual Property Law and International Trade Agreements Affect Access to Antiretroviral Therapy? 1233.

[140] Lisa Forman, "Trading Health for Profit: The Impact of Bilateral and Regional Free Trade Agreements on Domestic Intellectual Property Rules on Pharmaceuticals," in *The Power of Pills: Social, Ethical and Legal Issues in Drug Development, Marketing and Pricing* ed. Jillian Clare Cohen, Patricia Illingworth and Udo Schuklenk (London: Pluto Press, 2006), 190.

[141] Forman, "Trading Health for Profit: The Impact of Bilateral and Regional Free Trade Agreements on Domestic Intellectual Property Rules on Pharmaceuticals," 192.

1986 and 2000.[142] Furthermore, it has negotiated numerous regional trade agreements involving about 50 countries, such as the Andean Free Trade Agreement (FTA), the Free Trade Area of the Americas (FTAA), the Central American Free Trade Agreement (CAFTA), the North American Free Trade Agreement (NAFTA), and the South African Customs Union Free Trade Agreement (SACU FTA).[143] These agreements have extensive effects for affordable access to drugs not only in these regions but globally, since a number of developing countries with generic local production capabilities, such as Argentina, Brazil, and Mexico, will be obliged to comply with TRIP-plus IP rules.[144]

The import of TRIP-plus measures enshrined in FTAs is understood not just as a matter of international trade law but of international human rights law. These agreements considerably restrict government ability to achieve the human rights to health and life of their population.[145] Forman argues "Given the urgent need for increased access to essential patented medicines (particularly for HIV/AIDS), there is no logical or palatable way to justify trading off the instrumental value of patent protection (to provide rewards and incentives for future innovation) at the present cost of the lives of millions of people."[146] Similarly, the Commission on Intellectual Property Rights (CIPR) articulates that, "there are no circumstances in which the most fundamental human rights should be subordinated to the requirements of IP protection."[147] It is pertinent to point out that IP rights are given by states for limited durations especially for patents and copyrights while in contrast human rights are inalienable and universal.[148]

The United States attempts to procure or has already procured the inclusion of various detrimental intellectual property provisions in its regional and bilateral trade agreements.[149] TRIPS-plus measures in FTAs usually comprise "limits on compulsory licensing; prohibitions on parallel imports; limiting market approval for generic

---

[142] Peter Drahos, "BITS and BIPS: Bilateralism in Intellectual Property," *Journal of World Intellectual Property* 4, no. 6 (November 2001): 791.

[143] Forman, "Trading Health for Profit: The Impact of Bilateral and Regional Free Trade Agreements on Domestic Intellectual Property Rules on Pharmaceuticals," 192.

[144] Forman, "Trading Health for Profit: The Impact of Bilateral and Regional Free Trade Agreements on Domestic Intellectual Property Rules on Pharmaceuticals," 192.

[145] Forman, "Trading Health for Profit: The Impact of Bilateral and Regional Free Trade Agreements on Domestic Intellectual Property Rules on Pharmaceuticals," 192.

[146] Forman, "Trading Health for Profit: The Impact of Bilateral and Regional Free Trade Agreements on Domestic Intellectual Property Rules on Pharmaceuticals," 192–193.

[147] Commission on Intellectual Property Rights, *Integrating Intellectual Property Rights and Development Policy,* 6.

[148] United Nations Sub-Commission on the Promotion and Protection of Human Rights, *Intellectual Property Rights and Human Rights*, (Geneva: United Nations, 2001), http://www.unhchr.ch/Huridocda/Huridoca.nsf/(Symbol)/E.CN.Sub.2.2001.12.En?Opendocument. (Accessed February 5, 2014).

[149] Medecins Sans Frontieres, *Access to Medicines at Risk Across the Globe: What to Watch Out for in Free Trade Agreements with the United States*, (Geneva, Switzerland: MSF, 2004), http://www.accessmed-msf.org/documents/ftabriefingenglish.pdf. (Accessed February 5, 2014)

drugs; data exclusivity; extended patent terms and evergreening provisions.[150] A brief discussion on TRIPS-plus provisions and the likely impact on affordable access to essential drugs for developing countries follow.

Compulsory Licensing

The US FTAs restrict the bases or circumstances for issuing compulsory licenses on pharmaceuticals to national emergencies or other conditions of extreme urgency, the compensation of practices considered to be anti-competitive and use for public non-commercial use.[151] This implies exclusion of any other bases for issuing compulsory licenses, comprising "the denial of a voluntary license, or as a measure to protect public health under TRIPS Article 8 that fell short of a national emergency."[152] While FTAs as well restrict the recipients of licenses to government entities or legal entities, functioning under the government's authority, TRIPS on the other hand, requires no such limitations, since licenses can be granted to independent private entities for commercial purposes.[153] Restrictions on compulsory licenses would imply that countries would not be able to use their basic right to grant a compulsory license in order to alleviate high prices that limit access to essential drugs and to promote generic competition in the private sector to increase affordable access to essential drugs.[154] Compulsory licensing is as well being undercut through restrictions on data exclusivity and marketing approval for generic drugs.[155]

Parallel Importing

FTAs such as with Singapore, Morocco and Australia authorize patent holders to block parallel importation. Similarly, the FTAA orders regional exhaustion within 5 years of being signed, and essentially excludes parallel importation from outside countries. These measures will prevent countries from importing cheap patented drugs sold in other countries.[156]

[150] Forman, "Trading Health for Profit: The Impact of Bilateral and Regional Free Trade Agreements on Domestic Intellectual Property Rules on Pharmaceuticals," 193.

[151] Medecins Sans Frontieres, *Access to Medicines at Risk Across the Globe: What to Watch Out for in Free Trade Agreements with the United States,* (Geneva, Switzerland: MSF, 2004), http://www.accessmed-msf.org/documents/ftabriefingenglish.pdf. (Accessed February 5, 2014)

[152] Forman, "Trading Health for Profit: The Impact of Bilateral and Regional Free Trade Agreements on Domestic Intellectual Property Rules on Pharmaceuticals," 194.

[153] Forman, "Trading Health for Profit: The Impact of Bilateral and Regional Free Trade Agreements on Domestic Intellectual Property Rules on Pharmaceuticals," 194.

[154] Medecins Sans Frontieres, *Access to Medicines at Risk Across the Globe: What to Watch Out for in Free Trade Agreements with the United States*, (Geneva, Switzerland: MSF, 2004), http://www.accessmed-msf.org/documents/ftabriefingenglish.pdf. (Accessed February 5, 2014)

[155] Forman, "Trading Health for Profit: The Impact of Bilateral and Regional Free Trade Agreements on Domestic Intellectual Property Rules on Pharmaceuticals," 194.

[156] Forman, "Trading Health for Profit: The Impact of Bilateral and Regional Free Trade Agreements on Domestic Intellectual Property Rules on Pharmaceuticals," 194.

Intellectual Property and Regulatory Authorities: Transforming Drug Regulatory Authorities into Patent Police

The US has created a new role for national drug regulatory authorities (NDRAs) through negotiating measures in FTAs that will entrust enforcement of drug patents to NDRAs. They would be prohibited from approving or registering a generic drug that is still under patent in a country unless the patent holder gives permission. This implies that registration should not be granted to generic producer before the expiration of the patent.[157] MSF argues that, "Linking a drug's registration (also known as its marketing approval) to its patent status is an underhanded way of preventing generic competition."[158] It also undercuts the utilization of compulsory licenses, since a generic company that has been awarded a license would not be able to register that drug, essentially making the license useless.[159]

Data Exclusivity: Preventing Competition to Non-Patented Drugs

Data exclusivity creates a new type of monopoly by blocking the registration of generic drugs even for a drug that is non-patented. It blocks NDRA from utilizing data offered by originator Company to approve the use of an equivalent generic drug, consequently providing as such the original manufacturer's monopoly.[160] Most FTAs make provisions for the protection of manufacturers' drug testing data, that is, data exclusivity to pharmaceutical products for 5 years from the date of the originator's approval.[161] The implication is that generic companies are discouraged from pursuing registration for their drugs due to the tedious task of generating their own test data. Generic manufacturers in developing countries would not be able to foot the bill for their test data due to the exorbitant costs and low margins of generic production. Furthermore, data exclusivity essentially blocks the use of compulsory licenses, because it prevents the registration of a generic drug equivalent for the duration of exclusivity.[162] Some trade agreements such as the U.S.-Morocco and U.S.-Bahrain FTAs make provision for an extra 3 years of data exclusivity when

---

157 Medecins Sans Frontieres, *Access to Medicines at Risk Across the Globe: What to Watch Out for in Free Trade Agreements with the United States,* (Geneva, Switzerland: MSF, 2004), http://www.accessmed-msf.org/documents/ftabriefingenglish.pdf. (Accessed February 5, 2014)

158 Medecins Sans Frontieres, *Access to Medicines at Risk Across the Globe: What to Watch Out for in Free Trade Agreements with the United States,* (Geneva, Switzerland: MSF, 2004), http://www.accessmed-msf.org/documents/ftabriefingenglish.pdf. (Accessed February 5, 2014)

159 Medecins Sans Frontieres, *Access to Medicines at Risk Across the Globe: What to Watch Out for in Free Trade Agreements with the United States,* (Geneva, Switzerland: MSF, 2004), http://www.accessmed-msf.org/documents/ftabriefingenglish.pdf. (Accessed February 5, 2014).

160 Medecins Sans Frontieres, *Access to Medicines at Risk Across the Globe: What to Watch Out for in Free Trade Agreements with the United States,* (Geneva, Switzerland: MSF, 2004), http://www.accessmed-msf.org/documents/ftabriefingenglish.pdf. (Accessed February 5, 2014).

161 Forman, "Trading Health for Profit: The Impact of Bilateral and Regional Free Trade Agreements on Domestic Intellectual Property Rules on Pharmaceuticals," 195.

162 Medecins Sans Frontieres, *Access to Medicines at Risk Across the Globe: What to Watch Out for in Free Trade Agreements with the United States,* (Geneva, Switzerland: MSF, 2004), http://www.accessmed-msf.org/documents/ftabriefingenglish.pdf. (Accessed February 5, 2014).

patent holders pursue marketing approval for already unapproved uses of registered drugs, including older generic drugs with expired patents.[163]

Extending Patent Life Beyond 20 Years
FTAs extend the protection of patents beyond the 20-year period guaranteed under TRIPS to compensate for delays in the process of awarding patents or marketing approval by a national drug regulatory authority, as well as for unreasonable delays. Some trade agreements such as the FTAA, U.S.-Singapore, U.S.-Australia, and U.S.-Morocco FTAs as well make provisions for extending the patent life when delays in granting the patent exceeds 5 years from filling, or 3 years after a request for examination of the application, whichever is later. The idea of unreasonable delays is contentious particularly considering that NDRAs and patent offices in developing countries have resource constraints.[164]

New Use or Evergreening Provisions
New use patent is a mechanism for prolonging the monopoly of pharmaceutical companies. Patent holders as well could utilize new use patents to harass competitors by arguing that they violated patent. TRIPS Agreement does not make any provision for granting patents on new uses of existing drugs, whereas FTAs do make it possible for pharmaceutical companies to extend patent durations on existing drugs, thus prolonging or evergreening their monopolies,[165] and endlessly block generic competition. Apart from patent protection, there are also several non-patent factors that significantly impede affordable access to essential drugs particularly antiretroviral drugs for developing countries.

3.2.2.4 Non-Patent Factors

It is pertinent to recognize that so many factors unconnected to the IP system play a critical role in determining affordable access to essential drugs in developing countries, including sub-Saharan Africa.[166] Schuklenk and Ashcroft contend that, "It would be wrong to paint a simplistic picture of the evil industry versus the brave governments of developing countries trying to save the lives of their suffering peoples."[167] The pharmaceutical industry argued that significant limitations to

[163] Forman, "Trading Health for Profit: The Impact of Bilateral and Regional Free Trade Agreements on Domestic Intellectual Property Rules on Pharmaceuticals," 195.

[164] Medecins Sans Frontieres, *Access to Medicines at Risk Across the Globe: What to Watch Out for in Free Trade Agreements with the United States*, (Geneva, Switzerland: MSF, 2004), http://www.accessmed-msf.org/documents/ftabriefingenglish.pdf. (Accessed February 5, 2014).

[165] Medecins Sans Frontieres, *Access to Medicines at Risk Across the Globe: What to Watch Out for in Free Trade Agreements with the United States*, (Geneva, Switzerland: MSF, 2004), http://www.accessmed-msf.org/documents/ftabriefingenglish.pdf. (Accessed February 5, 2014).

[166] Attaran and Gillespie-White, "Do Patents for Antiretroviral Drugs Constrain Access to AIDS Treatment in Africa?," 1886–1892.

[167] Udo Schuklenk and Richard Ashcroft, "Affordable Access to Essential Medication in Developing Countries: Conflicts between Ethical and Economic Imperatives", in *Ethics & AIDS in Africa: The*

affordable access to drugs in developing countries are not strong legal protection of patent, but the lack of spending in developing countries, and the lack of adequate health infrastructure to administer drugs safely and effectively.[168] Similarly, a report by the US pharmaceutical industry association succinctly articulates, "Handicapped by limited financial resources, these nations' ability to contain AIDS and address a host of other killer diseases is compromised by inadequate infrastructure, cultural barriers to care, and mismanaged health care systems. Some developing countries also are hampered by political leadership that lacks the will to confront or even acknowledge their nation's health care needs.".[169]

The lack of political will contributes significantly to the denial of affordable access to essential drugs.[170] Most African countries grapple with the failure of political leadership. The implication of this is usually the inability of their political leaders to step up to the responsibility of making critical decisions in order to identify and fight for public healthcare needs. However, there are some exceptions. For example, Senegal stood out regarding the existence of strong political will, which was contributory to the early identification of the HIV/AIDS crisis, the organization of financial resources, the utilization of mass media campaign to counteract cultural and religious taboos, the support for the use of condom, and the providing of universal access to anti-retroviral treatment. The significant result was that Senegal had one of the lowest rates of HIV infection in sub-Saharan Africa by the end of the 1990s.[171]

Poverty is also a critical factor limiting the ability of developing countries including sub-Saharan Africa to afford even the basic essential healthcare needs. Their poverty stems more from a combination of poor political leadership and an uneven international political-economic structure.[172] For example, in 1991 28 out of 48 African countries had an average per capita income of less than $1 per day compared to 19 out of 36 countries in 1981.[173] Furthermore, most developing countries in sub-Saharan Africa spend less than an average of US$ 10 per person every year

---

Challenge to Our Thinking ed. Anton A. Van Niekerk and Loretta M. Kopelman (Walnut Creek, California: Left Coast Press Inc., 2005), 127.

[168] Commission on Intellectual Property Rights, *Integrating Intellectual Property Rights and Development Policy*, 38.

[169] Pharmaceutical Research and Manufacturers of America, *Health Care in the Developing World*, (Washington D.C.: PhRMA, 2002), http://world.phrma.org/ip.access.aids.drugs.html (Accessed February 5, 2014).

[170] Donald, "The Political Economy of Technology Transfer," 1298.

[171] UNAIDS, *Acting Early to Prevent AIDS: The Case of Senegal*, (Geneva, Switzerland: UNAIDS, 1999), 23, http://data.unaids.org/publications/irc-pub04/una99-34_en.pdf (Accessed February 5, 2014).

[172] Chris Simms, Mike Rowson, and Siobhan Peattie, *The Bitterest Pill of All: The Collapse of Africa's Health Systems*, (London: Save the Children, 2001),2–3, http://www.savethechildren.org.uk/sites/default/files/docs/The_Bitterist_Pill_of_All_1.pdf (Accessed February 5, 2014).

[173] UNICEF, *The State of World's Children 2001*, (New York: UNICEF, 2001), 100–101, http://www.unicef.org/sowc/archive/ENGLISH/The%20State%20of%20the%20World'%20Children%202001.pdf (Accessed February 5, 2014).

on healthcare, which is far less than $60 per capita recommended by WHO as the minimum level of expenditure on basic healthcare services, as well as, about 20–40% below the World's Bank recommended minimum level of healthcare services.[174]

The International Monetary Fund (IMF) and the World Bank imposed reforms known as Structural Adjustment Programmes (SAPs) on African Governments as conditions for giving them loans in the 1980s, which exacerbated their poverty level and further impeded the ability of the populations to procure essential drugs including HIV/AIDS drugs. This scenario made them very vulnerable to deadly diseases, such as HIV/AIDS, tuberculosis and malaria.[175]

Large debt interest payments made by developing countries especially sub-Saharan Africa also negatively impacted affordable access to essential drugs. For example, sub-Saharan African countries by 1997 were previously remitting to Western creditors more than four times the amount invested in their domestic healthcare systems. A typical instance was that only Senegal expended more than five times the amount expended on health in loan repayments.[176] The total debt of Africa as of 2000 was US $230 billion, with annual repayments of US $15 billion, which amounts to about 5% of its income and 15% of its export earnings.[177] This financial squash disordered the healthcare systems of Africa, and resulted in people gradually more vulnerable to diseases. The situation deteriorated with the privatization of healthcare supported by the World Bank, which resulted in the commercialization of health services, consequently impeding access to essential healthcare to more people.[178]

The recent global financial crisis worsened the situation, as it compelled governments to reduce their already insufficient health budgets, notwithstanding the epidemic of infection from major diseases such as HIV/AIDS, tuberculosis, malaria and sleeping sickness. The global financial crisis as well undercut the activities of international donor agencies. For example, the global fund for AIDS, TB and Malaria in 2009 reported about US$ 4 billion deficit in what was required to adequately fund essential services for these diseases in 2010. This shortfall was in

---

[174] Chris Simms, Mike Rowson, and Siobhan Peattie, *The Bitterest Pill of All: The Collapse of Africa's Health Systems,* 3.

[175] Medact and the World Development Movement, *Deadly Conditions? Examining the Relationship between Debt Relief Policies and HIV/AIDS*, (1999), http://www.wdm.org.uk/campaigns/cambriefs/debt/aids.htm (Accessed February 5, 2014).

[176] Ann-Louise Colgan, "Hazardous to Health: The World Bank and IMF in Africa," *Africa Action* 1/2, (April 2002): http://www.africa.upenn.edu/Urgent_Action/apic041802.html (Accessed February 5, 2014).

[177] International Intellectual Property Institute, *Patent Protection and Access to HIV/AIDS Pharmaceuticals in Sub-Saharan Africa*, (Washington, DC: IIPI, 2000), 5, http://www.wipo.int/export/sites/www/about-ip/en/studies/pdf/iip_hiv.pdf (Accessed February 5, 2014).

[178] Colgan, "Hazardous to Health: The World Bank and IMF in Africa," *Africa Action* 1/2, (April 2002): http://www.africa.upenn.edu/Urgent_Action/apic041802.html (Accessed February 5, 2014).

addition to a US$ 10.7 billion deficit in the funding for the implementation of the Global Plan to Stop TB at regional levels.[179]

Poor international media coverage of diseases prevalent in developing countries is another factor identified as impeding affordable access to essential drugs. HIV/AIDS in sub-Saharan Africa only got international spotlight in the 1990's when it was viewed as a threat to U.S. national security.[180] However, the media attention diminished when the fears of catastrophic impact of the disease both in the US and other Western world dispelled. The lack of international media coverage could significantly prevent or constrain the organization of the resources required to guarantee a timely and constant supply of drugs to assist people afflicted with the disease in sub-Saharan Africa.[181]

Finally, other non-patent factors militating against affordable access to essential drugs in sub-Saharan include the brain drain of qualified medical professionals from sub-Saharan Africa to overseas countries,[182] tariffs and other forms of indirect taxation,[183] which could be as high as 30% in some circumstances.[184] It is significant that national tax systems function in a way that promotes public health policies, in the same way that the patent system should.[185] The adverse effects of the TRIPS Agreement, TRIPS-plus provisions and non-patent factors in limiting access to antiretroviral drugs for developing countries create a context to address the issue of the social responsibility of the pharmaceutical companies.

3.3 Social Responsibility and Access Strategies

3.3.1 The Social Responsibility of Pharmaceutical Companies

Two dominant perspectives emerge in the discussion of scholars about the social responsibility of pharmaceutical companies. The leading proponent of the first perspective known as Resnik argues that pharmaceutical companies are like moral

---

[179] Colgan, "Hazardous to Health: The World Bank and IMF in Africa," *Africa Action* 1/2, (April 2002): http://www.africa.upenn.edu/Urgent_Action/apic041802.html (Accessed February 5, 2014).

[180] Barton Gellman, "The Belated Global Response to AIDS in Africa," *Washington Post* (July 5, 2000): A1.

[181] Gellman, "The Belated Global Response to AIDS in Africa," A1.

[182] Kristine Novak, "The WTO's Balancing Act," *The Journal of Clinical Investigation* 112, no. 9 (November 2003): 1269–1273.

[183] Commission on Intellectual Property Rights, *Integrating Intellectual Property Rights and Development Policy*, 38.

[184] International Intellectual Property Institute, *Patent Protection and Access to HIV/AIDS Pharmaceuticals in Sub-Saharan Africa,* (Washington, DC: IIPI, 2000), 5, http://www.wipo.int/export/sites/www/about-ip/en/studies/pdf/iip_hiv.pdf (Accessed February 5, 2014).

[185] Commission on Intellectual Property Rights, *Integrating Intellectual Property Rights and Development Policy*, 38.

agents who have obligations to avoid causing harm and to promote social welfare. They have social responsibilities and moral obligations to meet the health needs of the populations in developing countries.[186]

On the other hand, the second perspective propounded by Brock argues that corporations are unlike moral agents as long as their responsibilities are to their shareholders. He appeals to an argument in support of role differentiation. He argues that corporations do not have similar moral obligations like individuals due to the fact that they serve a different social role that entails shareholder primacy.[187] The shareholder primacy view of corporations emphasizes maximization of shareholders wealth.[188]

An inclusive view of the corporation's responsibilities is currently supported by the enactment of corporate constituency statutes.[189] The implication is that corporations have both shareholder and social responsibilities. A brief discussion of the theoretical approaches to the social responsibility of the pharmaceutical companies will be pertinent.

3.3.1.1 Theoretical Approaches

The classical theory of the primacy of shareholder invoked by Brock maintains the existence of a fiduciary relationship between directors and shareholders that prioritizes the interests of shareholders.[190] This position is supported by persuasive statements from legal cases, scholars, and the economist Milton Friedman.[191] The central legal argument supporting the shareholder primacy is focused on agency law, which considers shareholders as the owners or the principals of the corporation and managers as the agents. In this context, the agent must always act for the interest of the principal, excluding the interests of other constituencies, including the manager himself. The application of this line of thought to the modern cooperation falls short of a remote possibility. The modern corporation significantly departs from the principal-agent model because it is grounded on the separation of ownership and

---

[186] David B. Resnik, "Developing Drugs for the Developing World: An Economic, Legal, Moral, and Political Dilemma," *Developing World Bioethics* 1, no. 1 (May 2001): 18.

[187] Dan W. Brock, "Some Questions about the Moral Responsibilities of Drug Companies in Developing Countries," *Developing World Bioethics* 1, no. 1 (May 2001): 34.

[188] Milton Friedman, "The Social Responsibility of Business is to Increase its Profits," *The New York Times Magazine*, September 13, 1970, http://www.colarado.edu/studentgroups/lidertarians/issues/friedman-soc-resp-business.html (Accessed February 05, 2014).

[189] Eric W. Orts, "Beyond Shareholders: Interpreting Corporate Constituency Statutes," *George Washington Law Review* 61, no. 1 (November 1992): 14–135.

[190] Cohen and Illingworth, "The Dilemma of Intellectual Property Rights for Pharmaceuticals: The Tension Between Ensuring Access of the Poor to Medicines and Committing to International Agreements," 41.

[191] Adolf A. Berle, Jr. "Corporate Powers as Powers in Trust," *Harvard Law Review* 44, no. 7 (May 1931):1049–1074.

control of the corporation, with managers shouldering a very active role and shareholders a comparatively passive one.[192]

The idea of the primacy of the shareholder has as well been defended on the basis that the shareholders own the corporation, that they are the principals, and that the directors are obliged to maximize their wealth.[193] This notion of the modern corporation known as the property conception was originally articulated by authors such as Adolph A. Berle and Milton Friedman. Berle argues that "all powers granted to a corporation or the management of a corporation … are necessarily and at all times exercisable only for the ratable benefit of all the shareholders as their interest appears."[194] He further argues that corporations were solely mediums for advancing and protecting the interests of shareholders and that corporate law should be interpreted within the context of this principle. He contends that any other account of the function and purpose of corporations would "defeat the very object and nature of the corporation itself.".[195]

Similarly, Friedman argues that in a free enterprise, private-property system, the primary responsibility of a corporate executive is to conduct business in conformity with the interests of the owners, which usually entails making as much money as possible while complying with the basic rules of the society.[196] The idea of maximization of stockholders' wealth stems from the fact that they are considered as owners of the corporation. Nevertheless, the myth of shareholder primacy is debunked, because even if they are considered as owners of the corporation, there is nothing as such that validates that corporation must focus on shareholder profit, more especially when shareholders make very minimal contribution of only money. More so, the popular practice of extending stock options definitely undermines any assertion of shareholder ownership.[197]

A newer concept with a significant change to the argument in support of shareholder primacy focuses on the agency costs' view. This view broadens a corporation's constituencies and consequently, its priority interests. The agency costs' view emphasizes that "in the best of all possible worlds, it would be preferable if managers could consider the interests of all of a corporation's constituencies (all those who affect the organization and are affected by it), including among others employees,

[192] Cohen and Illingworth, "The Dilemma of Intellectual Property Rights for Pharmaceuticals: The Tension Between Ensuring Access of the Poor to Medicines and Committing to International Agreements," 42.

[193] Cohen and Illingworth, "The Dilemma of Intellectual Property Rights for Pharmaceuticals: The Tension Between Ensuring Access of the Poor to Medicines and Committing to International Agreements," 42.

[194] Berle, "Corporate Powers as Powers in Trust," 1049.

[195] Berle, "Corporate Powers as Powers in Trust," 1074.

[196] Friedman, "The Social Responsibility of Business is to Increase its Profits," *The New York Times Magazine*, September 13, 1970, http://www.colarado.edu/studentgroups/lidertarians/issues/friedman-soc-resp-business.html (Accessed February 05, 2014).

[197] Fischer Black and Myron Scholes, "The Pricing of Options and Corporate Liabilities," *The Journal of Political Economy* 81, no. 3 (May–June 1973): 637.

clients, and the community."[198] In contrast, there is a conception that deviating from shareholder primacy would involve giving managers who are also mere humans too much discretion to the point that they become too opportunistic. A similar problem encountered in the principal-agent model could also be experienced when managers are given such discretion, because they would not necessarily act in the interests of society but instead act in their own interest.[199] Roe articulates, "… a stakeholder measure of managerial accountability could leave mangers so much discretion that managers could easily pursue their own agenda, one that might maximize neither shareholder, employer, consumer, nor national wealth, but only their own.".[200]

Another perspective known as the social entity conception views the corporation as a social construction, with social purposes.[201] A leading scholar who originally articulated this idea in response to Berle's shareholder primacy position was E. Merrick Dodd, a professor at Harvard Law School. He argues that, "there is in fact a growing feeling not only that business has responsibilities to the community but that our corporate managers who control business should voluntarily and without waiting for legal compulsion manage it in such a way as to fulfill those responsibilities."[202] He cited the heads of some major corporations, such as General Electric, to buttress his argument that business leaders had come to acknowledge that corporate managers are required to consider social responsibility when running their companies.[203]

Dodd offered some interpretations of the social entity view in relation to corporate law. He argued that if social responsibility entailed that corporate managers focused more on the needs of their employees and consumers, this would in the long run result in the shareholders' benefit. His reasoning was based on the fact that employee satisfaction results in greater productivity and eventually increased profits. The implication is that managers could actually increase profits by concentrating on the needs of groups other than shareholders.[204]

---

[198] Cohen and Illingworth, "The Dilemma of Intellectual Property Rights for Pharmaceuticals: The Tension Between Ensuring Access of the Poor to Medicines and Committing to International Agreements," 42–43.

[199] Cohen and Illingworth, "The Dilemma of Intellectual Property Rights for Pharmaceuticals: The Tension Between Ensuring Access of the Poor to Medicines and Committing to International Agreements," 43.

[200] Mark J. Roe, "Symposium Norms and Corporate Law: The Shareholder Wealth Maximization Norm and Industrial Organization," *University of Pennsylvania Law Review* 149, (September 2001): 2063, 2065.

[201] Cohen and Illingworth, "The Dilemma of Intellectual Property Rights for Pharmaceuticals: The Tension Between Ensuring Access of the Poor to Medicines and Committing to International Agreements," 43.

[202] E. Merrick Dodd, "For Whom are Corporate Managers Trustees*? Harvard Law Review* 45, no. 7 (May 1932): 1145–1163.

[203] Lainie Rutkow, "Should Corporations Serve Shareholders or Society?: The Origins of the Debate," Corporations and Health Watch: Tracking the Effects of Corporate Practices on Health, http://www.corporationsandhealth.org/2011/04/06/should-corporations-serve-shareholders-or-society-the-origins-of-the-debate/ (Accessed February 20, 2014).

[204] Dodd, "For Whom are Corporate Managers Trustees? 1156.

Dodd further argued that courts had offered enough leeway to corporate managers, permitting them "a wide range of discretion as to what policies will best promote the interests of the stockholders …"[205] For example, he indicated that corporate charitable giving, although may not directly increase shareholder wealth, but could engender good will in the community.[206] This good will could lead to shareholders' benefit, because consumers would be more likely to think positively of the corporation and purchase its products. By this logic, he thinks that corporations are "affected not only by the laws which regulate business but by the attitude of public and business opinion as to the social obligations of business."[207] He asserted that the opinion of the society about corporation as a purely private enterprise was shifting, and that corporate managers should "recognize that the attitude of law and public opinion towards business was changing….".[208]

He thinks the case for social responsibility is even more compelling with regard to companies that have strong public dimensions, such as railways and public utilities. Pharmaceutical companies may be classified under this category because of the strong public health dimension. The social entity conception takes into account that corporations have much broader social purposes and duties than merely maximizing the wealth of shareholders.[209] A review of the three major theories of the corporation indicates that the duties of pharmaceutical companies will be determined by the particular theory one adopts.

Allen highlights that the courts and legislatures have recognized the social entity view and the social obligations it supports.[210] The legal system realized this objective with the enactment of corporate constituency statutes. These statutes have undercut shareholders' primacy, in support of other constituencies, such as employees and the community. Currently, about 29 states have adopted corporate constituency statutes in the United States.[211] For example, the New York statute's provisions of 4 and 5 authorize directors to consider other constituencies when they act, and to act on behalf of these other constituencies.[212] Corporate constituency statutes significantly modify a corporation's duty of care to encompass others in addition to shareholders. This implies that managers may have a legal right to consider interests of other constituencies in addition to shareholders without infringing on their obligations to shareholders.[213] The responsibilities of pharmaceutical companies to their

[205] Dodd, "For Whom are Corporate Managers Trustees? 1157.

[206] Dodd, "For Whom are Corporate Managers Trustees? 1159.

[207] Dodd, "For Whom are Corporate Managers Trustees? 1161.

[208] Dodd, "For Whom are Corporate Managers Trustees? 1163.

[209] Dodd, "For Whom are Corporate Managers Trustees? 1149–1150.

[210] William T. Allen, "Our Schizophrenic Conception of the Business Corporation," *Cardozo Law Review* 14 (1992): 276.

[211] Orts, "Beyond Shareholders: Interpreting Corporate Constituency Statutes," 14–135.

[212] N.Y. Bus. Corp. Law Section 717 (b).

[213] Cohen and Illingworth, "The Dilemma of Intellectual Property Rights for Pharmaceuticals: The Tension Between Ensuring Access of the Poor to Medicines and Committing to International Agreements," 44.

shareholders do not protect them from increased moral responsibilities for ensuring affordable access to anti-retroviral drugs for developing countries.[214]

3.3.1.2 Global Pharmaceutical Companies

Assigning social responsibilities to global pharmaceutical companies has been hotly contested by several authors. Resnik argues strongly in support of ascribing social responsibilities to global pharmaceutical companies, appealing to the view that business only functions well within the context of social values such as honesty, integrity, fidelity, diligence and fairness. Business would be undercut by corruption, theft, fraud and disloyalty without allegiance to such social values.[215] In contrast, Daniels argues that Resnik's argument falls short, because the specific obligations or responsibilities assigned to global pharmaceutical companies cannot be deduced from the list of social values identified by him.[216]

Furthermore, Resnik argues that businesses have other social responsibilities because they exist within societies where people are concerned about the environment, public safety, public health and other values. Ignoring such social responsibilities grounded in what society cares about may incur the wrath of the public and eventually stringent regulations.[217] On the other hand, Daniels argues that the specific responsibilities or obligations ascribed to businesses within a society result from a societal negotiation "in which the protection of business incentives and productivity are weighed against the consequences to the public of failing to impose specific – legal and administrative – duties and obligations.[218]

Daniels postulates moral responsibilities that conform to and possibly justify the legal duties and obligations that result from such a negotiation. However, he went further to argue that the specific nature of any of these duties or responsibilities should result from a kind of social contract that establishes them. They cannot be deduced simply from what society is concerned about or what it impends. Daniels contends that the duties and obligations we may impose on global corporations, internationally also result from both domestic and international negotiation. He thinks we can have more clarity and specificity regarding the social responsibilities that are ascribed to corporations in either the domestic or international case only when we have executed the appropriate negotiation within the appropriate social or

[214] Cohen and Illingworth, "The Dilemma of Intellectual Property Rights for Pharmaceuticals: The Tension Between Ensuring Access of the Poor to Medicines and Committing to International Agreements," 44.

[215] David Resnik, "Access to Affordable Medication in the Developing World: Social Responsibility VS. Profit," in *Ethics & AIDS in Africa: The Challenge to Our Thinking* ed. Anton A. Van Niekerk and Loretta M. Kopelman (Walnut Creek, California: Left Coast Press Inc., 2005), 114.

[216] Norman Daniels, "Social Responsibility and Global Pharmaceutical Companies," *Developing World Bioethics* 1, no1 (May 2001): 39.

[217] Resnik, "Access to Affordable Medication in the Developing World: Social Responsibility VS. Profit," 114.

[218] Daniels, "Social Responsibility and Global Pharmaceutical Companies," 39.

inter-societal contract setting.[219] Similarly, Brock disagrees with Resnik argument of facing the public's wrath if pharmaceutical companies ignore their social responsibilities but for a different reasoning distinct from Daniels. He thinks that this is not the reason why they have any moral responsibilities, rather, it simply indicates that it is in their self-interest, not a moral obligation, to execute some responsibilities.[220]

Resnik offered second reason for ascribing social responsibilities to pharmaceutical companies, articulating that they are like moral agents in that their decisions have significant consequences for people affected by them. Daniels agrees partially with him that we all get involved in moral blame of some of these consequences and some of these actions. However, he doubts the clarity of responsibilities that derive from the fact that corporate decisions have consequences on people outside the corporations. He thinks Resnik does not provide sufficient details regarding the derivation. Furthermore, he contends that both in different nations and globally, societies and their legal institutions take these consequences into account, engage in evaluating of benefits in contrast to consequences, including the consequences of corporate decisions on the rights of other parties, and establish a legal and possibly moral framework for discussing social responsibilities. Noteworthy, is that the specifics do not derive from the nature of agency but from the kind of discussion he has identified.[221]

On the other hand, Brock argues that corporations are unlike moral agents in many other ways, pointing to the fact that various social institutions are established for specific purposes and functions, which make them unlike persons and influence their responsibilities. More so, he contends that many people believe that the responsibilities of corporations go to their shareholders and that they do not have moral responsibilities of beneficence and justice similar to those of individuals. He thinks that Resnik too hurriedly supposes that corporations have the moral obligations like those of individuals.[222]

Resnik argues that global pharmaceutical companies have social responsibilities or duties of beneficence and justice to developing countries. He appeals to such social or moral responsibilities as a solution for providing affordable access to essential drugs in developing countries. There are various ways that global pharmaceutical companies can exercise their social or moral responsibilities to developing countries, including sponsoring research and development for diseases that affect people in developing countries such as malaria, tuberculosis and HIV/AIDS, providing free medications to them, and offering substantial discounts on drug prices. He argues that these social responsibilities are not absolute requirements and may

---

219 Daniels, "Social Responsibility and Global Pharmaceutical Companies," 39.

220 Dan W. Brock, "Some Questions About the Moral Responsibilities of Drug Companies in Developing Countries," *Developing World Bioethics* 1, no1 (May 2001): 34.

221 Daniels, "Social Responsibility and Global Pharmaceutical Companies," 39–40.

222 Brock, "Some Questions About the Moral Responsibilities of Drug Companies in Developing Countries," 34.

be weighed against other obligations and commitments with respect to economic, social, political, legal and other relevant conditions.[223]

Furthermore, Resnik argues that the degree to which a global pharmaceutical company may exercise social responsibility in developing countries extensively hinges on two major factors: (1) the expectations of a reasonable profit and (2) the expectations of a good business environment.[224] Developing countries can either assist or hinder the efforts of pharmaceutical companies to execute social responsibility through several policies and practices. They could guarantee a reasonable profit for pharmaceutical companies by honoring pharmaceutical patents. If they do not comply with the patents, the pharmaceutical companies will significantly lose some profits which will take away money that could be invested in projects or programs aimed at promoting affordable access to essential drugs in developing countries. Guaranteeing a good business environment for pharmaceutical companies entails that developing countries should make honest efforts "to promote the rule of law, ethical business practices, stable currencies, reliable banking systems, free and open markets, democracy, and other social, economic, legal and political conditions conducive to business.".[225]

Conversely, Brock thinks that more argument is required to buttress that global pharmaceutical companies have a moral obligation to conduct business and exercise social responsibilities in developing countries.[226] He thinks that fulfilling basic needs of individuals in a society is the responsibility of the government, citing Rawls' concept of the function of basic social institutions. For example, fulfilling the basic needs of food and shelter is not a special responsibility of the food and real estate industries, but a governmental responsibility. He asserts that providing prescription drugs coverage for the elderly acknowledged as a major political issue in the United States, has never been argued by any of the parties in the debate as the pharmaceutical companies' social responsibility of beneficence and justice to fulfil the need.[227]

Furthermore, Brock points to large income inequalities between developed and poorer developing countries, which impede the latter's affordability of the prices of patented drugs, and is acknowledged as one of the most grave injustices in the world currently. He argues that when developing countries fail to honor pharmaceutical patents, in order to provide drugs essential to save lives and protect their citizens' health, it is debatably a step towards ensuring greater justice between developed and

[223] Resnik, "Access to Affordable Medication in the Developing World: Social Responsibility VS. Profit," 123.

[224] Resnik, "Access to Affordable Medication in the Developing World: Social Responsibility VS. Profit," 123.

[225] Resnik, "Access to Affordable Medication in the Developing World: Social Responsibility VS. Profit," 123.

[226] Brock, "Some Questions About the Moral Responsibilities of Drug Companies in Developing Countries," 34.

[227] Brock, "Some Questions About the Moral Responsibilities of Drug Companies in Developing Countries," 35.

developing countries, as well as, a situation where the threat of overruling pharmaceutical patents may have greater influence on the voluntary efforts of pharmaceutical companies than arguments regarding their social responsibility.[228]

Daniels also expresses some doubts regarding Resnik's focus on social or moral responsibilities as a solution to the problem of affordable access to essential drugs in developing countries. He acknowledges that developing countries have endured for a long time, a situation in which global pharmaceutical companies have hardly exercised the social responsibilities that Resnik assigns to them. He contends that though debatable, developing countries do a better job of fulfilling the health needs of their populations by engaging in local manufacture of drugs as well as refusing to honor intellectual property rights. Resnik's contention that such action by developing countries takes away some profits from pharmaceutical companies which they need in order to be competitive, falls short because it cannot be buttressed from existing facts, and Resnik himself indicates that they make a substantial profit regardless of renegade states.[229]

Daniels challenges Resnik's argument that in the long run, the global pharmaceutical companies would better fulfill the health needs of developing countries if both global pharmaceutical companies and developing nations reciprocally fulfilled their social responsibilities. He indicates that Resnik may be right, but also recognizes that there is no assurance to developing countries that global pharmaceutical companies will fulfill their social responsibilities on a consistent basis, regardless of the abundant evidence that they fulfill other commitments. In the absence of resolving the problem of assurance which may rest on some reliable strategies for enforcement of those responsibilities, there is no way to effectively deal with the behavior of developing countries eager to resolve the problem of affordable access to essential drugs by their local production. In fact, the international agreements on intellectual property, such as TRIPS Agreement and TRIPS-plus provisions that establish international property rights fall short of implementing the social responsibilities of global pharmaceutical companies and concentrate only on the responsibilities of developing countries to honor property rights. Daniels forcefully argues that, "Without the quid pro quo, there is no solution to the assurance problem and so no basis for appealing to moral commitment to solve the problem of public goods lurking in the background."[230] Finally, Daniels contends that the solution to affordable access to essential drugs in developing countries rests on domestic and international action to control global pharmaceutical companies and to standardize their contributions towards fulfilling the needs of developing countries.[231]

---

[228] Brock, "Some Questions About the Moral Responsibilities of Drug Companies in Developing Countries," 37

[229] Daniels, "Social Responsibility and Global Pharmaceutical Companies," 40.

[230] Daniels, "Social Responsibility and Global Pharmaceutical Companies," 40.

[231] Daniels, "Social Responsibility and Global Pharmaceutical Companies," 41.

3.3.2 *Access Strategies*

Global pharmaceutical companies, international agencies and national governments exercise their social and moral responsibilities to ensure affordable access to essential drugs for developing countries with the following strategies: private donations, price reductions and differential pricing, International collaborative initiatives and public-private partnerships and compulsory licensing. A brief discussion of the strategies follows.

3.3.2.1 Private Donations

Several pharmaceutical companies embark on various programs to donate AIDS drugs free of charge to developing countries. For example, Boehringer-Ingelheim pledged to provide Nevirapene, an effective drug for significantly reducing the mother-to-child transmission of HIV, free of charge for a limited period of time. Pfizer also offered to provide Fluconazole free of charge to the people of South Africa affected by cryptococcal meningitis.[232]

Various pharmaceutical organizations also have donated funds to developing countries aimed at ensuring affordable access to essential drugs, including antiretroviral drugs. For example, Merck donated US$3 million to the Harvard AIDS Institute for developing and implementing a care program in Senegal and Brazil. Similarly, Bristol-Myers Squibb funded its own "Secure the Future Programme" with US$100 million, in order to establish a large number of programs in African Countries, as well as for training of health care professionals from Africa at US tertiary institutions.[233]

Most authors have questioned the sustainability and effectiveness of donations by pharmaceutical companies in resolving the perennial issue of affordable access to essential drugs for developing countries. Schuklenk and Ashcroft articulate, "The problem with these handouts is, of course, that such offers are fraught with conditions, time and quantity-based limitations and a continuing dependence of the developing country's health care planning on the generosity of commercial organizations.".[234]

Philanthropic approaches to the issue of affordable access to essential drugs for developing countries have both advantages and disadvantages. In a way, they depict an acknowledgement by corporations that they possess the capacity to act morally and, probably, possess an obligation to help those affected by disaster. The positive aspects of such donations comprise that they take place, they are not coerced and

[232] Schuklenk and Ashcroft, "Affordable Access to Essential Medication in Developing Countries: Conflicts between Ethical and Economic Imperatives", 132.

[233] Schuklenk and Ashcroft, "Affordable Access to Essential Medication in Developing Countries: Conflicts between Ethical and Economic Imperatives", 132.

[234] Schuklenk and Ashcroft, "Affordable Access to Essential Medication in Developing Countries: Conflicts between Ethical and Economic Imperatives", 132.

they depict an assumption of moral agency and moral obligations by corporations. Conversely, there is recognition of the existence of moral distinction between charitable giving considered as voluntary and honorable and acting on duty. In this context, duty implies duty to prevent avoidable deaths where it is possible for one, is not optional but obligatory. Charity entails the freedom to turn away one's giving elsewhere if one considers it appropriate, and to assume the right to cease if the recipient is not grateful or not deserving.[235]

Another case against charity is that "it morally degrades the individual by fostering dependence, promoting an attitude of humility toward the giver, and depriving the recipient of the ability to set terms and negotiate the terms of receipt."[236] Counting on this argument presents serious problem because it may portray aid as such as wrong, instead of particular type of aid, such as supererogatory, discretionary, and conditional charity. However, donation of drugs has been argued as an improvised solution, which may partly solve the problem of affordable access to essential drugs for developing countries, but it comes with moral problem. This implies that on the one hand, from aretaic point of view, it deals with the motives and character of donor and recipient, and on the other hand, from consequentialist viewpoint, the solution is not sustainable and it ignores the perspective in which pharmaceutical companies and states have a duty to prevent avoidable deaths when they possess the power to do so.[237]

3.3.2.2 Price Reductions and Differential Pricing

Price reductions and differential pricing or equity pricing are also ad hoc solutions used by pharmaceutical companies to tackle the issue of affordable access to patented essential drugs for people in developing countries. Initial efforts to reduce the exorbitant price of AIDS drugs were made by UNAIDS. In 1997, UNAIDS started a collaborative effort that engaged three pharmaceutical companies and health officials in Chile, Cote d'Ivoire, Uganda, and Vietnam. However, prices of antiretroviral drugs continued to be very exorbitant for most people in these countries, despite this plan.[238]

Pharmaceutical companies also embarked on reducing the price of life-prolonging, health preserving anti-retroviral drugs for poor developing countries. For example, in February 2001, Oxfam launched a campaign to pressure multinational drug companies to cut prices in poor countries. The charity initiative was not restricted to HIV/AIDS, but encompasses other drugs such as effective antibiotics.

[235] Schuklenk and Ashcroft, "Affordable Access to Essential Medication in Developing Countries: Conflicts between Ethical and Economic Imperatives", 132–133.

[236] Schuklenk and Ashcroft, "Affordable Access to Essential Medication in Developing Countries: Conflicts between Ethical and Economic Imperatives", 133.

[237] Schuklenk and Ashcroft, "*Affordable Access to Essential Medication in Developing Countries: Conflicts between Ethical and Economic Imperatives*", 133.

[238] Macklin, *Double Standards in Medical Research in Developing Countries*, 168.

More so, Oxfam challenged the patent laws that have hindered poor countries from importing inexpensive, generic drugs from other countries without fear of reprisal.[239]

Merck, a big pharmaceutical company in March 2001 also offered to sell two of the AIDS drugs it produces to developing countries at much reduced prices than it charges in the United States. For example, Crixivan and Stocrin were sold for $600 and $500 per patient per year respectively, at a time when the US prices for these drugs were $6000 and $4700, respectively. Merck also offered to reduce the prices of both drugs in Brazil, but at higher prices than it had provided other developing countries, $1029 for Crixivan and $920 for Stocrin. Merck explained that its decision was based on the countries that would qualify for its lowest price on the United Nations Human Development index. The Brazilian government as well exerted pressure on Hoffmann-La Roche, which manufactures another AIDS drug to reduce its price.[240]

Novartis, another pharmaceutical company based in Swiss in May 2001 also offered to cut the price of Riamet, a powerful drug to treat malaria. The price at the time in industrialized countries was about $12, but Novartis decided to sell the drug to WHO for $2 for a complete treatment. Overall, between March and May 2001, the companies that chose to sell their AIDS drugs at considerably reduced prices in developing countries were Merck, GlaxoSmithKline, Bistol-Myers Squibb, and Abbott Laboratories.[241]

Various pharmaceutical companies are also effectively advancing towards a solution to affordable access that entails differential pricing or equity pricing in developed and developing countries. Differential pricing or equity pricing is defined as "setting the price of essential drugs in a way that reflects countries' ability to pay, as measured by their level of income."[242] Differential pricing uses a tiered pricing system with emphasis on market segmentation based on the economic profile of a country. Pharmaceutical companies provide countries with a differential pricing or an equity pricing scheme, based on the economic profile of the poorest buyer in a country. This utilizes the notion of price discrimination which allows a pharmaceutical company to sell the same drug to different buyers at different prices. Prices are specifically not based on the costs of production but on what the buyer would be able to pay.[243]

---

[239] Donald G. McNeil Jr., "Oxfam Joins Campaign to Cut Drug Prices for Poor Nations," *New York Times* (February 13, 2001): A6.

[240] Melody Petersen and Larry Rohter, "Maker Agrees to Cut Prices of 2 AIDS Drugs in Brazil," *New York Times* (March 31, 2001): A4.

[241] Melody Petersen, "Novartis Agrees to Lower Price of a Medicine Used in Africa," *New York Times* (May 3, 2001): C1.

[242] Macklin, *Double Standards in Medical Research in Developing Countries*, 166.

[243] Cohen and Illingworth, "The Dilemma of Intellectual Property Rights for Pharmaceuticals: The Tension Between Ensuring Access of the Poor to Medicines and Committing to International Agreements," 36.

The notion of differential pricing is distinct from the circumstance seen in various countries where they price the same drugs differently. The latter condition stems the policies adopted by individual countries that help them to control the drug market for their own people. Differences are also traceable to taxes, import duties, whole-sale and mark-ups, and several other factors. Variations in prices of the same drugs in different countries are not usually accounted for by a deliberate and systematic international policy, but differential pricing focuses on that purpose and structure.[244]

The purpose of differential pricing is to enable poor developing countries to achieve access to essential drugs for their populations.[245] In most poor developing countries, both the government and the majority of the people are not able to afford essential drugs that are required for various treatable diseases. On the other hand, in affluent developed countries, public funding is the chief source for financing healthcare services, varying from over 95% in the U.K., more than 90% in Norway, to a low of less than 50% in the US. This implies that out of packet payments, instead of prepaid insurance, are the major means for financing healthcare services, including buying of essential drugs in most poor developing countries.[246] Both price reductions and differential pricing present serious challenges to pharmaceutical companies by opening avenues for parallel trade or parallel importing, which undercuts their profits.[247] They are also ineffective solutions to the issue of affordable access to essential drugs for developing countries. The failure of the market mechanism to guarantee affordable access to essential drugs, as well as the limitations imposed by intellectual property rights resulted in the establishment and growth of public-private partnerships.[248]

3.3.2.3 International Collaborative Initiatives and Public-Private Partnerships

An astronomical growth in the establishment of public-private partnerships, and numerous collaborations among international agencies was experienced in the 1990s.[249] Pharmaceutical companies, international agencies, developed and developing countries' governments were more readily willing to collaborate in order to ensure affordable access to essential drugs for people in developing countries. Notable among those collaborative initiatives were the Global Alliance for Vaccines and Immunization (GAVI), the International AIDS Vaccine Initiative (IAVI), and

[244] Macklin, *Double Standards in Medical Research in Developing Countries*, 166.

[245] WHO Secretariat, "More Equitable Pricing for Essential Drugs: What Do We Mean and What Are the Issues?" Background Paper for WHO-WTO Secretariat Workshop on Differential Pricing and Financing of Essential Drugs, Hosbjor, Norway, April 8–11, 2001.

[246] Macklin, *Double Standards in Medical Research in Developing Countries*, 166.

[247] Macklin, *Double Standards in Medical Research in Developing Countries*, 169.

[248] Macklin, *Double Standards in Medical Research in Developing Countries*, 182.

[249] Macklin, *Double Standards in Medical Research in Developing Countries*, 174.

the Global Fund to Fight AIDS, Tuberculosis, and Malaria. These initiatives function by utilizing funds to support research and development aimed at manufacturing new drugs that are badly needed for people in developing countries who cannot afford them. The pharmaceutical companies involved in these efforts benefit by utilizing patents that stem from their collaboration to develop drugs they can market more profitably in developed countries. In return for engaging in the partnership, they usually make a commitment to offer drugs to developing countries at reasonable prices.[250]

GAVI was established in 1999 to guarantee the protection of children against diseases that can be avoided by vaccines. GAVI supports new vaccine development, organizes current immunization programs, and operates at international, regional, and national levels. Its special focus is to expedite research and development of vaccines for developing countries.[251]

GAVI has broadened its intiative by attracting several public and private partners, comprising "the United Nations agencies, WHO, the World Bank, and UNICEF; private foundations, the Bill and Melinda Gates Children's Vaccine Program and the Rockefeller Foundations; the industry group, International Federation of Pharmaceutical Manufacturers Association as well as public health and research institutions and national governments."[252] GAVI created the Vaccine Fund, which offers direct support to countries in two ways.The first is providing new and underused vaccines, in addition to safe immunization equipment; the second is providing funds in order to help governments in fortifying their immunization services.[253]

The Global Fund is one of the most recent intiatives by various United Nations agencies to establish private-partnerships for the purpose of ensuring affordable access to essential drugs in poor developing countries. The Global fund was established for the purpose of fighting AIDS, Tuberculosis and Malaria prevalent in development countries, where the local populations are unable to afford drugs for treating these diseases. The Global fund initiative was set in motion by the call from both Gro Harlem Bruntland, the former Director General of the World Health Organization, and Kofi Annan, the former Secretary General of the United Nations, for establishment of a large fund to combat diseases that kill or disable millions of people in poor developing countries. Both leaders envisioned the necessity for obligations from affluent and poor countries' governments, and also from private foundations, nongovernmental agencies, and the private sector, to embark on this initiative.[254] The Global Fund was structured to raise funds for broad objectives, including the purchase of drugs from manufacturers, the launch of better educational and prevention programs, the construction of new clinics or improvement of

[250] Macklin, *Double Standards in Medical Research in Developing Countries*, 175.

[251] Macklin, *Double Standards in Medical Research in Developing Countries*, 177.

[252] Macklin, *Double Standards in Medical Research in Developing Countries*, 177.

[253] Macklin, *Double Standards in Medical Research in Developing Countries*, 177.

[254] Macklin, *Double Standards in Medical Research in Developing Countries*, 179.

current ones, training of healthcare workers, and fortifying the infrastructure in other ways.[255]

The Global Fund devised the strategy of supporting poor nations to buy generic drugs rather than more exorbitant brand-name drugs currently under patent protections by the global pharmaceutical companies. This strategy made it possible for countries such as India and Brazil, which are large manufacturers of generic drugs to market their products in other poor countries. Because, the Global Fund offers grants to countries that apply for them, such countries are required to purchase the lowest priced drugs of guaranteed quality, which is a more efficient and effective utilization of the fund's money.[256] One of the major challenges which hampered the progress of the Global Fund was meager funding from voluntary donations, which falls far short of its anticipated goal and the US was blamed for setting a poor example for other countries.[257] The TRIPS Agreement also continued to impede the improved access to generic drugs by poor countries, making it more imperative for such countries to use compulsory licensing in order to address their public health emergency needs.

3.3.2.4 Complusory Licensing

The TRIPS Agreement makes provision for countries to respond effectively to public health emergency situations such as HIV/AIDS epidemics by issuing compulsory licensing. This provision authorizes countries to bypass patent protection and manufacture or import copies or versions of patented drugs in a case of national health emergency. The procedure for accomplishing this is to issue a compulsory license in order to produce a generic copy of a drug, and the patent-holder is paid an affordable royalty under this arrangement.[258] Countries lacking manufacturing capacity were also allowed under this provision to import a generic copy of a patented drug.

A major contention among WTO member countries was the definition of what constitutes a national emergency. Countries were invested with the right to determine what constitutes a national emergency and there were express indications that public health crises, such as HIV/AIDS,malaria, tuberculosis and other epidemics will be considered as national emergencies.[259]

[255] Barbara Crossette, "A Wider War on AIDS in Africa and Asia," *New York Times* (April 30, 2001): A6.

[256] Donald G. McNeil Jr., "UN Disease Fund Opens Way to Generics," *New York Times* (October 16, 2002): A6.

[257] Sheryl Gay Stolberg, "AIDS Fund Falls Short Goal and US is Given Some Blame," *New York Times* (February 13, 2002): http://www.nytimes.com/2002/02/13/health/13AIDS.html (Accessed February 26, 2014).

[258] Macklin, *Double Standards in Medical Research in Developing Countries*, 183.

[259] Divya Murthy, "The Future of Compulsory Licensing: Deciphering the Doha Declaration on the TRIPS Agreement and Public Health," *American University International Law Review* 17, no. 6 (2002): 1324–1325.

Another obstacle to countries issuing compulsory licensing was the pressure from big multinational pharmaceutical companies, US and other western governments who are opposed to invoking the provision. Their pressure sometimes resulted in litigations and retaliatory sanctions against such countries as evident in the South African case.[260]

The validity of complusory licensing approach has been defended by several authors on moral and pragmatic bases by elimination of other alternatives such as donation, ad hoc price reduction, and public-private partnerships.[261] Schuklenk and Ashcroft argue that, "the effective prevention of avoidable deaths, the operation of efficient competitive markets through lowering of artificial barriers to entry, and the assertion of legitimate national sovereignty in the international arena are conclusive prima facie justifications of compulsory licensing."[262]

Some risks associated with compulsory licensing were also highlighted. Pharmaceutical companies may have less capacity to shoulder the risks of research and development (R&D), if their right to market their products with a substantial profit, in order to recoup research and development costs is significantly undermined by the risk of compulsory licensing. This scenario may also adversely impact their choice of drugs to develop. For example, they may focus on luxury high-cost drugs for lifestyle conditions, instead of essential drugs for life-threatening and chronic disease.[263] Nevertheless, Compulsory licensing has been argued as the most effetive means available to developing countries' governments to provide essential drugs, including anti-retroviral drugs to their people in a time-efficiently manner.[264]

3.4 Conclusion

Concluding remarks on post-trial access to drugs in developing countries stress the compelling need and urgency for development of cheaper generic copies of anti-retroviral drugs for addressing HIV/AIDS epidemics. Conducting international clinical trials was considered as a primary strategy for providing affordable access to essential drugs in developing countries, but available evidence shows that participants and host populations usually do not share in the benefits that result from such trials

---

[260] Macklin, *Double Standards in Medical Research in Developing Countries*, 183–184.

[261] Schuklenk and Ashcroft, "Affordable Access to Essential Medication in Developing Countries: Conflicts between Ethical and Economic Imperatives", 137.

[262] Schuklenk and Ashcroft, "Affordable Access to Essential Medication in Developing Countries: Conflicts between Ethical and Economic Imperatives", 138.

[263] Schuklenk and Ashcroft, "Affordable Access to Essential Medication in Developing Countries: Conflicts between Ethical and Economic Imperatives", 138.

[264] Schuklenk and Ashcroft, "Affordable Access to Essential Medication in Developing Countries: Conflicts between Ethical and Economic Imperatives", 137.

Most drugs that result from international clinical trials conducted in developing countries are not marketed in that setting, because they are unable to afford them, rather they are usually sold in developed countries, where pharmaceutical companies recoup substantial profits on R&D.

Three major factors were identified as contributing significantly to the problem of lack of access to essential drugs in developing countries. First, is the meager investment on R&D by pharmaceutical companies on diseases of poverty prevalent in developing countries. Second, is the adverse effects of strict enforcement of patent protection. Third, is poverty and inadequate health infrastructure.

Post-trial access to drugs in developing countries was discussed within the broader contexts of intellectual property law, international trade agreements and non-patent factors. The tension between enforcement of strict patent protection and affordable access to essential drugs, including anti-retroviral drugs for the poor people in developing contries was acknowledged. A discussion on the analysis of the issue of post-trial access to drugs established that both patent and non-patent factors adversely impede access to affordable drugs in developing countries.

The severe impact of international trade agreements currently negotiated by United States, which further exacerbate impeded access to drugs for people in developing countries was clearly recognized. Two key aspects of TRIPS agreement was discussed, including strict patent protection in order to promote incentives for innovation and promotion of public health interests, and maintaining a delicate balance between them was considered imperative. Compulsory licensing and parallel importation were encouraged in order to assist countries with dealing effectively with national health emergencies.

The social responsibility of pharmaceutical companies was also argued. Pharmaceutical companies were concluded to have broader responsibilities both to shareholders and to the society.

Finally, current strategies for dealing with the issue of post-trial access to drugs for developing countries were discussed, including private donations, price reductions and differential pricing, international collaborative initiatives and public-private partnerships and compulsory licensing. Private donations and price reductions were viewed as improvised solutions that are not effective and sustainable. Compulsory licensing was argued to be most effective stategy for countries to exercise their duty of providing essential drugs, including anti-retroviral drugs to their citizens, as well as for preventing avoidable deaths from treatable conditions for millions of people in developing countries. The combined problems of clinical research protocols in developing nations (Chap. 2) and of affordable access to resulting drugs for host populations (Chap. 3) establish the context for the ethical analysis in the subsequent chapters: the issue of global justice to address post-trial access to drugs in developing countries.

Chapter 4
National Responsibility and Humanitarian Assistance

4.1 Introduction

There has been a contentious debate about the issue of global justice to address post-trial access to drugs in developing countries. Two major approaches have emerged in this regard, cosmopolitan and statist.[1] Millum acknowledges that, "a central question that divides theorists writing about global justice, and that affects most directly problems in international bioethics, concerns what people and governments of rich countries owe to those outside their borders."[2] Cosmopolitans such as Pogge argue that distributive justice principles that apply in the domestic realm apply equally globally. The implication is that if Rawls' difference principle were considered the right way to distribute primary goods within a country, there should as well be a global difference principle distributing primary goods among all people in the world.[3] In contrast, statists such as Rawls argue that the principle of distributive justice applies only within domestic society, that is, individual nations.[4] Underlying the approach of Rawls is the view that justice deals with the basic structure of the society.[5] The implication of the statist approach of Rawls is to rely merely upon the duty of humanitarian assistance from the perspective of global justice to provide affordable access to drugs in developing nations. Perspectives of participants in international clinical trials on the issue of justice in contrast to Rawls' statist approach would be discussed. The notion of justice for local participants in international clinical trials is committed to global distributive justice principle.

---

[1] Joseph Millum, "Global Bioethics and Political Theory," in *Global Justice and Bioethics*, ed. Joseph Millum and Ezekiel J. Emmanuel (New York: Oxford University Press, 2012), 20.

[2] Millum, "Global Bioethics and Political Theory," 20.

[3] Millum, "Global Bioethics and Political Theory," 20.

[4] Millum, "Global Bioethics and Political Theory," 20.

[5] John Rawls, *A Theory of Justice*, Revised Edition (Cambridge, Massachusetts: The Belknap Press of Harvard University Press, 1999), 6.

E.C. Obi, *Post-Trial Access to Drugs in Developing Nations*, Advancing Global Bioethics 7, DOI 10.1007/978-3-319-60028-4_4

This chapter focuses on Rawls's statist approach to the issue of post-trial access to drugs in developing countries in contrast to the view of participants on the issue of justice in international clinical trials. The first part of this chapter deals with Rawls major contribution to political liberalism which begins with his landmark book titled A Theory of Justice (TOJ), first published in 1971 and later revised in 1999 that focuses on domestic justice, that is, justice within societies. He presents a conception of justice which he refers to as "justice as fairness."[6] Rawls's idea of social justice focuses on establishing criteria for evaluating the distribution of the primary social goods in a society.

The second part deals with just health care within nations which focuses on Rawls's view about the classification of health as well as Daniels' different conception of the position of health in Rawls's theory. The third part of this chapter focuses on Rawls's statist approach to post-trial access to drugs in developing countries which restricts the principle of justice to the domestic society.

The fourth part tackles Rawls's duty of assistance and post-trial access to drugs in contrast to the idea of justice for participants in international clinical trials. Rawls's notion of international responsibilities focuses on assisting burdened societies to attain well-ordered societies. Providing affordable access to drugs as a transition strategy is argued as an important component of Rawls's duty of assistance to burdened societies in order to attain well-ordered societies. On the other hand, the notion of justice for local participants in international clinical trials is strongly committed to distributive justice. A more detailed analysis of Rawls's position on post-trial access to drugs in developing countries begins with a discussion of justice as fairness.

4.2 Fairness and Just Health Care Within Nations

4.2.1 Justice and Fairness: John Rawls

4.2.1.1 A Fair System of Social Cooperation

Justice as fairness evolves from a political tradition that supports the public culture of a democratic society which emphasizes a basic idea of society as a fair system of cooperation.[7] Rawls articulates that, "one practical aim of justice as fairness is to provide an acceptable philosophical and moral basis for democratic institutions and thus to address the question of how the claims of liberty and equality are to be

[6] Rawls, *A Theory of Justice*, Revised Edition, 3.

[7] John Rawls, *Political Liberalism* (New York: Columbia University Press, 1993), 14–15.

understood."[8] The notion of society as a fair system of cooperation establishes a foundation for developing a political conception of justice for a democratic regime.[9]

In conjunction with the fundamental idea of society as fair system of cooperation, two other ideas are also considered central in understanding the concept of justice as fairness. These ideas comprise of the idea of citizens which denotes people involved in cooperation as free and equal persons; as well as the idea of well-ordered society, which indicates a society controlled effectively by a public conception of justice.[10]

Rawls identifies three important features of the notion of social cooperation. First, social cooperation is distinguished from an activity simply socially organized. The implication is that social cooperation is regulated by publicly acknowledged rules and procedures which cooperating members recognize as suitable to govern their conduct. Second, the notion of cooperation involves the idea of fair terms of cooperation, which implies the terms every cooperating member accepts as reasonable in all cases. Fair terms of cooperation stipulate an idea of reciprocity, and mutuality which entails that all those who played their role as required by the publicly acknowledged rules would benefit as stipulated by a public and consensus standard. Third, the notion of social cooperation involves an idea of each participant's rational advantage, or good. This idea of rational advantage stipulates what those involved in cooperation are looking to accomplish from the perspective of their own good.[11]

The idea of reciprocity in social cooperation rests between "the idea of impartiality, which is altruistic (being moved by the general good), and the idea of mutual advantage understood as everyone's advantaged with respect to each person's present or expected future situations as things are."[12] In the context of justice as fairness, reciprocity refers to a relation between citizens articulated by principles of justice that govern a social institution in which everyone benefits with regard to a suitable benchmark of equality delineated with respect to that institution.[13] Furthermore, Rawls highlights that "reciprocity is a relation between citizens in a well-ordered society expressed by its political conception of justice."[14] For example, the idea of reciprocity is established between citizens when you consider the two principles of justice with the difference principle emphasizing equal division as a benchmark.[15] Noteworthy also is that the idea of reciprocity is distinct from the idea of mutual advantage.[16]

---

[8] John Rawls, *Justice as Fairness: A Restatement* (Cambridge, Massachusetts: Harvard University Press, 2001), 5.

[9] Rawls, *Justice as Fairness: A Restatement*, 5.

[10] Rawls, *Justice as Fairness: A Restatement*, 5.

[11] Rawls, *Justice as Fairness: A Restatement*, 6.

[12] Rawls, *Political Liberalism*, 16–17.

[13] Rawls, *Political Liberalism*, 17.

[14] Rawls, *Political Liberalism*, 17.

[15] Rawls, *Political Liberalism*, 17.

[16] Rawls, *Political Liberalism*, 17.

Rawls also makes a distinction between two fundamental and complementary ideas, reasonable and rational as they relate to the basic idea of society as a fair system of social cooperation. In the context of persons involved in social cooperation and positioned as equals in several regards, "reasonable persons are ready to propose, or to acknowledge when proposed by others, the principles needed to specify what can be seen by all as fair terms of cooperation."[17] There is a consensus that reasonable persons also comprehend that they are to respect these principles, even when it is detrimental to their own interests but with the caveat that other members engaged in the cooperation may be required to respect them. It is deemed unreasonable not to be prepared to propose such principles, or not to respect fair terms of cooperation that other cooperating members may reasonably be required to consent to. Furthermore, it is worse than unreasonable if the person simply fakes to propose or respect the principles but is prepared to infringe on them to one's advantage as the situation allows.[18]

Conversely, what may be deemed unreasonable, may in general, be deemed rational. For example, we suggest this distinction in a situation when we concur that among persons engaged in cooperation, certain people due to their superior bargaining position, their proposal is clearly rational, but all the same unreasonable.[19]

The critical role of the principles of justice is to stipulate the fair terms of social cooperation. These principles stipulate the basic rights and duties to be allocated by the key political and social institutions, and they control the allocation of benefits and burdens arising from social cooperation. The principles of a democratic conception of justice may be seen as stipulating the fair terms of cooperation between citizens in a democratic society, where citizens are considered as free and equal persons from the perspective of political conception.[20] Another critical feature of justice as fairness is the basic structure of the society which is considered as the primary subject of justice.

4.2.1.2 The Basic Structure of the Society

Rawls's idea of the basic structure applies to a well-ordered society, which is a society controlled effectively by a public conception of justice.[21] Rawls contends that "the primary subject of justice is the basic structure of society, or more exactly, the way in which the major social institutions distribute fundamental rights and duties and determine the division of advantages from social cooperation."[22] The major institutions of basic structure of a society consist of political constitution, legal protection of freedom of thought and liberty of conscience, legally recognized forms of

[17] Rawls, *Justice as Fairness: A Restatement*, 6–7.

[18] Rawls, *Justice as Fairness: A Restatement*, 7.

[19] Rawls, *Justice as Fairness: A Restatement*, 7.

[20] Rawls, *Justice as Fairness: A Restatement*, 7.

[21] Rawls, *Justice as Fairness: A Restatement*, 8–10.

[22] Rawls, *A Theory of Justice*, Revised Edition, 6.

private property, the structure of the economy in the form of competitive markets and the monogamous family.[23] The activities of associations and individuals occur within the context of the basic structure which is the background social framework. Thus, "a just basic structure secures what we may call background justice".[24]

The major institutions of the basic structure of a society specify rights and duties of the citizens and significantly impact their life prospects, aspirations and opportunities, as well as their ability to take advantage of them in order to be successful in the society.[25] Therefore, the basic structure is the primary subject of justice because its impacts are pervasive and present from the start of life.[26] Significant inequalities are acknowledged in the basic structure of the society, because of the impact of the natural and social lotteries. The purpose of the principles of justice applied to the basic structure of the society is to address these inequalities.[27]

The principles of justice as fairness are limited to the basic structure and consequently regulate this structure, but they do not apply directly to or control institutions and associations internally within a domestic society. Constraints from the principles of justice apply only indirectly to private associations and institutions such as firms, labor unions, churches, universities and the family. For example, the two principles of justice are not meant to regulate the internal organization of churches and universities. The difference principle does not regulate how parents should treat their children or distribute the wealth of the family among its members.[28]

The principles of justice as fairness which may be deemed reasonable and just for the basic structure may not also be usually considered reasonable and just for institutions, associations and social practices. Although, the principles of justice as fairness impose restrictions on these social arrangements within the basic structure, the basic structure and the associations and social forms within it are separately regulated by different principles relative to their goals and their distinctive nature and peculiar requirements.[29] Rawls articulates, "Justice as fairness is a political, not a general, conception of justice: it applies first to the basic structure and sees other questions of local justice and also questions of global justice (what I call the law of peoples) as calling for separate considerations on their merits."[30]

The principles of justice that apply directly to or regulate associations and institutions within the basic structure may be termed principles of local justice. Rawls identifies three levels of justice, first, local justice that entails principles that apply directly to institutions; second, domestic justice that implies principles that apply to the basic structure of society; and third, global justice that involves principles that

---

[23] Rawls, *A Theory of Justice*, Revised Edition, 6.

[24] Rawls, *Justice as Fairness: A Restatement*, 10.

[25] Rawls, *A Theory of Justice*, Revised Edition, 6–7.

[26] Rawls, *A Theory of Justice*, Revised Edition, 7.

[27] Rawls, *A Theory of Justice*, Revised Edition, 7.

[28] Rawls, *Justice as Fairness: A Restatement*, 10.

[29] Rawls, *Justice as Fairness: A Restatement*, 11.

[30] Rawls, *Justice as Fairness: A Restatement*, 11.

apply to international law.[31] Justice as fairness begins with domestic justice which is the justice of the basic structure. It moves outward to the law of peoples and then inward to local justice.[32]

4.2.1.3 The Elements of Theory of Justice

Original Position

The idea of original position is necessitated by the requirement to specify the fair terms of cooperation in a fair system of cooperation between free and equal persons. In the context of justice as fairness, the fair terms of cooperation are specified by an agreement arrived at by free and equal citizens engaged in cooperation, which is reached with the understanding of what they consider as their reciprocal advantage, or good.[33] Rawls indicates that, "the original position is the appropriate initial status quo which insures that the fundamental agreements reached in it are fair. This fact yields the name "justice as fairness."[34]

The original agreement must be reached under certain conditions that are ideally fair if it is to be a valid agreement from the perspective of political justice. Specifically, these conditions must position free and equal persons fairly and must not allow some to possess unfair bargaining advantages over others.[35] More so, certain impeding conditions such as "threats of force and coercion, deception and fraud must be excluded."[36]

The original position is a thought experiment, an imaginary condition where every real citizen is represented and all these representatives reach an agreement on the principles of justice that would regulate the political institutions of the real citizens.[37] In order to reach the most impartial situation, the parties who enter the agreement are deprived of salient information that might bias their choice of principles. This idea is captured by indicating that the parties are to choose under a veil of ignorance.[38] The conditions of original position executed under the veil of ignorance "define the principles of justice as those which rational persons concerned to advance their interests would consent to as equals when none are known to be advantaged by social and natural contingencies."[39] The veil of ignorance is introduced in order to nullify the influences of natural and social circumstances

---

[31] Rawls, *Justice as Fairness: A Restatement,* 11.

[32] Rawls, *Justice as Fairness: A Restatement*, 11.

[33] Rawls, *Justice as Fairness: A Restatement,* 15.

[34] Rawls, *A Theory of Justice*, Revised Edition, 15.

[35] Rawls, *Justice as Fairness: A Restatement*, 15.

[36] Rawls, *Political Liberalism*, 23.

[37] Stanford Encyclopedia of Philosophy, *John Rawls*, (March 25, 2008), http://plato.stanford.edu/entries/rawls/ (Accessed March 28, 2014).

[38] Rawls, *A Theory of Justice*, Revised Edition, 17.

[39] Rawls, *A Theory of Justice,* Revised Edition, 17.

which could be exploited by some citizens to their own advantage. In this context of the veil of ignorance, the parties do not know how numerous options will impact their own specific case and they are bound to assess principles exclusively on the basis of general considerations.[40]

Rawls argues that citizens should not be favored or disfavored by social institutions based on for example characteristics such as their race, class and gender. Each party in the original position is deprived of information regarding the race, class, and gender of the real citizen they represent. The parties in the original position are deprived of all particular facts regarding citizens that are irrelevant to the choice of principles of justice: not only their race, class, and gender but as well their age, natural assets and abilities, etc. The veil of ignorance also eliminates specific information regarding the society of the citizens in order to acquire a clearer opinion of the enduring features of a just social system.[41] However, the parties know that citizens in the society possess different comprehensive doctrines and plans of life. They know as well that all citizens are interested in more primary social goods.[42] The parties also know the general facts about human society and whatever general facts influence the choice of the principles of justice.[43]

The veil of ignorance is aimed at positioning the representatives of free and equal citizens fairly with regard to one another. Parties in the original position cannot push for agreement on principles that will randomly favor the particular citizen they represent, because they do not know the particular attributes of the citizen they represent. The condition of the parties therefore symbolizes reasonable conditions, within which the parties can enter a rational agreement. Each party makes honest efforts to agree to principles that will be most advantageous for the citizen they represent, which implies maximizing the share of the citizen's primary goods. The implication is that because the parties are fairly situated, they will reach an agreement that will be fair to all actual citizens.[44] The primary task of the parties in the original position is to choose the principles of justice that will regulate the social life and the basic structure of the society that are ideally fair.

Principles of Justice as Fairness

Rawls establishes principles that he thinks would be chosen in the original position. The principles are considered as ones that free and equal persons could accept as a fair basis for social cooperation. The principles are presented as guiding ideas of

[40] Rawls, *A Theory of Justice,* Revised Edition, 118.

[41] Stanford Encyclopedia of Philosophy, *John Rawls,* (March 25, 2008), http://plato.stanford.edu/entries/rawls/ (Accessed March 28, 2014).

[42] Stanford Encyclopedia of Philosophy, *John Rawls*, (March 25, 2008), http://plato.stanford.edu/entries/rawls/ (Accessed March 28, 2014).

[43] Rawls, *A Theory of Justice*, Revised Edition, 119.

[44] Stanford Encyclopedia of Philosophy, *John Rawls*, (March 25, 2008), http://plato.stanford.edu/entries/rawls/ (Accessed March 28, 2014).

justice as fairness thus: "… Each Person has the same indefeasible claim to a fully adequate scheme of equal basic liberties, which scheme is compatible with the same scheme of liberties for all (the equal liberty principle); … Social and economic inequalities are to satisfy two conditions: (a) They are to be attached to offices and positions open to all under conditions of fair equality of opportunity (fair equality of opportunity principle); (b) They are to be to the greatest benefit of the least-advantaged members of society (the difference principle)."[45] These principles mainly apply to basic structure of society and rule the assignment of rights and duties and control the distribution of social and economic advantages.[46]

The principles of justice as fairness in all consist of the equal liberty principle, fair equality of opportunity principle and the difference principle. The difference principle requires that any social and economic inequalities work to the greatest benefit of the least advantaged. Fair equality of opportunity principle requires that all citizens of a domestic society have equal opportunities for obtaining position of power.[47] Rawls's difference principle and fair equality of opportunity principle provide protection and compensation for people who are disadvantaged by natural and social lotteries.[48]

Rawls's general conception of justice is that, "all social primary goods - liberty and opportunity, income and wealth, and the social bases of self-respect – are to be distributed equally unless an unequal distribution of any or all of these goods is to the advantage of the least favored."[49] A brief discussion of the principles of justice as fairness follows.

Equal Liberty Principle

The equal liberty principle guarantees equal basic rights and liberties for all citizens, including "political liberty, (the right to vote and to hold public office) and freedom of speech and assembly; liberty of conscience and freedom of thought; freedom of the person, which includes… the right to hold personal property; and freedom from arbitrary arrest and seizure as defined by the concept of the rule of law."[50] The equal liberty principle requires equal rights for all in all normal conditions because unequal rights would not be to the advantage of those who would get a lesser share of rights.[51] The history of democratic thought reveals that emphasis

[45] Rawls, *Justice as Fairness: A Restatement*, 42–43.

[46] Rawls, *A Theory of Justice,* Revised Edition, 53.

[47] Rawls, *A Theory of Justice*, Revised Edition, 53.

[48] Tristram H. Engelhardt, *The Foundations of Bioethics* (New York: Oxford University Press, 1996), 393.

[49] John Rawls, *A Theory of Justice*, Original Edition (Cambridge, Massachusetts: Harvard University Press, 1971), 303.

[50] Rawls, *A Theory of Justice*, Revised Edition, 53.

[51] Stanford Encyclopedia of Philosophy, *John Rawls*, (March 25, 2008), http://plato.stanford.edu/entries/rawls/ (Accessed March 28, 2014).

has been on realizing certain specific rights and liberties as well as specific constitutional guarantees, as enshrined in various bills of rights and declarations of the rights of man. Justice as fairness conforms to this traditional view of human rights.[52]

A list of basic liberties can be established in two different ways, historical and analytical.[53] In the historical context, various democratic regimes are reviewed and a list of rights and liberties gathered that appear fundamental and firmly guaranteed in apparently more historically successful regimes. Obviously, cognizant of the veil of ignorance, the implication is that this type of specific information is not accessible to the parties in the original position. On the other hand, it is accessible to you and me in establishing justice as fairness.[54]

In the analytical context, the focus is on what liberties offer the political and social circumstances critical for the sufficient development and full exercise of the two moral powers of free and equal persons. In line with this thought, some conclusions are made: first, that equal political liberties and freedom of thought make it possible for citizens to cultivate and to exercise these powers in evaluating the justice of the basic structure of the society as well as its social policies. Second, that liberty of conscience and freedom of association make it possible for citizens to cultivate and exercise their moral powers in establishing and reviewing and in pursuing rationally their notions of the good either individually or collectively.[55] Rawls argues that, "those basic rights and liberties protect and secure the scope required for the exercise of the two moral powers in the two fundamental cases just mentioned…."[56] It is pertinent to point out that exercising our moral powers in these ways is important to us as free and equal citizens.[57]

Noteworthy also is that the equal liberty principle of justice applies not only to the basic structure of society but more precisely to either written or unwritten constitution. Some of these liberties, such as the equal political liberties and freedom of thought and association, are securely protected by a constitution.[58] Rawls argues that constituent power distinguished by Locke as people's power to constitute the legislative as the first and fundamental law of all commonwealths, in contrast to ordinary power is to be enshrined in the bill of rights, in the right to vote and to occupy office and in the measures for amending the constitution.[59]

Rawls stipulated the strict lexical priority of the equal liberty principle over fair equality of opportunity principle and the difference principle. This implies that one may not trade off one's basic liberties for gains either in difference principle or in fair equality of opportunity. It is also important to note that fair equality of opportu-

---

[52] Rawls, *Justice as Fairness: A Restatement*, 45.
[53] Rawls, *Justice as Fairness: A Restatement*, 45.
[54] Rawls, *Justice as Fairness: A Restatement*, 45.
[55] Rawls, *Justice as Fairness: A Restatement*, 45.
[56] Rawls, *Justice as Fairness: A Restatement*, 45.
[57] Rawls, *Justice as Fairness: A Restatement*, 45.
[58] Rawls, *Justice as Fairness: A Restatement*, 46.
[59] Rawls, *Justice as Fairness: A Restatement*, 46.

nity, the non-discrimination principle, has strict lexical priority over the difference principle.[60]

Fair Equality of Opportunity Principle

Fair equality of opportunity requires that not only that public offices and social positions be open in the proper sense, but that all should have a fair opportunity to achieve them.[61] Fair equality of opportunity further requires that citizens with the same talents and abilities and the same disposition to utilize them should have the same educational and economic opportunities regardless of whether they were born rich or poor.[62] Rawls argues that, "in all parts of the society there are to be roughly the same prospects of culture and achievement for those similarly motivated and endowed."[63]

The fair equality of opportunity principle specifies what types of inequalities are allowed in a domestic society. In this context, inequalities are regarded as "not any differences between offices and positions, but differences in the benefits and burdens attached to them either directly or indirectly, such as prestige and wealth, or liability to taxation and compulsory services."[64] Inequalities are understood here as differences stemming from distribution established by a practice or enabled by it, of the things citizens endeavor to achieve or avoid.[65]

Rawls identifies fair equality of opportunity with liberal equality. Fair equality of opportunity achieves its purpose by imposing certain requirements on the basic structure of the society. For example, a free market system must be regulated by political and legal institutions in order to avoid excessive concentrations of property and wealth, which may likely result in political domination.[66]

Fair equality of opportunity is linked to pure procedural justice.[67] From this perspective, issues regarding distributive shares are handled as pure procedural justice. The implication is to establish a social system that guarantees just result.[68] Rawls argue that, "pure procedural justice obtains when there is no independent criterion for the right result: instead there is a correct or fair procedure such that the outcome is likewise correct or fair, whatever it is, provided that the procedure has been prop-

[60] Rawls, *A Theory of Justice,* Revised Edition, 266.

[61] Rawls, *Justice as Fairness: A Restatement,* 43.

[62] Rawls, *Justice as Fairness: A Restatement*, 44.

[63] Rawls, *Justice as Fairness: A Restatement*, 44.

[64] Samuel Freeman, ed., *John Rawls: Collected Papers* (Cambridge, Massachusetts: Harvard University Press, 1999), 50.

[65] Freeman, ed., *John Rawls: Collected Papers*, 50.

[66] Rawls, *Justice as Fairness: A Restatement*, 44.

[67] Rawls, *A Theory of Justice*, Revised Edition, 73.

[68] Rawls, *A Theory of Justice*, Revised Edition, 74.

erly followed."[69] The function of the principle of fair equality of opportunity is to guarantee that the system of cooperation conforms to pure procedural justice.[70]

The social and economic process is established within the appropriate framework of political and legal institutions. The result of the distributive process will not be just without a suitable system of these background institutions. This implies that background fairness is absent.[71] The government attempts to guarantee equal opportunities of education and culture for those who are equally gifted and motivated either by funding private schools or by setting up a public-school system. Moreover, it implements and guarantees equality of opportunity in economic activities and in the free choice of occupation.[72] Inequalities of any sort can only be authorized by the government if they are to the greatest advantage of the least privileged.

Difference Principle

The difference principle requires that social institutions be arranged so that inequalities of wealth and income work to the greatest benefit of the least advantaged.[73] The emphasis here is that every party involved in the social cooperation must benefit from the inequality. The implication is that every person involved in the social cooperation "must find it reasonable to prefer his condition and prospects with the inequality to what they would be under the practice without it."[74] The principle in essence rules out the justification of inequalities on the basis that the disadvantages in one position are overshadowed by the greater advantages of those in another position.[75]

The difference principle is established within the context that "social cooperation is always productive, and without cooperation there would be nothing produced and so nothing to distribute."[76] A system of cooperation is established mainly by how its rules organize productive activity, stipulate the division of labor and allocate various roles to the members engaged in it. Beginning from an imagined baseline of equality, greater returns to the more advantaged can be produced by permitting inequalities in wages and salaries. In this context, higher wages can cover the costs of training and education, and can offer incentives to fill jobs that are in more demand.[77]

The difference principle requires that inequalities which increase the total product work to the advantage of everyone, and precisely to the greatest advantage of

[69] Rawls, *A Theory of Justice,* Revised Edition, 75.

[70] Rawls, *A Theory of Justice,* Revised Edition, 77.

[71] Rawls, *A Theory of Justice,* Revised Edition, 243.

[72] Rawls, *A Theory of Justice,* Revised Edition, 243.

[73] Stanford Encyclopedia of Philosophy, *John Rawls*, (March 25, 2008), http://plato.stanford.edu/entries/rawls/ (Accessed March 28, 2014).

[74] Freeman, ed., *John Rawls: Collected Papers*, 50.

[75] Freeman, ed., *John Rawls: Collected Papers*, 50.

[76] Rawls, *Justice as Fairness: A Restatement*, 61.

[77] Rawls, *Justice as Fairness: A Restatement*, 61.

those least favored. The difference principle does not permit the affluent in the society to get richer at the expense of the poor. The difference principle exemplifies "equality-based reciprocity: from an egalitarian baseline it requires inequalities that are good for all, and particularly for the worst-off."[78]

Rawls argues that inequalities of birth, natural endowment and historical circumstances are undeserved, and, in a fair system of cooperation where justice is promoted, every effort should be invested in compensating those who have been disadvantaged by the identified factors.[79] He further argues that advantages that people have over others that are the outcomes of accidents of biology and history appear arbitrary from the moral point of view, and should then be redressed as far as possible.[80] He emphasizes the notion of redressing the bias of contingencies emanating from the inequalities of birth and natural endowment in order to maximize equality for the least privileged in the fair system of cooperation.[81] Citizens endowed with different talents and abilities can use them for the benefit of everyone. In a society regulated by the difference principle citizens consider distribution of natural endowments as an asset that benefits all. Those better endowed are encouraged to utilize their talents and abilities to make themselves better off, but provided that they as well contribute to the good of those with less endowments.[82] Rawls contends that, "in justice as fairness men agree to share one another's fate."[83] Daniels extends Rawls's theory of justice to health care[84] in the domestic society which will be the focus of the discussion in the next section of the chapter.

4.2.2 *Just Health Care Within Nations*

4.2.2.1 Index of Primary Goods

The basic structure of society as contended by Rawls distributes certain primary goods which imply the things that every rational person is supposed to want. These goods usually play a significant role in a person's rational plan of life.[85] Rawls argues for index of primary social goods which is a truncated scale of well-being utilized by moral agents pursuing a hypothetical social contract.[86] His list of primary social goods includes rights, liberties and opportunities, income and wealth,

---

78 Stanford Encyclopedia of Philosophy, *John Rawls*, (March 25, 2008), http://plato.stanford.edu/entries/rawls/ (Accessed March 28, 2014).

79 Rawls, *A Theory of Justice*, Original Edition, 100–108.

80 Rawls, *A Theory of Justice*, Original Edition, 15.

81 Rawls, *A Theory of Justice*, Original Edition, 100–102.

82 Rawls, *A Theory of Justice*, Original Edition, 101.

83 Rawls, *A Theory of Justice*, Original Edition, 102.

84 Daniels, *Just Health Care*, 42–48.

85 Rawls, *A Theory of Justice*, Revised Edition, 54.

86 Daniels, *Just Health Care*, 42.

and social bases of self-respect.[87] Evidently absent from his list of primary social goods are health and health care. He classified them as natural goods that can be influenced by the basic structure in their possession but are not considered to be regulated by the difference principle.[88]

Rawls simplified the construction of his theory that he assumed individuals engaged in the social contract would be fully functional, active and normal over the course of their life span and that no one would become ill or die prematurely.[89] Daniels argues that "Rawls index of primary goods seems to be too truncated a scale, once we drop the idealizing assumption that all people are normal. People with equal indices will not be equally well-off once we allow them to differ in health care needs."[90] Similarly, Kenneth Arrow argued that Rawls's index of primary goods was inadequate because it fell short in articulating for us how to compare the ill rich with the well poor.[91] Amartya Sen also argued that the index is not sensitive to the way in which disease, disability, or other individual variations would produce inequalities in people's capabilities for those who had the same primary social goods.[92]

Arrow pointed out some shortcomings related to merely including health care to the list of primary goods. He argued that the import of Rawls's difference principle which requires inequalities to work to the benefit of the least advantaged individuals, would be to invest all social resources into fulfilling special needs of persons with excessive health care needs, probably to a situation where the rest of society is impoverished.[93] He further argued that including health care to the creation of the index, and permitting its exchange against income and wealth, would compel Rawls into interpersonal comparisons of utility he had intended his index would avoid.[94] Nevertheless, Daniels argues that extending Rawls's theory to include health care through the equal opportunity account undermines some of Arrow's and Sen's criticisms.[95]

Despite Rawls simplified idealization of his theory, it still provided a clue about how to extend it to the real world of illness and premature death. The objective of public health and medicine is to restore people as close as possible to the ideal of

[87] Rawls, *A Theory of Justice*, Revised Edition, 54.

[88] Rawls, *A Theory of Justice*, Revised Edition, 54.

[89] Norman Daniels, Bruce P. Kennedy and Ichiro Kawachi, "Why Justice is Good for our Health: The Social Determinants of Health Inequalities," *Daedalus* 128, no. 4 (Fall 1999): 228.

[90] Daniels, *Just Health Care*, 43.

[91] Kenneth J. Arrow, "Uncertainty and the Welfare Economics of Medical Care," *American Economic Review* 53, no.5 (December 1963): 941–973.

[92] Amartya K. Sen, *Inequality Reexamined* (Cambridge, Mass.: Harvard University Press, 1992), cited in Daniels, Kennedy and Kawachi, "Why Justice is Good for our Health: The Social Determinants of Health Inequalities," 233.

[93] Kenneth J. Arrow, "Some Ordinalist-Utilitarian Notes on Rawls's Theory of Justice," *Journal of Philosophy* 70, no. 9 (May 1973):251.

[94] Arrow, "Some Ordinalist-Utilitarian Notes on Rawls's Theory of Justice," 254.

[95] Norman Daniels, "Equality of What: Welfare, Resources, or Capabilities?" *Philosophy and Phenomenological Research* 50, supplement (Autumn 1990): 273–296.

normal functioning, within the reasonable constraints of the resources. Resources are essentially limited because maintaining health cannot be the society's only social good or objective.[96]

Daniels adopted a different conception of the place of health in Rawls's theory as articulated in his theory of just health care discussed in his 1985 work Just Health Care.[97] He argues that health care institutions should be regulated by a principle of fair equality of opportunity, but under two conditions: (1) an adequate general theory of justice encompasses a principle which obliges basic institutions to guarantee fair equality of opportunity, and (2) the fair equality of opportunity principle functions as a control on allowable economic inequalities.[98] Daniels strongly advocates for the fair equality of opportunity as a suitable principle to regulate macro decisions regarding the health care system's design. He further articulates that, "such a principle defines, from the perspective of justice, what the moral function of the health-care system must be – to help guarantee fair equality of opportunity.[99] A brief discussion of fair equality of opportunity and Daniels's just health care theory follows.

4.2.2.2 Fair Equality of Opportunity and Just Health Care

The most viable strategy for extending Rawls's theory in Daniels's just health care entails adding health care institutions and practices to the basic institutions engaged in guaranteeing fair equality of opportunity.[100] Daniels acknowledges that "meeting the health needs of all persons, viewed as free and equal citizens, is of comparable and special moral importance."[101] Daniels contends that since meeting health care needs has a critical influence on the distribution of opportunity, the health care institutions are governed by a fair equality of opportunity principle. He further argues for a special correlation between species functioning and the opportunity range open to an individual in a society.[102] Daniels clearly writes "since meeting health needs protects the range of opportunities people can exercise, then any social obligations we have to protect opportunity imply obligations to protect and promote health for all."[103]

Similarly, Beauchamp and Childress articulate that, "Daniels's thesis is that social institutions affecting health care distribution should be arranged, as far as

[96] Daniels, Kennedy and Kawachi, "Why Justice is Good for our Health: The Social Determinants of Health Inequalities," 228.

[97] Daniels, *Just Health Care*, 41.

[98] Daniels, *Just Health Care*, 41.

[99] Daniels, *Just Health Care*, 41.

[100] Daniels, *Just Health Care*, 45.

[101] Norman Daniels, *Just Health: Meeting Health Needs Fairly* (Cambridge: Cambridge University Press, 2008), 141.

[102] Daniels, *Just Health Care*, 45.

[103] Daniels, *Just Health: Meeting Health Needs Fairly*, 141.

possible, to allow each person to achieve a fair share of the normal range of opportunities present in that society."[104] The normal range of opportunity entails the range of life plans that a reasonable person could pursue, taken into account his or her talents and skills in a given society. Daniels's just health care theory, like Rawls's acknowledges a positive obligation of the society to eliminate or reduce obstacles that prevent fair equality of opportunity, an obligation that encompasses programs to rectify or compensate for numerous disadvantages.[105] Daniels writes "just health requires that we protect people's shares of the normal opportunity range by treating illness when it occurs, by reducing the risks of disease and disability before they occur, and by distributing those risks equitably."[106]

Daniels contends that if it is critical to utilize resources to compensate for the advantages in opportunity some people suffer in the natural lottery, it is equally critical to utilize resources to compensate for the natural disadvantages caused by disease.[107] Disease etiology has been noted as significantly impacted by social conditions which varies with class and which refutes the conception that disease is a product of the natural lottery.[108] Daniels et al. argue that health status is principally determined by choices regarding what are termed the social determinants of health. They articulate, "properly understood, justice as fairness tells us what justice requires in the distribution of all socially controllable determinants of health."[109] There is no intention to engage in a futile goal of eradicating all differences between people. Daniels highlights, "health care has normal functioning as its goal: it concentrates on a specific class of obvious disadvantages and tries to eliminate them. That is its limited contribution to guaranteeing fair equality of opportunity."[110] Disease and disability are seen as unjustified constraints on persons' opportunities to fulfill basic goals. Health care is then necessary to attain, restore or maintain adequate or "species-typical" levels of functioning, in order to accomplish basic goals.[111]

Another important point noted by Daniels in extending Rawls's theory to health care is that Rawls's contractarian theory requires a thick veil of ignorance in order to ensure the impartiality of free and equal moral agents. However, Daniels advocates for a thinner veil of ignorance in selecting principles to regulate health-care resource allocation decisions. This is important because in this context, parties involved in the negotiations must know about some essential features of the society,

---

[104] Tom L. Beauchamp and James F. Childress, *Principles of Biomedical Ethics*, 5th ed. (New York: Oxford University Press, 2001), 234.

[105] Beauchamp and Childress, *Principles of Biomedical Ethics*, 5th ed., 234.

[106] Daniels, *Just Health: Meeting Health Needs Fairly*, 141.

[107] Daniels, *Just Health Care*, 46.

[108] Daniels, *Just Health Care*, 46.

[109] Daniels, Kennedy and Kawachi, "Why Justice is Good for our Health: The Social Determinants of Health Inequalities," 232.

[110] Daniels, *Just Health Care*, 46.

[111] Beauchamp and Childress, *Principles of Biomedical Ethics*, 5th ed., 234.

for instance, its resource limitations.[112] It would be a futile effort for individuals to negotiate behind the veil of ignorance for benefits in the realm of health care that end up being completely unaffordable in real life. Health care is one of those contemporary societal benefits that cannot be available in any society without limitations.[113]

Daniels identifies four levels of health care institutions that should be provided in order to reflect the original idealization under which Rawls's theory was constructed. The idealization entails the ideal to enable normal, fully functioning persons to complete their normal life span. The four levels of health care institutions include: (1) Preventive health-care institutions that reduce the prospect of departures from the normality assumption. (2) Institutions that deliver personal medical and rehabilitative services that restore normal functioning. (3) Institutions that provide more extended medical and social support services for people who are moderately chronically ill or disabled, comprising the frail elderly. (4) Institutions that provide health care and related social services to people who are seriously ill and cannot be restored closer to the idealization comprising terminally ill people and mentally and physically disabled people.[114] The implication is that forms of health care that have a substantial influence on preventing, limiting and compensating for declines in normal species functioning should be prioritized in designing health care institutions and distributing health care.[115] Just health emphasizes the priority of preventive measures, since it is preferable to avoid the burdens of disease than to decrease them when they happen. Daniels argues that "it is more effective to prevent disease and disability than it is to cure them (or to compensate individuals for loss of function when cure is not possible)."[116] Rawls proposes basic health care which encompasses essential drugs,[117] as one of the five guarantees of any constitutional democracy.[118] A related issue of Rawls's view of the nature and scope of international responsibilities in providing affordable access to drugs in developing nations is discussed in the next part of this chapter.

---

[112] Daniels, *Just Health Care*, 47.

[113] Anton A van Niekerk, "Principles of Global Distributive Justice and the HIV/AIDS Pandemic: Moving Beyond Rawls and Buchanan," *in Ethics & AIDS in Africa: The Challenge to Our Thinking* ed. Anton A. van Niekerk and Loretta M. Kopelman (Walnut Creek, California: Left Coast Press Inc., 2005), 88.

[114] Daniels, *Just Health Care*, 47–48.

[115] Daniels, *Just Health Care*, 34–58, cited in Beauchamp and Childress, *Principles of Biomedical Ethics*, 5th ed., 234.

[116] Daniels, *Just Health: Meeting Health Needs Fairly*, 141.

[117] WHO, *WHO Medicines Strategy: Framework for Action in Essential Drugs and Medicines Policy 2000–2003,* Geneva: WHO/EDM/2000.1.2000.

[118] Rawls, *The Law of Peoples with "The Idea of Public Reason Revisited,"* 50.

4.3 Humanitarian Duty of Assistance Among Nations

4.3.1 The Concept of Statism

Rawls briefly outlined some principles of international justice in A Theory of Justice. In this context, he talked about deriving justice within the state and used a second hypothetical contract at which representatives of states choose principles to govern international relations from behind a veil of ignorance. Rawls contends that such representatives would not choose any principles of international distribution.[119] Rawls's extensive discussion about international relations and international justice was presented in The Law of Peoples (LOP). The central aim of The Law of Peoples is to explore how the content of a theory of international justice "might be developed out of a liberal idea of justice similar to, but more general than, the idea of justice as fairness."[120]

In the LOP, Rawls's first task which is the first stage of his global project was to extend the idea of the social contract from the domestic society discussed in TOJ to society of liberal peoples, deriving what he dubbed the "Law of Peoples."[121] The LOP is described as a "political conception of right and justice that applies to the principles and norms of international law and practice."[122] One of the key features in Rawls's argument is his typology of societies. He distinguishes between the five kinds of regime including liberal peoples, decent hierarchical peoples, outlaw states, societies with unfavorable conditions, and benevolent absolutisms.[123]

Rawls argues that representatives of liberal peoples ignore any knowledge of the people's comprehensive conception of the good, because a liberal society with a constitutional regime does not have a comprehensive conception of the good.[124] The first task of the parties in the second original position is "to specify the Law of Peoples – its ideals, principles, and standards – and how those norms apply to political relations among peoples.[125] Rawls argues that both liberal and decent hierarchical societies would be able to agree to eight principles of justice. He contends that just principles are those that liberal and decent societies will approve. Among the principles avowed here is the duty of humanitarian assistance, which clearly states that "peoples have a duty to assist other peoples living under unfavorable conditions that prevent their having a just or decent political and social regime."[126]

The second aspect of Rawls's ideal theory which relates to one of the principal ideas of the Law of Peoples, focuses on how and why representatives of certain

---

[119] Rawls, *A Theory of Justice, Revised Edition,* 331–333.

[120] Rawls, *The Law of Peoples with "The Idea of Public Reason Revisited,"* 3.

[121] Rawls, *The Law of Peoples with "The Idea of Public Reason Revisited,"* 3.

[122] Rawls, *The Law of Peoples with "The Idea of Public Reason Revisited,"* 3.

[123] Rawls, *The Law of Peoples with "The Idea of Public Reason Revisited,"* 4.

[124] Rawls, *The Law of Peoples with "The Idea of Public Reason Revisited,"* 34.

[125] Rawls, *The Law of Peoples with "The Idea of Public Reason Revisited,"* 40.

[126] Rawls, *The Law of Peoples with "The Idea of Public Reason Revisited,"* 37.

nonliberal but well-ordered societies would also approve the same set of principles. The nonliberal societies do not approve the standard range of liberal democratic rights such as the freedoms of expression and association, religious equality and the right to equal participation.[127] Furthermore, individuals in nonliberal societies are "not regarded as free and equal citizens, nor as separate individuals deserving equal representation (according to the maxim: one citizen, one vote)."[128] However, nonliberal societies respect basic human rights, including right to life, to the means of subsistence and security, to freedom from slavery, serfdom, and forced occupation, and are respectful of other peoples[129] as demanded by the law of peoples. Rawls points out that these nonliberal decent people qualify as "societies in good standing," and are, thus, to be tolerated by liberal societies. The implication is that liberal societies are "to recognize these nonliberal societies as equal participating members in good standing of the society of peoples," and not merely to "refrain from exercising political sanctions – military, economic, or diplomatic – to make a people change its ways."[130] Kok-Chor Tan notes that "nonliberal peoples are tolerated as a matter of liberal principle, and not merely accommodated on account of practicality."[131]

The Law of Peoples aims to attain a global stability with regard to justice, and not stability as a way of life, that is, stability as a balance of forces.[132] The first two aspects of Rawls's ideal theory is critical to understanding his Law of Peoples, because it tries to show that the global principles adopted by liberal peoples conform to the principles that can be endorsed independently by decent nonliberal peoples. More so, it is important to note that it is not the case that liberal peoples did not adapt their global principles precisely with respect to accommodating nonliberal peoples or existing global institutional arrangements.[133]

The first two aspects just discussed conclude the ideal theory part of the Law of Peoples. The goal of ideal theory is to recognize the principles that should regulate the relationship between societies with the necessary political and economic conditions to be well ordered and to conform to the Law of Peoples. This implies that the goals of justice and stability for the right reason between societies can be accomplished in this ideal situation.[134]

The Third part of the Law of Peoples focuses on societies without the economic resources to support well-ordered institutions or societies that deliberately refuse to conform to the principles of the law of Peoples. It grapples with the issues that arise from the "highly non-ideal conditions of our world with its great injustices and

[127] Rawls, *The Law of Peoples with "The Idea of Public Reason Revisited,"* 71–75.

[128] Rawls, *The Law of Peoples with "The Idea of Public Reason Revisited,"* 71.

[129] Rawls, *The Law of Peoples with "The Idea of Public Reason Revisited,"* 64–67.

[130] Rawls, The Law of Peoples with "The Idea of Public Reason Revisited," 59.

[131] Kok-Chor Tan, *Justice Without Borders: Cosmopolitanism, Nationalism and Patriotism* (Cambridge: Cambridge University Press, 2004), 64.

[132] Rawls, *The Law of Peoples with "The Idea of Public Reason Revisited,"* 12–13, 44–45.

[133] Tan, *Justice Without Borders: Cosmopolitanism, Nationalism and Patriotism*, 64.

[134] Tan, *Justice Without Borders: Cosmopolitanism, Nationalism and Patriotism*, 64.

widespread social evils."[135] The nonideal theory part of the Law of Peoples therefore tackles (1) the issue of noncompliance, with reference to conditions when outlaw societies "refuse to comply with a reasonable Law of Peoples,"[136] and (2) the issue of unfavorable conditions, which entails that burdened societies lack the basic resources to become well ordered.[137] A comprehensive approach in the Law of Peoples has to deal with these nonideal issues, and provide direction on how well-ordered peoples may protect themselves against outlaw regimes and assist in establishing needed reform within these regimes in the long run.[138] It needs to further address how they may assist burdened societies and help to bring them "into the society of well-ordered peoples."[139]

Rawls's focus on burdened societies and a duty of assistance clearly shows that he does not support an isolationist foreign policy which advocates for liberal and decent peoples' indifference to the concerns of burdened societies. He stresses that societies that are better off have a duty of assistance towards burdened societies so as to help them attain the required level of economic and social development to become well ordered. The duty of assistance would stem from the principle avowing basic human rights which consist of the right to subsistence.[140] The duty of assistance has been referred to as humanitarian duty because its goal is to fulfil individuals' basic needs and their collective capacity to sustain decent institutions. However, Rawls also emphasizes that this duty of humanitarian assistance is clearly different from, and does not involve a duty of distributive justice.[141] Tan also notes "so while a duty of humanitarian assistance is required by the Law of Peoples as part of its nonideal theory, a distributive principle has no place at all here."[142]

Elucidating this point further, it implies that the principles in Rawls's Law of Peoples are clearly principles of justice, but the LOP is a theory of justice exclusively for the society of peoples. Furthermore, the LOP cannot be regarded as merely advocating for the status quo, that is, the current state of affairs. The requirement of a duty of assistance by the LOP inevitably will result in a fundamental change in the contemporary world, where a fifth of the world's population live in absolute poverty, 1.2 million people lack access to clean water, about 17 million people yearly die from curable diseases of poverty, and millions more lack access to essential drugs, including anti-retroviral drugs especially in developing countries. The requirement in the LOP's nonideal theory that liberal and decent peoples assist burdened societies to attain the required level of economic and political developments to be well-ordered institutions, entails a substantial change. More so, the requirement that liberal and decent peoples honor and defend basic rights, which encom-

[135] Rawls, *The Law of Peoples with "The Idea of Public Reason Revisited,"* 89.

[136] Rawls, *The Law of Peoples with "The Idea of Public Reason Revisited,"* 90.

[137] Rawls, *The Law of Peoples with "The Idea of Public Reason Revisited,"* 90.

[138] Rawls, *The Law of Peoples with "The Idea of Public Reason Revisited,"* 92–93.

[139] Rawls, *The Law of Peoples with "The Idea of Public Reason Revisited,"* 106.

[140] Tan, *Justice Without Borders: Cosmopolitanism, Nationalism and Patriotism*, 65.

[141] Tan, *Justice Without Borders: Cosmopolitanism, Nationalism and Patriotism*, 65.

[142] Tan, *Justice Without Borders: Cosmopolitanism, Nationalism and Patriotism*, 65.

pass individuals' access to subsistence, constitute fundamental departures from how individuals that are better-off currently understand their global responsibilities towards the poor.[143] Nevertheless, Tan acknowledges that "what is lacking in Rawls's account of global justice is the commitment to distributive justice. That is there are no ongoing distributive principles regulating the inequalities between the rich and the poor of the world beyond the duty of the better-off to ensure that the badly-off are able to meet a certain threshold level of basic needs."[144] There is a contention that Rawls's account of international justice discussed in the LOP made some progress in international relations and politics, but did not go far enough. The critical issue under consideration is "whether there should be distributive principles to regulate global relations, as many cosmopolitans think, or whether Rawls is right that there is no place for distributive principles in the global setting."[145] A brief discussion of the distinctive features of humanitarian duty and duty of justice is the task of the next section of this chapter.

4.3.2 Humanitarian Duty and Duty of Justice

Rawls provides two arguments for rejecting the concept of global distributive justice. The first is that global principles of distributive justice would be redundant, since a duty of humanitarian assistance is presently required by the Law of peoples as an integral part of nonideal theory. The second is that more so, global distributive principles would produce unacceptable results.[146] The redundancy argument is defended in the first instance by Rawls's acknowledgement of gross injustices, huge inequality and dismal poverty in our nonideal world and the need for well-ordered societies to assist burdened societies to bring them into the society of well-ordered peoples.[147] Furthermore, he argues these "goals of attaining liberal and decent institutions, securing human rights, and meeting basic needs … are (adequately) covered by the duty of assistance."[148] The implication is that a global distributive principle does not have any additional role to play in this context.

Rawls's redundancy argument obfuscates an essential distinction between duties of humanity and duties of justice, which is not merely a distinction in semantic. The implication is that if we agree that affluent countries have only duty of humanity to poorer countries, we are as well agreeing that the current criterion for resource and wealth distribution is fair, and that the global basic institutions established around and justifying the existing allocation of wealth and resources are satisfactory. In this context, duties to assist each other entail duties that occur within an institutional

[143] Tan, *Justice Without Borders: Cosmopolitanism, Nationalism and Patriotism*, 65.
[144] Tan, *Justice Without Borders: Cosmopolitanism, Nationalism and Patriotism*, 65.
[145] Tan, *Justice Without Borders: Cosmopolitanism, Nationalism and Patriotism*, 66.
[146] Tan, *Justice Without Borders: Cosmopolitanism, Nationalism and Patriotism*, 66.
[147] Rawls, *The Law of Peoples with "The Idea of Public Reason Revisited*," 106.
[148] Rawls, *The Law of Peoples with "The Idea of Public Reason Revisited*," 116.

framework that is fair. Duties of humanity focus on how states should interact with one another without paying attention to the global basic structure including the norms regulating the allocation and ownership of resources and wealth where the interactions take place. On the other hand, duties of justice focus directly on the basic structure, hence, justice is related to the criterion for distribution of wealth and resources, and the basic institutions and principles that justify and rationalize this distribution.[149] Tan writes, "to put it perspicuously, while duties of humanity aim to redistribute wealth, duties of justice aim to identify what counts as a just distribution in the first place."[150] Put succinctly, the goal of justice is not to transfer wealth per se, which entails taking it from its just owners and redistributing it to others, but, rather to establish the conditions of just ownership, to reformulate "what justly belongs to a country."[151] Duties of justice would require us to reevaluate our current global basic structure, whereas duties of humanity regard this to be more or less sensible, and merely urge countries to do more within this particular framework. Brian Barry argues that justice is prior to humanity in that "we cannot sensibly talk about humanity unless we have a baseline set by justice. To talk about what I ought, as a matter of humanity, to do with what is mine makes no sense until we have established what is mine in the first place."[152] Barry also distinguishes the obligations of humanity and those of justice based on goals and rights respectively. He argues that "the obligations of humanity are goal-based, whereas those of justice are rights-based."[153]

On the other hand, the long-term goals of humanity and justice are entirely distinct, not only in their objective or duration, as indicated by Rawls, but as well in their scope and focus. The long-term goals of humanity require greater humanitarianism between countries within the present institutional framework, whereas the long-term goals of justice require a critical assessment of that framework. This distinction is critical because tackling issues of global dimension such as poverty, inequality, access to essential drugs etc. requires reforming global institutions and arrangements.[154]

Furthermore, the distinction in focus has some implications for foreign policy. For example, if foreign aid is viewed from the perspective of humanitarian aid, it could be exposed to criteria compelled by donor countries. However, if foreign aid is seen as a matter of justice, it would not be subject to a redistribution which denotes a transfer from the rightful owner to poorer ones, but to an alteration of an initial

---

[149] Tan, *Justice Without Borders: Cosmopolitanism, Nationalism and Patriotism*, 66–67.

[150] Tan, *Justice Without Borders: Cosmopolitanism, Nationalism and Patriotism*, 67.

[151] Brian Barry, "Humanity and Justice in Global Perspective," in *Nomos XXIV: Ethics, Economics and the Law* ed. J. Roland Pennock and John W. Chapman (New York: New York University Press, 1982), 248.

[152] Barry, "Humanity and Justice in Global Perspective," 249.

[153] Brian Barry, "Humanity and Justice in Global Perspective," in *Global Justice: Seminal Essays, Global Responsibilities* Vol. 1 ed. Thomas Pogge and Darrel Moellendorf (St. Paul: Paragon House, 2008), 202.

[154] Tan, *Justice Without Borders: Cosmopolitanism, Nationalism and Patriotism*, 67.

unjust distribution.[155] Barry notes that discussing global inequality as a matter of humanity obfuscates the fundamental point, "that if some share of resources is justly owed to a country, then it is (even before it has been actually transferred) as much that country's as it now normally thought that what a country produces belongs to that country."[156]

There is a significant difference when distribution of wealth between countries is viewed as a matter of humanitarian assistance or as a matter of justice. Discussing duties between countries as a matter of justice emphasizes the proper place to be concerned about, which are the institutions and their fundamental norms. It also highlights that the critical issue is eventually the issue of rightful ownership instead of humanitarian contribution. In the context of nonideal case of burdened societies, it makes a significant normative distinction whether we are assisting from the point of view of humanitarian concern, or whether we are assisting as a result of acknowledging the current injustices in our global arrangements.[157]

It is pertinent to note that humanitarian assistance within the context of the current global arrangement merely deals with the symptoms of injustice rather than deals with the fundamental cause of it.[158] Tan writes, "Humanitarian assistance applies as long as there are burdened societies, but principles of justice would push us to assess the framework within which such assistance is being rendered."[159] Furthermore, justice focuses on structural equality of some kind, whereas humanitarianism emphasizes mainly fulfilling basic needs.[160]

It is evident that Rawls's concept of domestic egalitarianism aims at structural transformation of the basic structure of the society in such a way that his second principle offers a framework for evaluating and criticizing the basic institutions of society. In this context, institutional arrangements which preserve and justify inequality of opportunity between citizens are disallowed. Therefore, it will appear that for Rawls to be consistent with his basic philosophical ideals that he should evaluate the basic structure of the society of peoples against his principles of justice, rather than take it as a given.[161]

Therefore, Rawls is cognizant of the significant distinction between humanitarian duties and duties of justice. His contention is that the global distribution of resources and wealth is not an issue of justice. He argues that a global distribution of wealth that does not fulfil the egalitarian requirement of his difference principle is acceptable provided that assistance is offered to help burdened societies.[162]

Rawls's second argument rejecting global distributive principles throws some light with respect to his position above regarding a global distribution of resources

[155] Tan, *Justice Without Borders: Cosmopolitanism, Nationalism and Patriotism*, 67.

[156] Barry, "Humanity and Justice in Global Perspective," 248.

[157] Tan, *Justice Without Borders: Cosmopolitanism, Nationalism and Patriotism*, 68.

[158] Tan, *Justice Without Borders: Cosmopolitanism, Nationalism and Patriotism*, 68.

[159] Tan, *Justice Without Borders: Cosmopolitanism, Nationalism and Patriotism*, 68.

[160] Tan, *Justice Without Borders: Cosmopolitanism, Nationalism and Patriotism*, 68.

[161] Tan, *Justice Without Borders: Cosmopolitanism, Nationalism and Patriotism*, 68.

[162] Tan, *Justice Without Borders: Cosmopolitanism, Nationalism and Patriotism*, 69.

and wealth. He believes that in contrast to domestic distributive principle, global distributive principles would have unacceptable results. He argues that a duty of humanitarian assistance is a "principle of transition… (it) holds until all societies have achieved just liberal or decent basic institutions. (It is) defined by a target beyond which (it) no longer hold(s)."[163] The implication is that the duty of assistance is accomplished when all societies have achieved the basic level of development adequate for establishing and maintaining decent institutions. On the other hand, distributive "principles do not have a defined goal, aim, or cut-off point, beyond which aid may cease."[164] Therefore, whereas, the duty of humanitarian assistance is aimed at improving the circumstance of societies burdened by unfavorable conditions, such assistance is not needed as part of ideal theory for societies that have achieved the basic level of development required for a decent society. Conversely, distributive justice principle is an essential component of ideal theory, and therefore would apply so long as there are inequalities, excessive injustices and impairing poverty between societies, even "after the duty of assistance is fully satisfied."[165]

Rawls's central argument is that upholding global distributive principle would have unacceptable results because we should not be able to distinguish between societies which have increased their wealth through foresight and prudence[166]; or, societies which have succeeded to curtail their population growth through sound population policies and consequently boosted their wealth, and societies which have failed to control their population and consequently remain worse-off.[167] He argues that in both cases identified, a global egalitarian principle without target would maintain that resources be transferred from the more affluent societies to the poor ones, despite the fact that both may have begun with an equal amount of wealth and resources. He contends that this is unacceptable because it would imply punishing some societies for their sound domestic policies so as to reward other societies for their imprudent policies.[168]

Rawls is repudiating a situation where "distributive principles would insist on redistribution as long as there is inequality between peoples no matter what the cause of this inequality."[169] The underlying implication of Rawls's argument is that there is a distinction between inequality due to choice and inequality attributable to circumstance. In the same way that we don't want a domestic scheme to compensate individuals for their poor decisions by taking from those who have made good decisions, we also don't want a global scheme to compensate societies for their poor governance by punishing other societies for their good governance.[170] The goal of

---

[163] Rawls, *The Law of Peoples with "The Idea of Public Reason Revisited,"* 118.

[164] Rawls, *The Law of Peoples with "The Idea of Public Reason Revisited,"* 106.

[165] Rawls, *The Law of Peoples with "The Idea of Public Reason Revisited,"* 117.

[166] Rawls, *The Law of Peoples with "The Idea of Public Reason Revisited,"* 117.

[167] Rawls, *The Law of Peoples with "The Idea of Public Reason Revisited,"* 117–118.

[168] Rawls, *The Law of Peoples with "The Idea of Public Reason Revisited,"* 117–118.

[169] Tan, *Justice Without Borders: Cosmopolitanism, Nationalism and Patriotism*, 70.

[170] Tan, *Justice Without Borders: Cosmopolitanism, Nationalism and Patriotism*, 70.

distributive justice is to offset the influences of unchosen inequality due to circumstances on persons, rather than to compensate them for their poor decisions.[171] Rawls's concern is that a global distributive principle would not take into account the distinction between inequality as a result of either choice or circumstance, rather it would unfairly deal with citizens of properly managed economies by transferring their benefits to citizens of poorly managed economies unceasingly so long as global inequality persists.[172] Tan articulates, "the choices a people make about its domestic arrangements would not be respected if the gains or losses due to these choices were annulled by a distributive principle between peoples."[173]

The fundamental premise of Rawls's argument is that the reasons for a country's inability to espouse good social and economic policies are mainly internal, and thus freely espoused by governments of worse-off countries. Rawls alludes to several domestic factors that are instrumental to society's economic and social performance, consisting of its political culture and virtues (comprising a respect for basic human rights), its civic society, "its members' probity and industriousness," and its population policy.[174] However, this premise which was dubbed "explanatory nationalism" was refuted by Thomas Pogge on the basis of empirical fact.[175] Explanatory nationalism "present(s) poverty as a set of national phenomena explicable mainly as a result of bad domestic policies and institutions that stifle, or fail to stimulate, national economic growth and engender national economic injustice."[176] On the other hand, Pogge indicates that this explanation "leave(s) open important questions, such as why national factors (institutions, officials, policies, culture, natural environments, level of technical and economic development) have these effects rather than others"[177] by disregarding the causal influences of global factors, for example, trade practices, patterns of consumption by wealthy countries, international law, etc. on a country's domestic policies and their results.[178]

If the distributive goal is to counter the impacts of these unchosen global factors and not the influences of chosen national policies on a people's welfare, then, distributive arrangements between societies should not be indifferent to choice. A poorer country that benefits from a global distributive principle need not be viewed as a society that is being funded unfairly for the domestic decisions it has made, but is more accurately being compensated for the impacts of global factors imposed on it without its choice.[179] So worthy of note is that "global distributive justice and

[171] Will Kymlicka, *Contemporary Political Philosophy: An Introduction* (Oxford: Oxford University Press, 1990), 73–76.

[172] Tan, *Justice Without Borders: Cosmopolitanism, Nationalism and Patriotism,* 70.

[173] Tan, *Justice Without Borders: Cosmopolitanism, Nationalism and Patriotism,* 70.

[174] Rawls, *The Law of Peoples with "The Idea of Public Reason Revisited,"* 108–111.

[175] Thomas Pogge, "The Bounds of Nationalism," in *Rethinking Nationalism* ed. J. Couture, K. Nielsen and M. Seymour (Calgary: University of Calgary Press, 1998), 497–502.

[176] Pogge, "The Bounds of Nationalism," 497.

[177] Pogge, "The Bounds of Nationalism," 498.

[178] Pogge, "The Bounds of Nationalism," 498–499.

[179] Tan, *Justice Without Borders: Cosmopolitanism, Nationalism and Patriotism,* 71.

national self-determination, the latter being the underlying premise in Rawls's argument, are not incompatible goals."[180] Despite Rawls's lack of commitment to global distributive principles and his inherent flaws and inconsistency with his own moral individualism as argued in the domestic case, his statist approach implies relying merely upon humanitarian assistance from the perspective of global justice to provide affordable access to drugs in developing nations.

The analysis will focus on the account of international responsibilities as presented in the LOP. Rawls's notion of international responsibilities focuses on assisting burdened societies to attain well-ordered societies. Providing affordable access to drugs as a transition strategy is argued as a critical component of Rawls's duty of assistance to burdened societies in order to attain well-ordered societies.

4.3.3 Post-trial Access to Drugs

Acknowledging Rawls's imperative position that we have a duty to assist burdened societies, this part explores whether it is in consonant with a Rawlsian approach to extend these duties to health care, and specifically to provision of essential drugs especially anti-retroviral drugs. It examines three arguments regarding the claim that enhancing access to essential drugs is a desideratum for fulfilling Rawls's duty of assistance.

4.3.3.1 Defense of Minimal Human Rights

The first argument focuses on Rawls's defense of human rights. The human right to health is codified in the 1948 United Nations' Universal Declaration of Human Rights. Article 25 states that "Everyone has the right to a standard of living adequate for the health and well-being of himself and of his family, including food, clothing, housing and medical care and necessary social services…."[181] The right to essential medicines is increasingly receiving support as a sub-right of the right to health. In Montreal Statement on the Human Right to Essential Medicines, Pogge writes, "we have a responsibility to achieve a social and international order in which human rights – including the right to essential medicines – are fully realized."[182]

An initial consideration about the duty of assistance is that it would strongly favor policy interventions supporting the human right to health, comprising the sub-right to essential medicines, based on three reasons. The first reason focuses on the

---

[180] Tan, *Justice Without Borders: Cosmopolitanism, Nationalism and Patriotism*, 71–72.

[181] United Nations, *Universal Declaration of Human Rights*, 1948, http://www.un.org/Overview/rights.html (Accessed April 11, 2014).

[182] Thomas Pogge, "Montreal Statement on the Human Right to Essential Medicines," *Cambridge Quarterly of Healthcare Ethics* 15, no. 2 (2006): 10.

sixth Law of Peoples which affirms: "Peoples are to honor human rights."[183] Second, Rawls stresses the importance of policies that support human rights over economic transfers.[184] Third, Rawls emphasizes the importance of policy interventions supporting women's basic rights and interests.[185] Among the global poor, women and girl children disproportionately bear the burden of disease consisting of problems of unsafe abortion and childbirth. The health of women and children are significantly adversely affected by HIV/AIDS pandemic especially in sub-Saharan Africa. Data shows that they are increasingly among the victims of HIV/AIDS and disproportionately among many new HIV/AIDS infections especially in sub-Saharan Africa. In many poor countries, women and girl children lack access to treatment for various diseases because they are excluded due to gender discrimination.[186] Gender-based violence and gender inequality are increasingly mentioned as critical determinants of women's HIV risk.[187] Maternal mortality also continues to be astronomically high especially in many developing countries, as a result of limited health infrastructure, inadequately trained birth attendants, and women's unavoidable resort to illegal and unsafe abortions.[188] Cognizant of these vast inequalities affecting women, a policy favoring access to drugs can be certain to possess a strong pro-female effect.[189]

The argument in support of improving access to essential drugs based on Rawls's restrictive notion of human rights does not go through, because his use of the term human rights is equivocal. Rawls refused to favor an expansive definition of human rights that might encompass all rights codified in international treaties in support of a minimum that he views as more reasonable basis for international consensus.[190] The human rights supported by Rawls in LOP reveal a sub-set of human rights that he sees not only as widely supported by liberal democratic societies, but as well capable of receiving similar support in decent non-democratic societies. These human rights "express a special class of urgent rights, such as freedom from slavery and serfdom, liberty (but not equal liberty) of conscience, and security of ethnic

[183] Rawls, *The Law of Peoples with "The Idea of Public Reason Revisited,"* 37.

[184] Rawls, *The Law of Peoples with "The Idea of Public Reason Revisited,"* 108–109.

[185] Rawls, *The Law of Peoples with "The Idea of Public Reason Revisited,"* 109–110.

[186] World Health Organization, *World Health Report 2005: Make Every Mother and Child Count*, (Geneva, Switzerland: World Health Organization, 2005), http://www.who.int/whr/2005/whr2005_en.pdf.?us=1 (Accessed April 11, 2014).

[187] Kristin L Dunkle et al., "Gender-Based Violence, Relationship Power, and Risk of HIV Infection in Women Attending Antenatal Clinics in South Africa," *Lancet* 363, no. 9419 (May 2004): 1415–1421.

[188] Ruth Macklin, *Double Standards in Medical Research in Developing Countries* (Cambridge: Cambridge University Press, 2004), 74.

[189] Johri et al., "Sharing the Benefits of Medical Innovation: Ensuring Fair Access to Essential Medicines," 2006, http://www.lawweb.usc.edu/centers/paccenter/assets/docs/Ehrenreich_Johri_2006_04_15.pdf. (Accessed March 20, 2014).

[190] Johri et al., "Sharing the Benefits of Medical Innovation: Ensuring Fair Access to Essential Medicines," 2006, http://www.lawweb.usc.edu/centers/paccenter/assets/docs/Ehrenreich_Johri_2006_04_15.pdf. (Accessed March 20, 2014).

groups from mass murder and genocide."[191] Rawls argue that societies whose political institutions and legal order fulfil this special class of human rights are well-ordered, and cannot justifiably be subjected to the use of sanctions or military force. These urgent human rights establish a minimum framework for use among peoples that Rawls sees as non-ethnocentric.[192]

There is a charge that Rawls's account of minimal rights is ad hoc. Rawls's global minimum is not compatible with his views about domestic justice. It does not fulfil the condition of a theory of human rights – the criterion of domestic-compatibility. Rawls attempts to ground civil and political rights in persons' moral powers and interests but at the same contradictorily refute that all persons enjoy these very same civil and political rights.[193] Caney argues that "the force of the scope claim is that one cannot, as it were, apply these universalist arguments for citizens and not apply them to foreigners when the very terms of the arguments (the moral powers and interests of persons) do not justify this kind of domestic/international split."[194] One cannot logically support Rawls's domestic theory and stick to his international theory. It fails short of the criterion of domestic-compatibility.[195]

In short, Caney argues that "one cannot coherently both embrace the rights that Rawls does embrace and reject some of the rights that Rawls rejects. The claim is that they stand or fall as a package."[196] There is an apparent contradiction between Rawls's account of minimum rights and his domestic theory on the one hand, and a contradiction between his account of minimum rights and his rejection of other proposed human rights on the other hand. The charge is that his theory does not fulfil the criterion of coherence.[197] Caney also employs what he termed rights holism in his criticism of Rawls's minimal rights account. Rights holism "maintains that the acceptance of some specific rights implies the acceptance of some other specific rights. It claims that certain rights are interconnected."[198] Based on Rawls's minimal rights account, the policy supporting improving access to essential drugs including anti-retroviral drugs would not be favored by Rawls's duty of assistance.

[191] Rawls, *The Law of Peoples with "The Idea of Public Reason Revisited,"* 78–79.

[192] Rawls, *The Law of Peoples with "The Idea of Public Reason Revisited,"* 80.

[193] Simon Caney, *Justice Beyond Borders: A Global Political Theory* (New York: Oxford University Press, 2005), 82.

[194] Caney, *Justice Beyond Borders: A Global Political Theory*, 82.

[195] Caney, *Justice Beyond Borders: A Global Political Theory*, 82.

[196] Caney, *Justice Beyond Borders: A Global Political Theory*, 82–83.

[197] Caney, *Justice Beyond Borders: A Global Political Theory*, 83.

[198] Caney, *Justice Beyond Borders: A Global Political Theory*, 83.

4.3.3.2 Redress for the Unjustified Distributive Effects of Cooperative Organizations

The second argument is based on redress for the unjustified distributive effects of cooperative organizations. Rawls indicates that parties to the second original position in LOP would not only agree to the eight basic principles or laws, they would as well formulate guidelines for establishing cooperative organizations and criteria for fairness in trade. Rawls posits that three such organizations would be established: one to ensure fair trade among peoples, another to institute a cooperative banking system, and a third to play a diplomatic and coordinating role similar to that of the United Nations.[199] With regard to fair trade, Rawls contends that the parties to the original position negotiating behind the veil of ignorance would agree to fair standards of trade to keep the market fair and competitive as well as to everyone's mutual advantage in the long run, irrespective of whether its economy is large or small. He stressed that these standards must guarantee the fairness of market transactions, and guarantee that unjustified inequalities among people do not develop over time. Their function is therefore similar to that of the fair background structure in the domestic case discussed in TOJ.[200] In a situation where these cooperative arrangements should lead to unjustified distributive effects between peoples, Rawls argues that "…these would have to be corrected, and taken into account by the duty of assistance…."[201]

The pharmaceutical industry and its advocates claim that the TRIPS agreement established by WTO signifies "the optimal balance between stimulating innovation and promoting access."[202] The impact of the TRIPS agreement is non-symmetrical and the distributive benefits accrue mainly to the high-income nations where pharmaceutical industry is concentrated.[203] Rawls's duty of assistance would support compensating those who have suffered the unintended distributive consequences of this situation. Interventions for providing affordable access to essential drugs including anti-retroviral drugs for developing countries would constitute logical approaches for redress.[204]

Rawls also includes another condition to his explanation of fair trade: "A further assumption here is that the larger nations within the wealthier economies will not attempt to monopolize the market, or to conspire to form a cartel, or to act as an

[199] Rawls, *The Law of Peoples with "The Idea of Public Reason Revisited*," 42.

[200] Rawls, *The Law of Peoples with "The Idea of Public Reason Revisited*," 43.

[201] Rawls, *The Law of Peoples with "The Idea of Public Reason Revisited*," 43.

[202] Johri et al., "Sharing the Benefits of Medical Innovation: Ensuring Fair Access to Essential Medicines," 2006, http://www.lawweb.usc.edu/centers/paccenter/assets/docs/Ehrenreich_Johri_2006_04_15.pdf. (Accessed March 20, 2014).

[203] Johri et al., "Sharing the Benefits of Medical Innovation: Ensuring Fair Access to Essential Medicines," 2006, http://www.lawweb.usc.edu/centers/paccenter/assets/docs/Ehrenreich_Johri_2006_04_15.pdf. (Accessed March 20, 2014).

[204] Johri et al., "Sharing the Benefits of Medical Innovation: Ensuring Fair Access to Essential Medicines," 2006, http://www.lawweb.usc.edu/centers/paccenter/assets/docs/Ehrenreich_Johri_2006_04_15.pdf. (Accessed March 20, 2014).

oligopoly."[205] Violating this condition of fairness implies violation of the ideal of reciprocity among peoples and which calls into question the legitimacy of trade arrangements.[206] In the history of access to drugs debate, an abundance of evidence shows that government of wealthy countries and multinational pharmaceutical companies defend their own interests at the cost of access to essential drugs for the poor.[207]

4.3.3.3 Access to Drugs as a Transition Strategy Supporting Politically Well-Ordered Nations

The third and final argument is based on access to drugs as a transition strategy favoring the establishment of politically well-ordered nations. This implies considering whether a policy improving access to drugs can be viewed as an effective strategy that enables burdened societies to become politically well-ordered, and therefore as a policy that should be supported by Rawls's duty of assistance. The argument is presented in two aspects. The first focuses on the assertions that countries with a high burden of disease and severe lack of access to drugs do not fulfil Rawls's criteria for well-ordered societies. The second contends that improving access to essential drugs would help in the transition to attaining politically well-ordered.[208]

According to Johri et al. "the principal correlates of a high burden of disease and lack of access to medicines are economic."[209] More so, poor and middle-income countries accepted various types of governance structures, ranging from democracies and dictatorships.[210]

Clarifying the correlation between disease and being well-ordered requires a review of the characteristics of well-ordered societies. Well-ordered societies include liberal or decent societies. Rawls explains societies that satisfy a liberal conception of justice as fulfilling three characteristic principles. The first guarantees basic rights and liberties of the sort familiar to constitutional democracies.

---

[205] Rawls, *The Law of Peoples with "The Idea of Public Reason Revisited,"* 43.

[206] Johri et al., "Sharing the Benefits of Medical Innovation: Ensuring Fair Access to Essential Medicines," 2006, http://www.lawweb.usc.edu/centers/paccenter/assets/docs/Ehrenreich_Johri_2006_04_15.pdf. (Accessed March 20, 2014).

[207] Nathan Ford et al., "The Role of Civil Society in Protecting Public Health Over Commercial Interests: Lessons from Thailand," *Lancet* 363, no. 9408 (February 2004): 560.

[208] Johri et al., "Sharing the Benefits of Medical Innovation: Ensuring Fair Access to Essential Medicines," 2006, http://www.lawweb.usc.edu/centers/paccenter/assets/docs/Ehrenreich_Johri_2006_04_15.pdf. (Accessed March 20, 2014).

[209] Johri et al., "Sharing the Benefits of Medical Innovation: Ensuring Fair Access to Essential Medicines," 2006, http://www.lawweb.usc.edu/centers/paccenter/assets/docs/Ehrenreich_Johri_2006_04_15.pdf. (Accessed March 20, 2014).

[210] Johri et al., "Sharing the Benefits of Medical Innovation: Ensuring Fair Access to Essential Medicines," 2006, http://www.lawweb.usc.edu/centers/paccenter/assets/docs/Ehrenreich_Johri_2006_04_15.pdf. (Accessed March 20, 2014).

The second allocates a special priority to these rights, liberties and opportunities, with regard to claims of the general good and perfectionism values. The third guarantees for all citizens the required primary goods to assist them to make intelligent and effective use of their freedoms.[211] One preferred interpretation of these principles was presented by Rawls's in the domestic case as justice as fairness. However, other interpretations that represent liberal perspectives can be accepted provided that they fulfil conditions consistent with the idea of the social contract that supports the freedom and equality of all citizens, and of society as a fair system of cooperation over time.[212]

Rawls in exploring why democratic nations are peaceful explains in a nutshell the five features of the basic structure of society that he views as important to a reasonably just constitutional democracy that can endure over time. He contends that, peace is made more secure internally among citizens and externally among states to the degree that these features are fulfilled. He delineates five institutions, without which "excessive and unreasonable inequalities tend to develop."[213] (1) A guaranteed fair equality of opportunity, particularly in education and training. (2) A decent distribution of income and wealth fulfilling the third condition of liberalism which assures all citizens the necessary means for intelligent and effective use of their basic freedom. (3) Society playing the role of employer of final recourse through general or local government, or other social and economic policies. (4) Basic health care guaranteed for all citizens. (5) Public financing of elections and means of guaranteeing the ability of public information on issues of policy.[214] Furthermore, decent societies considered as well-ordered are required to fulfill strict conditions. Rawls sees them as jointly fulfilling the following two criteria: (1) lack of aggressive aims and means; and (2) a system of law guaranteeing human rights.[215] In addition to these human rights that are principally political, Rawls stresses the importance of basic economic entitlements. He indicates that the right to life includes a claim "to the means of subsistence and security."[216] Therefore, decent societies must show a respect for human rights which includes economic subsistence.

Based on these important preliminary clarifications, we will now discuss whether a high burden of disease constitutes a barrier to attaining a politically well-ordered society in Rawls's perspective. Rawls's account supports that all liberal societies have a domestic responsibility to ensure provision of basic health care that involves access to drugs. Providing these basic health care services varies among liberal societies. Furthermore, there are also clear indications that some low and middle-income societies that guarantee political rights, liberties and freedoms, but have not

[211] Rawls, *The Law of Peoples with "The Idea of Public Reason Revisited,"* 14.
[212] Rawls, *The Law of Peoples with "The Idea of Public Reason Revisited,"* 14.
[213] Rawls, *The Law of Peoples with "The Idea of Public Reason Revisited,"* 49.
[214] Rawls, *The Law of Peoples with "The Idea of Public Reason Revisited,"* 50.
[215] Rawls, *The Law of Peoples with "The Idea of Public Reason Revisited,"* 65.
[216] Rawls, *The Law of Peoples with "The Idea of Public Reason Revisited,"* 65.

been successful in providing basic health or access to essential drugs to all citizens. Countries like South Africa, India and Guatemala have been cited as examples.[217]

The social determinants of health indicate that poor health is correlated with lower socio-economic status.[218] Poverty is also established to have a correlation with health.[219] Based on the assertions, an inference could be made that health problems are concentrated excessively in population sub-groups that are disadvantaged and inability to provide access to essential drugs reveal and exacerbate social and economic inequalities between the members of these groups.[220]

The implication of this situation is that the principle of equality of opportunity as argued by Rawls in TOJ is violated. The prevalence of high burden of disease in any situation results in very considerable mortality differentials between the members of different social groups. This in turn results in a situation where the chance to survive to the adult stage of life when freedom, liberties and opportunities can be achieved varies considerably across social classes and other group separations, comprising gender.[221] Similarly, "morbidity differentials aggravate this situation by compounding inequalities in the ability to flourish."[222]

Rawls's explanation of the principle of equality of opportunity in LOP necessary to fulfil the criteria of a liberal society is less rigorous than that discussed in TOJ. Nevertheless, he still emphasizes some type of equality of opportunity particularly in education and training. Both childhood mortality and school performance are universally predicted to be significantly affected by preventable and treatable conditions such as malaria, pneumonia, diarrhea and nemotodes and parasites.[223]

Furthermore, LOP acknowledges the significance of fulfilling basic economic entitlements such as subsistence rights, in the absence of which one would have "not liberalism at all but libertarianism."[224] It is pertinent to note that "a high burden of disease contributes to the entrenchment of poverty and threatens subsistence

---

217 Johri et al., "Sharing the Benefits of Medical Innovation: Ensuring Fair Access to Essential Medicines," 2006, http://www.lawweb.usc.edu/centers/paccenter/assets/docs/Ehrenreich_Johri_2006_04_15.pdf. (Accessed March 20, 2014).

218 Michael Marmot, "Social Determinants of Health Inequalities," *Lancet* 365, no. 9464 (March 2005): 1099–1104.

219 Adam Wagstaff, "Poverty and Health Sector Inequalities," *Bulletin of the World Health Organization* 80, no. 2 (January 2002): 97–105.

220 Johri et al., "Sharing the Benefits of Medical Innovation: Ensuring Fair Access to Essential Medicines," 2006, http://www.lawweb.usc.edu/centers/paccenter/assets/docs/Ehrenreich_Johri_2006_04_15.pdf. (Accessed March 20, 2014).

221 World Health Organization, *World Health Report 2003: Shaping the Future* (Geneva, Switzerland: World Health Organization, 2003).

222 Johri et al., "Sharing the Benefits of Medical Innovation: Ensuring Fair Access to Essential Medicines," 2006, http://www.lawweb.usc.edu/centers/paccenter/assets/docs/Ehrenreich_Johri_2006_04_15.pdf. (Accessed March 20, 2014).

223 Johri et al., "Sharing the Benefits of Medical Innovation: Ensuring Fair Access to Essential Medicines," 2006, http://www.lawweb.usc.edu/centers/paccenter/assets/docs/Ehrenreich_Johri_2006_04_15.pdf. (Accessed March 20, 2014).

224 Rawls, *The Law of Peoples with "The Idea of Public Reason Revisited,"* 49.

rights."[225] Data shows that the influence of this burden is usually distributed among groups disadvantaged in other means, such as income, wealth, power and prestige.[226] This situation is exacerbated by lack of access to essential drugs. There is an indication that calamitous illness is a major cause of household poverty in developing countries, and expenditure on drugs account for the biggest out-of-pocket costs.[227]

Some authors argue that there are many societies in Europe, the Americas, Asia and Africa that have lively democracy and thus in the process of becoming well-ordered, but unfortunately lack of the social determinants of health and health care consisting of access to essential drugs continues to be an impediment to accomplishing this political goal.[228] Johri et al. further argue that societies confronting these problems fall short of not merely to correspond to the formal characteristics of liberal democracies articulated by Rawls, but as well to fulfil spirit of the criterion of reciprocity which entails not embodying principles of social organization that is reasonable to consent to as free and equal persons behind the veil of ignorance. These problems as well endanger the ability of societies to comply with the conditions argued by Rawls for decent societies, especially by jeopardizing subsistence rights.[229]

On the other hand, Johri et al. argue "that where the burden of disease is still high, guaranteeing effective access to medicines would speed the process of transition to well-ordered societies, by making it possible for individuals to enjoy real exercise of rights, liberties and opportunities and to avoid destitution."[230] More so, decreasing the burden of disease in low- and middle- income countries would provide these countries with an opportunity to advance economically which is a critical condition to be satisfied if a basic standard of living is to be offered to all, without external aid.[231] Policies supporting access to drugs and other health sector interven-

---

[225] Johri et al., "Sharing the Benefits of Medical Innovation: Ensuring Fair Access to Essential Medicines," 2006, http://www.lawweb.usc.edu/centers/paccenter/assets/docs/Ehrenreich_Johri_2006_04_15.pdf. (Accessed March 20, 2014).

[226] Davison R. Gwatkin and Michel Guillot, *The Burden of Disease Among the Global Poor: Current Situation, Future Trends, and Implications for Strategy* (Washington, D.C.: The World Bank, 2000), 1–44.

[227] World Health Organization, *WHO Medicines Strategy: Framework for Action in Essential Drugs and Medicines Policy 2000–2003* (Geneva: WHO/EDM, 2000), 1–70.

[228] Johri et al., "Sharing the Benefits of Medical Innovation: Ensuring Fair Access to Essential Medicines," 2006, http://www.lawweb.usc.edu/centers/paccenter/assets/docs/Ehrenreich_Johri_2006_04_15.pdf. (Accessed March 20, 2014).

[229] Johri et al., "Sharing the Benefits of Medical Innovation: Ensuring Fair Access to Essential Medicines," 2006, http://www.lawweb.usc.edu/centers/paccenter/assets/docs/Ehrenreich_Johri_2006_04_15.pdf. (Accessed March 20, 2014).

[230] Johri et al., "Sharing the Benefits of Medical Innovation: Ensuring Fair Access to Essential Medicines," 2006, http://www.lawweb.usc.edu/centers/paccenter/assets/docs/Ehrenreich_Johri_2006_04_15.pdf. (Accessed March 20, 2014).

[231] Johri et al., "Sharing the Benefits of Medical Innovation: Ensuring Fair Access to Essential Medicines," 2006, http://www.lawweb.usc.edu/centers/paccenter/assets/docs/Ehrenreich_Johri_2006_04_15.pdf. (Accessed March 20, 2014).

tions may then be more effective than monetary transfers in promoting sustainable economic growth and relieving poverty.[232]

This policy as well would deal with threats to international peace which can be significantly reduced by the international community. For example, situations such as the spread of HIV/AIDS pandemic may be linked with food security, drought and famine,[233] and act as a harbinger to war. The magnitude of this threat was acknowledged by the 2001 UN Special Session on HIV/AIDS, which indicated the first time that the General Assembly convened specifically for an issue related to a disease. This meeting led to the Declaration of Commitment on HIV/AIDS with the international community favoring provision of antiretroviral treatment for the first time, which as well improves prevention efforts in developing countries as one strategy to fight the pandemic, as well as a matter of social justice.[234]

Well-ordered societies have an obligation to assist burdened societies in becoming well-ordered. Johri et al. writes "…we believe that there are good empirical grounds for seeing a policy of improving access to medicines as an effective transition strategy that should be favored by Rawls's duty of assistance to burdened societies. Commitment to this policy is to be seen as a duty of charity, and is highly circumscribed."[235] This policy would be appreciated in view of its contribution to the aim of promoting a society of well-ordered peoples, and may not relate to individuals living in outlaw states, or benevolent absolutisms. The commitment will terminate as soon as burdened societies become well-ordered. Furthermore, improving access to drugs has a potentially critical role in attaining the goals established by Rawls in LOP. Essentially, this is the case even if obligation for present lack of success to ensure access to drugs or to guarantee a favorable distribution of the social and economic determinants of health is thought to rest at the national level as Rawls's account proposes.[236] A related issue is exploring local participants' perspectives of justice in international clinical trials as discussed in the next theme.

[232] World Health Organization, *Macroeconomics and Health: Investing in Health for Economic Development* (Canada: World Health Organization, 2001), 1–200.

[233] Zosia Kmietowicz, "Failure to Tackle AIDS Puts Millions at Risk of Starvation," *BMJ* 325, no. 7375 (November 2002): 1257.

[234] United Nations Special Session on HIV-AIDS, *Global Crisis – Global Action: Declaration on Commitment on HIV/AIDS*, 2001, http://www.un.org/ga/aids/coverage/FinalDeclarationHIVAIDS.html (Accessed April 13, 2014).

[235] Johri et al., "Sharing the Benefits of Medical Innovation: Ensuring Fair Access to Essential Medicines," 2006, http://www.lawweb.usc.edu/centers/paccenter/assets/docs/Ehrenreich_Johri_2006_04_15.pdf. (Accessed March 20, 2014).

[236] Johri et al., "Sharing the Benefits of Medical Innovation: Ensuring Fair Access to Essential Medicines," 2006, http://www.lawweb.usc.edu/centers/paccenter/assets/docs/Ehrenreich_Johri_2006_04_15.pdf. (Accessed March 20, 2014).

4.3.3.4 Local Participants' Perspectives of Justice in International Clinical Trials

The application of the principle of distributive justice to research which entails "a balancing of benefits and burdens … takes on different dimensions when research is conducted in developing countries because access to health benefits outside the context of research is so limited."[237] Obligations of justice in international clinical trials have been defined in various ways in leading international guidelines and in the bioethical frameworks as: conducting research that is responsive to the health needs and priorities of the host community or country[238]; the provision of fair benefits to participants during and after clinical trials, which may include access to treatments or practices developed by the study[239]; and ethical review and research capacity strengthening in host communities or countries.[240] These requirements of obligations of justice – responsiveness, fair benefits, and research capacity strengthening link international clinical trials to the promotion of global justice.[241] These obligations of justice on the macro-level connect international clinical trials to improving access to health care and to developing research systems in host countries, which links the research activity to a broader justice agenda to reduce health inequalities between nations.[242]

Provision of treatment to participants and host populations have been argued as a means to reducing inequalities in health care between collaborating nations,[243] as a fair balance of research-related risks (e.g. of acquiring HIV) and benefits (e.g. of gaining access to treatment)[244] or as reciprocity.[245]

Recent guidance stresses that engaging communities in international clinical research is a critical activity that helps with understanding local perceptions, norms and practices. It emphasizes that communities participating in the research should

---

[237] Macklin, *Double Standards in Medical Research in Developing Countries*, 77.

[238] World Medical Association, *Declaration of Helsinki, Ethical Principles for Medical Research Involving Human Subjects*, (Seoul: World Medical Association, 2008), Note 1, Guideline 10. http://www.wma.net/en/30publications/10policies/b3/index.html (Accessed July19, 2013).

[239] World Medical Association, *Declaration of Helsinki, Ethical Principles for Medical Research Involving Human Subjects*, (Seoul: World Medical Association, 2008), Note 1, Guideline 10. http://www.wma.net/en/30publications/10policies/b3/index.html (Accessed July19, 2013).

[240] World Medical Association, *Declaration of Helsinki, Ethical Principles for Medical Research Involving Human Subjects*, (Seoul: World Medical Association, 2008), Note 1, Guideline 20. http://www.wma.net/en/30publications/10policies/b3/index.html (Accessed July19, 2013).

[241] Bridget Pratt and Bebe Loff, "Justice in International Clinical Research." Developing World Bioethics 11, no. 2 (2011): 76.

[242] Pratt and Loff, "Justice in International Clinical Research." 76.

[243] K. Shapiro, and S.R. Benatar, "HIV Prevention Research and Global Inequality: Steps towards Improved Standards of Care." J Med Ethics 31, (2004): 39–47.

[244] C. Slack et al., "Provision of HIV Treatment in HIV Preventive Vaccine Trials: A Developing Country Perspective." Social Science & Medicine 60, no. 6 (March, 2005): 1197–1208.

[245] Ruth Macklin, "Changing the Presumption: Providing ART to Vaccine Research Participants." American Journal of Bioethics 6, no. 1 (2006): W1.

contribute to treatment and care decisions, which may include how treatment services are accessed and how services might be monitored.[246] These contributions from the communities can improve the quality and suitability of the research as well as counteract accusations of token community consultation.[247] The voice of the community can help in determining the most appropriate and least detrimental course of action for access to care and treatment. The views and expectations of communities and potential research participants are crucial in framing the appropriate ethical course of action.[248]

There is limited empirical research focusing on perspectives of medical research from developing countries,[249] and limited data are available explaining what communities in developing nations think regarding the issue of care and treatment in HIV prevention trials.[250] In a survey which explores community perceptions of research benefits and harms in Uganda,[251] the majority of the respondents supported that researchers should provide some benefit to the wider community following a trial. More than half of the respondents indicated that this benefit should be either general medical care or care for HIV- infected community members.[252] Many respondents perceived the direct provision of basic healthcare and referral of participants to off-site healthcare services as a fair means of fulfilling the health needs of trial participants.[253] On the other hand, issues were also identified regarding the sustainability of care beyond the life of the study and the adequacy of local healthcare facilities.[254] Respondents comprising clinical trial participants and researchers in a Kenyan study indicated that it would be unfair to stop antiretroviral therapy (ART) at the conclusion of a trial, and that researchers have a long term moral obligation to provide ART to participants after the conclusion of a trial.[255]

[246] D. Tarantola et al., "Meeting Report: Ethical Considerations Related to the Provision of Care and Treatment in Vaccine Trials." Vaccine 25 (2007): 4863–4874.

[247] AIDS Vaccine Advocacy Coalition (AVAC), "Vaccine Trials: Leaving Communities Better Off." In AIDS Vaccine Handbook: Global Perspectives, P. Kahn, ed. (New York, NY: AVAC, 2005), 137–144.

[248] Nichola Barsdorf et al., "Access to Treatment in HIV Prevention Trials: Perspectives from A South African Community." Developing World Bioethics 10, no. 2 (August, 2010): 81.

[249] N. Barsdorf & D. Wassenar, "Racial Differences in Public Perceptions of Voluntariness of Medical Research Participants in South Africa." Social Science & Medicine 60, no. 6 (March, 2005): 1087–1098.

[250] Christine Grady et al., "Research Benefits for Hypothetical HIV Vaccine Trials: The Views of Ugandans in the Rakai District." IRB: Ethics & Human Research 30, no. 2 (March – April, 2008): 1–7.

[251] Christine Grady et al., "Research Benefits for Hypothetical HIV Vaccine Trials: The Views of Ugandans in the Rakai District." 1–7.

[252] Nichola Barsdorf et al., "Access to Treatment in HIV Prevention Trials: Perspectives from A South African Community." 81.

[253] Nichola Barsdorf et al., "Access to Treatment in HIV Prevention Trials: Perspectives from A South African Community." 81–82.

[254] Nichola Barsdorf et al., "Access to Treatment in HIV Prevention Trials: Perspectives from A South African Community." 82.

[255] Nichola Barsdorf et al., "Access to Treatment in HIV Prevention Trials: Perspectives from A South African Community." 82.

The leading narrative regarding respondents' perspectives on the issue of access to treatment and care in HIV prevention trials was that researchers should offer some type of help to participants who need treatment.[256] They argued that researchers should help because they are in a relationship with participants. They view the participant-researcher relationship as supportive and caring.[257] Some respondents indicated that the idea of a supportive and caring relationship implies that researchers should help as a form of reciprocity for these participants' contribution to the research.[258] More than half of all respondents considered researchers' assistance in referral for accessing ART vital. The respondents indicated that researchers should make provision for appropriate referrals or arrangements for participants' access to treatment.[259] While, some respondents asserted that researchers should make provision for lifelong access to ART, others indicated that researchers should continue to help until participants can access treatment on their own.[260]

An ethical model for formulating researchers' obligations for providing health care to participants was endorsed by respondents. The model emphasizes that participants by implication entrust certain aspects of their health to researchers when they enroll for a study. When participants give permission to researchers to monitor disease under study, they as well entrust caring for that disease to researchers. From this perspective of entrustment, researchers have special responsibilities to participants even beyond the requirement of sound scientific and safe conduct of the trial.[261] The scope of these responsibilities is affected by the extent of the researcher-participant relationship, and the participants' vulnerability, dependence and uncompensated risks or burdens.[262] This perspective is echoed by a recent Kenyan study where 11 focus groups indicated the theme of researchers' moral obligations to participants and the community in clinical trials.[263] Kenyan patients, clinician researchers, and administrators argue that it would be unfair to discontinue

---

[256] Nichola Barsdorf et al., "Access to Treatment in HIV Prevention Trials: Perspectives from A South African Community." 83.

[257] Nichola Barsdorf et al., "Access to Treatment in HIV Prevention Trials: Perspectives from A South African Community." 84.

[258] Nichola Barsdorf et al., "Access to Treatment in HIV Prevention Trials: Perspectives from A South African Community." 84.

[259] Nichola Barsdorf et al., "Access to Treatment in HIV Prevention Trials: Perspectives from A South African Community." 85.

[260] Nichola Barsdorf et al., "Access to Treatment in HIV Prevention Trials: Perspectives from A South African Community." 85.

[261] Nichola Barsdorf et al., "Access to Treatment in HIV Prevention Trials: Perspectives from A South African Community." 86.

[262] Leah Belsky and Henry S. Richardson, "Medical Researchers' Ancillary Clinical Care Responsibilities." BMJ 328, no. 7454 (June, 2004): 1494–1496.

[263] D N Shaffer et al., "Equitable Treatment for HIV/AIDS Clinical Trial Participants: A Focus Group Study of Patients, Clinician Researchers and Administrators in Western Kenya." Journal of Medical Ethics 32, no. 1 (January,2006): 55–60.

antiretroviral therapy following an HIV/AIDS clinical trial and that researchers have a long term obligation to participants.[264]

Several other studies conducted in developing countries with local researchers indicated that majority of the respondents about 67% who expressed that their intervention was, or might be effective insisted that it would be provided to study participants or to any other host country residents at the end of the study. Nevertheless, researchers who did not know whether their intervention would be effective were less likely to indicate that it would be provided.[265]

In a similar study, 53% of researchers contend that "research to test an intervention should not be carried out in a developing country unless the intervention, if found to be successful, will be made available to that country at the conclusion of the study."[266] Some respondents about 27% of the researchers insisted that "international policy regarding research should require researchers to establish a mechanism for continuing delivery of medical care after completion of the study."[267] Regarding the issue of future provision of interventions, one researcher indicated: " I agree with the notion…that formal steps be taken to insure certain actions after the study is completed (i.e., availability of drug or procedure studied to the study population/host population if benefit shown; reporting of findings to community; attempt to maintain or continue health services that were introduced to the community as part of the study, etc.)."[268]

Another researcher contends that benefits should be enjoyed by the host countries, but indicated that capacity building should take place in all cases, in case identified interventions are not provided.[269] Cowboy research which implies researchers going into a community to collect data, and leave is not acceptable.[270] Kass and Hyder articulate, "They (study participants) need to be informed and something needs to go back to them. You can't just do the research and take the results and run. Intervention has to come through, and this is happening now."[271] Some respondents indicated that effective interventions should be implemented in

[264] D N Shaffer et al., "Equitable Treatment for HIV/AIDS Clinical Trial Participants: A Focus Group Study of Patients, Clinician Researchers and Administrators in Western Kenya." 55–60.

[265] Kass and Hyder, "Attitudes and Experiences of U.S. and Developing Country Investigators Regarding U.S. Human Subjects Regulations," B-39.

[266] Kass and Hyder, "Attitudes and Experiences of U.S. and Developing Country Investigators Regarding U.S. Human Subjects Regulations," B-39.

[267] Kass and Hyder, "Attitudes and Experiences of U.S. and Developing Country Investigators Regarding U.S. Human Subjects Regulations," B-39.

[268] Kass and Hyder, "Attitudes and Experiences of U.S. and Developing Country Investigators Regarding U.S. Human Subjects Regulations," B-39.

[269] Kass and Hyder, "Attitudes and Experiences of U.S. and Developing Country Investigators Regarding U.S. Human Subjects Regulations," B-39.

[270] Kass and Hyder, "Attitudes and Experiences of U.S. and Developing Country Investigators Regarding U.S. Human Subjects Regulations," B-97.

[271] Kass and Hyder, "Attitudes and Experiences of U.S. and Developing Country Investigators Regarding U.S. Human Subjects Regulations," B-97.

the host community after the conclusion of the study.[272] One respondent insisted that a critical part of a research contract between the government and the researchers should be to make the research intervention available to the participants if proven effective.[273]

4.4 Conclusion

Concluding remarks recapitulate the analysis about national responsibility and humanitarian assistance in relation to post-trial access of participants and host populations to drugs in developing nations. The analysis began with a controversial debate between two major approaches, including cosmopolitan and statist in the issue of global justice to address post-trial access to drugs in developing countries. The focus of the chapter was on Rawls's statist approach in dealing with the issue of post-trial access to drugs in developing countries especially for participants and host populations.

Rawls's two different approaches to justice both in the domestic society and in the international arena were discussed. His account of domestic justice was dubbed justice as fairness which emphasizes the idea that fundamental agreements of the parties to the original position were fair. A constellation of ideas critical to understanding justice as fairness were discussed.

The idea of the basic structure of the society which applies to a well-ordered society was considered fundamental because it is the primary subject of justice. The basic structure of the society accommodates significant inequalities stemming from natural and social lotteries. Principles of justice were considered imperative in order to address the inherent inequalities in the basic structure of the society.

Essential elements of theory of justice including original position and two principles of justice were highlighted and discussed. Rawls identified two principles of justice as fairness. The first principle is called equal liberty principle. The second principle is divided into two parts, including the difference principle and fair equality of opportunity principle. These principles primarily were applied to the basic structure of the society and they governed the allocation of rights and duties and control the distribution of social and economic advantages.

In the analysis about just health care within nations, Daniels extended Rawls's theory of justice to health care in the domestic society. He argues that health care institutions should be regarded as basic institutions that have exclusive responsibility of guaranteeing fair equality of opportunity.

Rawls's statist approach discusses the two aspects of Rawls's account of international relations and international justice, including ideal theory and non-ideal the-

---

[272] Kass and Hyder, "Attitudes and Experiences of U.S. and Developing Country Investigators Regarding U.S. Human Subjects Regulations," B-97.

[273] Kass and Hyder, "Attitudes and Experiences of U.S. and Developing Country Investigators Regarding U.S. Human Subjects Regulations," B-98.

ory. However, Rawls's account of global justice lacks a commitment to distributive justice.

In analyzing Rawls's duty of assistance and post-trial access to drugs, three arguments were explored, including argument based on Rawls's defense of human rights, argument based on redress for the unjustified distributive effects of cooperative organizations and argument based on access to drugs as a transition strategy favoring the establishment of politically well-ordered nations. Furthermore, it was argued that improving access to drugs would help as an effective transition strategy that would enable burdened societies to become politically well-ordered.

In contrast to limited approach to post-trial access to drugs for participants and host populations argued by Rawls, local participants' perspectives of justice in international clinical trials was discussed with emphasis on its commitment to distributive justice. Obligations of justice in international clinical trials was linked to post-trial access to successful interventions and to a broader justice implications to reduce health inequalities between nations. There was a consensus among most local participants that successful interventions should be provided to participants and host populations in developing countries after the end of the clinical research. Therefore, Rawls relies upon the duty of humanitarian assistance from the perspective of global justice to provide post-trial access to drugs. This reliance is merely a transition strategy until the nation can develop its own resources as a well-ordered society. In contrast, as discussed in the next chapter, Pogge's cosmopolitan approach adopts a more robust and expansive international perspective to global justice to justify post-trial access to drugs in developing countries.

Chapter 5
International Responsibility and the Health Impact Fund

5.1 Introduction

Post-trial access to drugs in developing countries is argued also from the perspective of another dominant approach to global justice, cosmopolitanism.[1] Pogge argued for a stronger interpretation of global responsibilities for providing affordable access to drugs for participants and host populations in developing countries. His work establishes that we have a critical duty of justice to take action on the issue of affordable access to essential drugs. This duty is grounded on human rights, and extends universally to all individuals.[2]

In his landmark work on World Poverty and Human Rights: Cosmopolitan Responsibilities and Reforms published in 2002, Pogge develops a perspective on global justice that challenges Rawls's account on several dimensions. He contends that severe poverty and global inequality persist because citizens of affluent countries do not consider their eradication compelling. Pogge challenges the thesis of explanatory nationalism propounded by Rawls, which indicates that the persistence of world poverty and inequality is adequately explained by appeal to local factors.[3] He argues that "the institutional order perpetuates harm, and so violates negative rights or human rights."[4] He develops a theory motivated by human rights that emphasized negative rights and duties not to be harmed or not to harm.[5]

---

[1] Thomas Pogge, "Cosmopolitanism and Sovereignty," in *Global Justice: Seminal Essays*, ed. Thomas Pogge and Darrel Moellendorf (St. Paul, MN: Paragon House, 2008), 356.

[2] Mira Johri et al., "*Sharing the Benefits of Medical Innovation: Ensuring Fair Access to Essential Medicines*," 2006, http://www.lawweb.usc.edu/centers/paccenter/assets/docs/Ehrenreich_Johri_2006_04_15.pdf. (Accessed March 20, 2014).

[3] Pogge, *World Poverty and Human Rights: Cosmopolitan Responsibilities and Reforms*, 145–146.

[4] Franklin Tennant Gairdner, "*A Defence of Thomas Pogge's Argument for a Minimally Just Institutional Order*" (MA thesis, Queens's University, 2009), 4.

[5] Pogge, "Assisting the Global poor," 263.

E.C. Obi, *Post-Trial Access to Drugs in Developing Nations*, Advancing Global Bioethics 7, DOI 10.1007/978-3-319-60028-4_5

In his paper, Human Rights and Global Health: A Research Program, Pogge presents a very bleak picture of massive incidence of mortality and morbidity especially in many poor countries.[6] He presents two different strategies for dealing with the increasing global burden of disease particularly in many poor countries which results in massive mortality and morbidity rates. The First approach emphasizes the eradication of severe poverty.[7] The second strategy identified by Pogge for tackling the massive mortality and morbidity rates is guaranteeing improved access to medical treatments, including preventive and remedial.[8]

The focus of this chapter is on the second strategy and in line with this, Pogge delineates that significant reduction of global disease burden can be attained by providing medical innovators with stable and reliable financial incentives to tackle the medical conditions of the poor.[9] Pogge argues that his primary goal is to "develop a concrete, feasible, and politically realistic plan for reforming current national and global rules for incentivizing the search for new essential drugs."[10] Pogge argues that the reformed plan will be cost effective and fairly distribute the cost of global health-care spending among countries, generations, and between healthy and unhealthy people.

This chapter focuses on Pogge's cosmopolitan approach to the issue of access to drugs in developing countries. The first part of the chapter deals with meaning of cosmopolitanism and four different approaches of cosmopolitanism with their nuances on the application of global distributive principles. The second part deals with current rules for incentivizing pharmaceutical research which focuses on seven problems identified by Pogge in the current global patent system. The third part of this chapter focuses on Pogge's new rules for reforming and incentivizing pharmaceutical research in which he proposes two basic reform strategies for dealing with monopoly pricing issues of the current patent system, a differential pricing strategy and a public good strategy.

The fourth part tackles the critical role of the Health Impact Fund (HIF), a global institution proposed by Pogge for the implementation of his new plan for reforming and incentivizing pharmaceutical research. A more detailed analysis of Pogge's position on post-trial access to drugs in developing countries begins with a discussion about cosmopolitanism and pharmaceutical research.

---

[6]Thomas W. Pogge, "Human Rights and Global Health: A Research Program," *Metaphilosophy* 36, no. 1/2, (January 2005): 182–183.

[7]Pogge, "Human Rights and Global Health: A Research Program," 183.

[8]Pogge, "Human Rights and Global Health: A Research Program," 183.

[9]Pogge, "Human Rights and Global Health: A Research Program," 184.

[10]Pogge, "Human Rights and Global Health: A Research Program," 184.

5.2 Cosmopolitanism and Pharmaceutical Research

5.2.1 Justice and Cosmopolitanism: Thomas Pogge

5.2.1.1 Meaning of Cosmopolitanism

Cosmopolitanism emphasizes that everyone should be treated as equals regardless of nationality and citizenship. Tan articulates, "from the cosmopolitan perspective, principles of justice ought to transcend nationality and citizenship, and ought to apply equally to all individuals of the world as a whole."[11] Succinctly put, "cosmopolitan justice is justice without borders".[12]

One of the major interpretations of cosmopolitan justice is that this impartiality with regard to nationality and citizenship as well relates to distributive justice in such a way that a person's lawful material entitlements would be independently regulated by his or her national and state membership.[13] From this perspective, Tan explained that Charles Beitz and Thomas Pogge took their inspiration from John Rawls's A Theory of Justice to contend that Rawls's arguments for social and economic equality should apply as well to the global setting.[14] Tan writes, "Just as Rawls considers a person's race, gender, talents, wealth, and other natural and social particularities to be "arbitrary from a moral point of view,"[15] so too, they argue, are factors like a person's nationality and citizenship morally arbitrary."[16] Furthermore, it was argued that in the same way that the influences of contingencies of the natural and social lotteries on a person's life chances were invalidated in the domestic realm by specific principles of distributive justice, conversely, the influences of global contingencies should be diminished by specific principles of global distributive justice. The implication is that Rawls's principles of justice comprising the principle regulating social and economic equality "should apply between individuals across societies and not just within the borders of a single society".[17]

Cosmopolitan accounts of distributive justice defend some basic claims. Pogge claims that all cosmopolitan views share three essential features. The first is individualism which emphasizes that the ultimate units of moral concern are human beings, or persons, rather than, units such as family lines, tribes, or ethnic, cultural, or religious communities, nations, or states. The Second is universality which stresses that the status of ultimate unit of concern ascribes equally to every living human being, not simply to some subgroup, such as men, aristocrats, whites, or

---

[11] Kok-Chor Tan, *Justice Without Borders: Cosmopolitanism, Nationalism and Patriotism* (New York: Cambridge University Press, 2004), 1.

[12] Tan, *Justice Without Borders: Cosmopolitanism, Nationalism and Patriotism*, 1.

[13] Tan, *Justice Without Borders: Cosmopolitanism, Nationalism and Patriotism*, 1.

[14] Tan, *Justice Without Borders: Cosmopolitanism, Nationalism and Patriotism*, 62.

[15] John Rawls, *A Theory of Justice*, Original Edition (Cambridge, Massachusetts: Harvard University Press, 1971), 15.

[16] Tan, *Justice Without Borders: Cosmopolitanism, Nationalism and Patriotism*, 62.

[17] Tan, *Justice Without Borders: Cosmopolitanism, Nationalism and Patriotism*, 62.

Muslims. The third is generality which emphasizes that the special status identified has global force. Here, the ultimate units of moral concern for everyone are persons.[18] Similarly, Charles Jones points out that the cosmopolitan perspective is "impartial, universal, individualist, and egalitarian."[19] It is pertinent to note that there are seemingly some extreme positions among cosmopolitans. For example, some authors liken giving priority to members of one's nation to racism[20] or to "bad faith."[21] More so, some authors contend that we have a duty to engage in getting rid of nations and national identification completely.[22] Some cosmopolitans as part of advancing the aim of decreasing the significance of national identities have argued for the establishment of a world state, or at least strengthening of the power of international political structures.[23] There is an acknowledgement by some cosmopolitans that one severe flaw of the cosmopolitan perspective is its apparent inability to recognize and appropriately explain the significance of the special ties and commitments that typify the lives of ordinary men and women.[24]

On the other hand, other cosmopolitans advocate for the strengthening of international political institutions without getting rid of national attachments and loyalties.[25] In this context, more moderate cosmopolitans acknowledge that in some instances there are special duties owed to co-nationals and citizens provided that they promote realization of the global justice. Robert E. Goodin endorsed this idea in his efficiency argument. He argues "that we all have general duties to all persons, but these duties may be effectively fulfilled through a system of special responsibilities towards compatriots."[26] He discussed assigned responsibility model which entails that special responsibilities are in his own perspective "merely devices whereby the moral community's general duties get assigned to particular agents."[27] In a similar vein, Jones articulates "It can be morally permissible, even required, that one be patriotic and loyal to one's country, but such permissions and requirements can never override the demands of impartial justice."[28] Moral obligations are recognized beyond the requirements of justice. There is a consideration that various

[18] Pogge, "*Cosmopolitanism and Sovereignty*," 356.

[19] Charles Jones, *Global Justice: Defending Cosmopolitanism* (New York: Oxford University Press, 1999), 15.

[20] Paul Gomberg, "Patriotism is like Racism," *Ethics* 101, no. 1 (October 1990): 144–150.

[21] Simon Keller, "Patriotism as Bad Faith," *Ethics* 115, no. 3 (April 2005): 563–592.

[22] Jon Mandle, *Global Justice* (Malden, MA: Polity Press, 2006), 42.

[23] Mandle, *Global Justice*, 42.

[24] Charles R. Beitz, "International Liberalism and Distributive Justice: A Survey of Recent Thought," *World Politics* 51, no. 2 (January 1999): 291.

[25] Mandle, *Global Justice*, 42.

[26] Robert E. Goodin, "What Is So Special about Our Fellow Countrymen," in *Global Justice: Seminal Essays*, ed. Thomas Pogge and Darrel Moellendorf (St. Paul, MN: Paragon House, 2008), 255.

[27] Goodin, "What Is So Special about Our Fellow Countrymen," 269.

[28] Jones, *Global Justice: Defending Cosmopolitanism*, 134.

local attachments may engender some of these supplementary moral obligations, but that they are not the basic foundation of any obligations of justice.[29]

Another important basic claim of cosmopolitans deals with who is owed the goods transferred. There is a consensus among most contemporary cosmopolitans that obligations of justice are owed to individuals and not states.[30] Addressing global inequality between states without any focus on the well-being of the individuals falls short of an effective global justice theory. Tan writes, "It is myopic to think that the problems of global injustices that impact on individuals can be settled by focusing solely on justice between states. A "morality of states" approach does not go far enough if we are interested in improving individual lives."[31] A contrary perspective was at a time espoused by Brian Barry who argued that states were entitled to receive resources.[32] However, in his later works he discards this earlier view and aligns his position with cosmopolitanism's individualist claims.[33]

Another important point worth noting regarding cosmopolitanism is the various classifications in the literature. Millum provides two distinctions of cosmopolitan perspectives based on the foundations that are considered to motivate them. The first is humanitarian cosmopolitans who assert that obligations of justice do not stem from associations like the state, but from persons' characteristics per se, irrespective of their associations with other persons. He cites an example with any utilitarian theory of global justice, which emphasizes that the ultimate justification for justice principles is their contribution to aggregate utility, and it is immaterial in whom that utility is situated.[34] The second is political cosmopolitans who claim that some associations' characteristics like those exemplified by the state are the bases for applying justice requirements, but contend that these characteristics are actually located as well in international associations.[35] A typical example is to argue that mutually beneficial cooperation between individuals is necessary and adequate to establish justice requirements in the distribution of the cooperation's products. The global trade based on its magnitude and significance has been cited as an indication that mutually beneficial cooperation goes completely beyond national borders. The implication is that the requirements of justice go beyond national borders as well.[36]

---

[29] Darrel Moellendorf, *Cosmopolitan Justice Reconsidered* (Boulder, Colo.: Westview Press, 2002), 35.

[30] Thomas Pogge, "An Egalitarian Law of Peoples", *Philosophy and Public Affairs* 23, no. 3 (Summer 1994): 202.

[31] Tan, *Justice Without Borders: Cosmopolitanism, Nationalism and Patriotism*, 37.

[32] Brian Barry, "Justice as Reciprocity", in *Liberty and Justice: Essays in Political Theory*, Volume 2 (Oxford: Clarendon, 1991), 239–240.

[33] Brian Barry, "International Society from a Cosmopolitan Perspective" *in International Society: Diverse Ethical Perspectives* ed. David Mapel and Terry Nardin (Princeton: Princeton University Press, 1998), 159–160.

[34] Joseph Millum, "Global Bioethics and Political Theory" in *Global Justice and Bioethics* ed. Joseph Millum and Ezekiel J. Emmanuel (New York: Oxford University Press, 2012), 20–21.

[35] Millum, "Global Bioethics and Political Theory", 21.

[36] Charles Beitz, *Political Theory and International Relations* (Princeton, NJ: Princeton University Press, 1999), 182–183.

Pogge also offered two distinctions of cosmopolitan perspectives. He distinguishes first between legal and moral cosmopolitanism. Legal cosmopolitanism focuses on a global order in the political realm which grants equal legal rights and duties to all persons considered to be fellow citizens of a universal republic. On the other hand, moral cosmopolitanism claims that all persons are in moral associations with one another, and in this context, there is a requirement to respect one another's status as ultimate units of moral concern, which imposes restrictions on our conduct and on our attempts to form institutional schemes. The dominant notion of moral cosmopolitanism entails that every human being possess a global stature as an ultimate unit of moral concern.[37]

Pogge's second distinction of cosmopolitan positions is between institutional cosmopolitanism and interactional cosmopolitanism. Institutional cosmopolitanism holds that principles of justice apply to institutions which encompass schemes of trade, communication, and interdependence largely. On the other hand, interactional cosmopolitanism holds that principles of justice apply still without a common institutional setting.[38] In this context, principles of justice apply directly to the conduct of persons and groups.[39] Pogge argues that "interactional cosmopolitanism assigns direct responsibility, for the fulfilment of human rights to other individual and collective agents, whereas institutional cosmopolitanism assigns such responsibility to institutional schemes."[40] The institutional approach also establishes a shared responsibility for all persons not only to refrain from cooperating in imposing a harmful institutional order that impedes fulfillment of human rights, but also to promote institutional reform. Pogge argues that "our negative duty not to cooperate in the imposition of unjust social institutions triggers obligations to promote feasible reforms that would enhance the fulfillment of human rights."[41] A discussion of the four different approaches of cosmopolitanism is the next task of this chapter.

5.2.1.2 Four Approaches of Cosmopolitanism

The four different approaches of cosmopolitanism with their nuances on the application of global distributive principles comprise contractarian, consequentialist, rights based and duty based. A detailed discussion of the four approaches follows.

Contractarian Approach

The two major proponents of the contractarian approach to cosmopolitanism are Charles Beitz and Thomas Pogge. Beitz and Pogge applied Rawls's Original Position to the world stage. Martha C. Nussbaum writes, "for both of them, the right

[37] Pogge, "Cosmopolitanism and Sovereignty," 356.
[38] Pogge, "Cosmopolitanism and Sovereignty," 357–364.
[39] Pogge, "Cosmopolitanism and Sovereignty," 357.
[40] Pogge, "Cosmopolitanism and Sovereignty," 357.
[41] Pogge, "Cosmopolitanism and Sovereignty," 359.

way to use Rawlsian insights in crafting a theory of global justice is to think of the Original Position as applied directly to the world as a whole."[42] Pogge and Beitz contend that the only way to adequately recognize the individual as a subject of justice, within the framework of Rawls's perspective, is to envisage that the whole global scheme is available, and that the parties are bargaining for a just global structure as individuals. Both argue in different ways that the outcome of a global original position will be a global structure that maximizes the advantage of the least well off.[43] Different specific arguments offered by Beitz and Pogge in defense of cosmopolitan view of justice are presented as follows.

Beitz unlike Rawls proposes principles of distributive justice that apply globally rather than within states. He presents two arguments to support his defense of cosmopolitan principles of distributive justice. The first argument focuses on the arbitrary distribution of natural resources. Beitz defended the view that the distribution of natural resources is morally arbitrary and that the representatives at the global original position will adopt a condition that will favor equal distribution of the natural resources. He supports a global principle of distributive justice which outlines the criteria for the distribution of natural resources.[44] He forcefully argues, "not knowing the resource endowments of their own societies, the parties would agree on a resource redistribution principle that would give each society a fair chance to develop just political institutions and an economy capable of satisfying its members' basic needs."[45]

Beitz's second argument defends the existence of a global system of cooperation, drawing on an extensive amount of empirical research.[46] He argues that "international economic interdependence constitutes a scheme of social cooperation like those to which requirements of distributive justice have often been thought to apply."[47] Beitz discounts the importance of domestic original position, but forcefully defends only the global original position, where the participants will adopt a global difference principle. Caney writes, "So whereas Rawls's domestic contract delivers a difference principle with a domestic scope, Beitz's global contract delivers a global difference principle."[48] The strength of Beitz's argument is based on the premise that global distributive principles apply to schemes of social cooperation.

Beitz's position was criticized by some authors. Barry refuting Beitz's position argues that there is no global interdependence of the suitable type. Barry writes, "Beitz's argument for extending the Rawlsian difference principle is in essence that the network of international trade is sufficiently extensive to draw all countries

[42] Martha C. Nussbaum, "Beyond the Social Contract: Capabilities and Global Justice", *Oxford Development Studies* 32, no. 1 (March 2004): 10–11.

[43] Nussbaum, "Beyond the Social Contract: Capabilities and Global Justice", 11.

[44] Beitz, *Political Theory and International Relations*, 136–143.

[45] Beitz, *Political Theory and International Relations*, 141.

[46] Beitz, *Political Theory and International Relations*, 144–152.

[47] Beitz, *Political Theory and International Relations*, 154.

[48] Simon Caney, *Justice Beyond Borders: A Global Political Theory* (New York: Oxford University Press, 2005), 109.

together in a single cooperative scheme. But it seems to be that trade, however multilateral, does not constitute a cooperative scheme of the relevant kind."[49] While, Barry's major contention is that principles of distributive justice apply only to schemes of cooperation that are mutually beneficial,[50] Beitz on the other hand, argues that they apply to groups of people who are engaged in interdependence of some kind, even if their interdependence is not mutually advantageous or cooperative.[51] Beitz clearly notes that "everyone need not be advantaged by the cooperative scheme in order for requirements of justice to apply".[52]

Another criticism of Beitz's position comes from David A.J. Richards who defends an original position that encompasses all persons in the world not on the bases of social cooperation but merely on the bases that persons are entitled to be involved in the contract as a result of their rights and interests as human beings. Richards argues that fair principles are the ones that would be adopted in a contract that involves all persons and in which people are behind a veil of ignorance.[53] He validates the utilization of such a global hypothetical contract on the bases that "ones membership in one nation as opposed to another and the natural inequality among nations may be as morally fortuitous as any other natural fact."[54] The implication is that all persons as such are entitled to be involved in the hypothetical contract. A consideration of everyone being a member of a common institutional system is basically irrelevant.[55]

Beitz reformulated his position following Richards' criticism. He maintains that the morally pertinent aspects of persons are the "two essential powers of moral personality—a capacity for an effective sense of justice and a capacity to form, revise, and pursue a conception of the good."[56] This entails that all those who possess these universal properties are eligible to be represented in a global original position.[57]

The second exponent of contractarian approach to cosmopolitanism is Pogge who defends the assertion that the global distributive principles apply to a scheme of international cooperation. Pogge who is usually considered as an unrestricted institutionalist argues that all principles of justice apply only within schemes of

[49] Brain Barry, "Humanity and Justice in Global Perspective" in *Global Justice: Seminal Essays*, ed. Thomas Pogge and Darrel Moellendorf (St. Paul, MN: Paragon House, 2008), 191.

[50] Barry, "Humanity and Justice in Global Perspective" 191–192.

[51] Beitz, *Political Theory and International Relations*, 131.

[52] Beitz, *Political Theory and International Relations*, 150.

[53] David A.J. Richards, "International Distributive Justice" in *Ethics, Economics, and the Law: NOMOS XXIV* ed. J. Roland Pennock and John W. Chapman (New York: New York University Press, 1982), 278–282.

[54] Richards, "International Distributive Justice" 290.

[55] Caney, *Justice Beyond Borders: A Global Political Theory*, 115.

[56] Charles Beitz, "*Cosmopolitan Ideals and National Sentiment*", The Journal of Philosophy 80, no. 10 (October 1983): 595.

[57] Charles R. Beitz, "Social and Cosmopolitan Liberalism", *International Affairs* 75, no. 3 (July 1999): 521–524.

cooperation.[58] An institutional perspective maintains that persons have obligations to defend the civil and political rights of people who are members of the same scheme, rather than those who are not part of the cooperation.[59] Darrel Moellendorf argues within the framework of an institutionalist when he insists both that justice applies within schemes of cooperation[60] as well as that there is a global scheme.[61]

Pogge distinguishes between positive and negative duties in his institutional approach to cosmopolitan justice. He contends that principles of justice require that persons possess a negative duty not to support an unjust socio-economic structure. He clearly articulates that persons have a "negative duty not to uphold injustice, not to contribute to or profit from the unjust impoverishment of others."[62] Membership of institutions is imperative because each member is bound by a negative duty not to support unjust institutions. In interpreting justice as requiring a negative duty not to uphold unjust institutional structures to which one belongs we come to the inference that duties of justice apply to, and among, members of institutions. On the other hand, if we assert that there are global institutional structures we can also come to the inference that persons are bound by a negative duty not to impose unjust global institutional structures on the rest of the world.[63] Critics argue that Pogge's unrestricted institutionalist perspective focuses a lot more on duty bearers without adequate attention to entitlement bearers which include the needy, the hungry and the sick.[64]

Pogge proposes what he calls a global resource dividend (GRD) which requires that people should be taxed for utilizing the resources in their territory and the proceeds expended for improving the poor all over the world.[65] He proposes a global resource tax (GRT) as a way for controlling global inequality. He maintains that the proceeds from the global resource tax will be invested in alleviating the global poverty. Pogge argues that proceeds from GRD are to be utilized for improving the lives of the global poor through provision of "access to education, health care, means of production (land) and/or jobs to sufficient extent to be able to meet their own basic needs with dignity and to represent their rights and interests effectively against the rest of humankind: compatriots and foreigners."[66] Revenues from GRD would assist

---

[58] Caney, *Justice Beyond Borders: A Global Political Theory*, 110.

[59] Caney, *Justice Beyond Borders: A Global Political Theory*, 111.

[60] Darrel Moellendorf, *Cosmopolitan Justice* (Boulder, Colorado: Westview Press, 2002), 30–36.

[61] Moellendorf, *Cosmopolitan Justice*, 36–38.

[62] Thomas Pogge, "Eradicating Systematic Poverty: Brief for a Global Resources Dividend" in *World Poverty and Human Rights: Cosmopolitan Responsibilities and Reforms* (Cambridge: Polity Press, 2002), 197.

[63] Thomas W. Pogge, *Realizing Rawls* (Ithaca, NY: Cornell University Press, 1989), 276–278.

[64] Henry Shue, *Basic Rights: Subsistence, Affluence, and U.S. Foreign Policy*, Second Edition (Princeton: Princeton University Press, 1996), 164–166.

[65] Thomas Pogge, "A Global Resources Dividend" in *Ethics of Consumption: The Good Life, Justice, and Global Stewardship* ed. David A. Crocker and Toby Linden (Lanham: Rowman and Littlefield, 1998), 501–536.

[66] Pogge, "An Egalitarian Law of Peoples", 201.

poorer governments in providing people with education, health care, microloans, infrastructure, and maintaining lower tax rates and higher tax exemptions.[67] A discussion of the outcome-based approach to cosmopolitan justice is the task of next section of this chapter.

Consequentialist Approach

The second approach to cosmopolitanism is consequentialist approach to cosmopolitan justice. Consequentialist approach to global justice focuses on the consequences and results of actions and structures. The major proponents are utilitarians. Consequentialists assert that "the present global economic order has stark consequences: it leaves hundreds of millions in profound poverty, with all its associated insecurities, ill-health and powerlessness."[68] Consequentialists support cosmopolitan principles of distributive justice. Some proponents of consequentialist approach to cosmopolitan justice include Peter Singer, Robert Goodin and Martha Nussbaum. A brief discussion of their different perspectives follows.

Singer proposes a utilitarian approach to global inequality in his article titled Famine, Affluence and Morality published in 1972. He argues that the richer nations in the developed world have obligations to aid the poorer nations in the developing world.[69] Singer begins with the assertion that poverty is evidently a bad.[70] He argues that we all have an obligation to prevent bad things from happening. Singer posits two versions of this claim. The first claim, which is pretty strong states, "if it is in our power to prevent something bad from happening, without thereby sacrificing anything of comparable moral importance, we ought, morally, to do it."[71] The second claim which is a weaker claim states, "if it is in our power to prevent something very bad from happening, without sacrificing anything morally significant, we ought, morally, to do it"[72] In the second claim which is a weaker one, a person can be relieved of the duty if it imposes a substantial moral cost on him or her. On the other hand, in the first claim which is a stronger one, a person can only be relieved of the duty if that moral cost is comparable to the cost to the poor person that the person involved would otherwise assist. Singer acknowledges only an important point that our rendering assistance need not make a bad situation worse or bring about a comparable harm.[73] Based on these premises, Singer proposes that wealthy persons have obligations to aid the needy regardless of where they live and what

[67] Pogge, "An Egalitarian Law of Peoples", 201–202.

[68] Onora O'Neill, *Bounds of Justice* (New York: Cambridge University Press, 2000), 122.

[69] Peter Singer, "Famine, Affluence and Morality", *Philosophy and Public Affairs* 1, no. 3 (Spring 1972): 229–230.

[70] Peter Singer, "Famine, Affluence and Morality", 231.

[71] Singer, "Famine, Affluence and Morality", 231.

[72] Singer, "Famine, Affluence and Morality", 231.

[73] Singer, "Famine, Affluence and Morality", 231.

country they come from.[74] The implication is that special attachments like nationality and citizenship are irrelevant when we consider our obligations to others. Singer further argues that the obligation of the affluent towards the poor is not minimized by the physical distance between rich and poor, or by the fact that there are many other people likewise able to assist.[75] He argues that giving to the distant poor generally regarded as an act of charity and/or supererogatory, is rather a matter of duty or obligation.[76]

Singer's claim regarding our obligation to assist has been criticized on the grounds that it is unrealistically overdemanding.[77] Deborah Zion citing John Arthur, for instance, proposes that "Singer's formulation produces a duty for healthy people to donate one eye or one kidney, on the grounds that the inconvenience caused to the donating agent is seriously outweighed by the good such organs might do to the blind or dying."[78] Singer's obligation to assist which is characterized by its "overdemandingness" has been strongly criticized by Michael Slote when he proposes that persons should not have their major life plans disturbed by the obligation to help others.[79] The important point established by critics is that there is need to set limits to beneficence.[80] Singer was also criticized for equating actions with omissions. Caney acknowledges that someone may contend that not saving a person's life is not the same as killing them. However, it is still pertinent to argue in consonant with Singer's perspective that we are bound by an obligation to distribute resources to the needy.[81]

To mitigate the overdemanding claim of Singer, Robert Goodin proposes a modest consequentialist theory of cosmopolitan distributive justice which does not oblige the sacrifice of people's own commitments. Goodin insists that we have an obligation to aid the vulnerable[82] and he further argues that this principle authorizes international aid.[83] He contends that this is required by justice,[84] even though he thinks it can be explained a lot better in terms of humanity.[85] The implication is that Goodin argues for a global application of his duty to aid the vulnerable and rebukes

[74] Singer, "Famine, Affluence and Morality", 232.

[75] Singer, "Famine, Affluence and Morality", 231–234.

[76] Singer, "Famine, Affluence and Morality", 235–237.

[77] John Arthur, "Rights and the Duty to Bring Aid" in *World Hunger and Morality* ed. W. Aiken and H. LaFollette 2nd Edition (Englewood Cliffs, NJ: Prentice Hall, 1996), 43–44.

[78] Deborah Zion, "HIV/AIDS Clinical Research and the Claims of Beneficence, Justice and Integrity", *Cambridge Quarterly of Healthcare Ethics* 13, no. 4 (October 2004): 407.

[79] Michael Slote, "The Morality of Wealth" in *World Hunger and Moral Obligation* ed. W. Aiken and H. LaFollette (Englewood Cliffs, NJ: Prentice Hall, 1997), 125–127.

[80] Zion, "HIV/AIDS Clinical Research and the Claims of Beneficence, Justice and Integrity", 407.

[81] Caney, *Justice Beyond Borders: A Global Political Theory*, 116–117.

[82] Robert E. Goodin, *Protecting the Vulnerable: A Reanalysis of our Social Responsibilities* (Chicago: University of Chicago Press, 1985), 33–144.

[83] Goodin, *Protecting the Vulnerable: A Reanalysis of our Social Responsibilities,* 161–169.

[84] Goodin, *Protecting the Vulnerable: A Reanalysis of our Social Responsibilities,* 159–161.

[85] Goodin, *Protecting the Vulnerable: A Reanalysis of our Social Responsibilities*, 161–164.

the present world order but does not endorse Singer's overdemandingness.[86] Goodin also acknowledges the significance of collective action in his discussion of foreign aid and world hunger.[87] He indicates that personal donations to schemes that focus on individuals, such as sponsor a child, do not account for massive restructuring required in needy communities. Therefore, Goodin argues that when contemplating aid to the severely impoverished, giving money is not adequate, but rather individuals must as well engage in political action to organize effective schemes.[88] Zion articulates, "the main advantages of collective action are, therefore, efficacy and an easing of the burden on individual donors, thus once again answering to some degree the "overdemandingness" objection.".[89]

Another exponent of consequentialist approach to cosmopolitanism is Martha Nussbaum who proposes an Aristotelian and modest version of consequentialism. Her version is termed Capabilities approach, which is outcome based. Nussbaum writes, "Some theories, such as Rawls, begin with the design of a fair procedure. My capabilities approach begins with outcomes: with a list of entitlements that have to be secured to citizens, if the society in question is a minimally just one."[90] She argues that we have a collective obligation to ensure adequate protection of human entitlements. Her emphasis is to promote capabilities that will enhance living a fulfilled and dignified life for every human person irrespective of one's nationality or citizenship. She writes, "Humanity is under a collective obligation to find ways of living and cooperating together so that all human beings have decent lives."[91] Caney noted that Nussbaum convincingly advocates for global principles of distributive justice that defends persons' capacity to flourish and lead fulfilling lives.[92] Nussbaum clearly articulates, "We insist that a fundamental part of the good of each and every human being will be to cooperate together for the fulfillment of human needs and the realization of fully human lives."[93] She describes fully human life as comprising of "adequate nutrition, education of the faculties, protection of bodily integrity, liberty for speech and religious self-expression and so forth."[94]

Nussbaum argues that justice requires that we have entitlements to a minimum of each of these basic goods.[95] Capabilities approach defends providing people with the necessary conditions to lead lives with human dignity. Consequentialist accounts of cosmopolitan justice fall short in providing a criterion for distributing the benefits and burdens of the society.[96] It is pertinent to note that an emphasis on capabilities

[86] Caney, *Justice Beyond Borders: A Global Political Theory*, 117.

[87] Goodin, *Protecting the Vulnerable: A Reanalysis of our Social Responsibilities*, 163.

[88] Goodin, *Protecting the Vulnerable: A Reanalysis of our Social Responsibilities*, 164.

[89] Zion, "HIV/AIDS Clinical Research and the Claims of Beneficence, Justice and Integrity", 407.

[90] Nussbaum, "Beyond the Social Contract: Capabilities and Global Justice", 13.

[91] Nussbaum, "Beyond the Social Contract: Capabilities and Global Justice", 13.

[92] Caney, *Justice Beyond Borders: A Global Political Theory*, 118.

[93] Nussbaum, "Beyond the Social Contract: Capabilities and Global Justice", 13.

[94] Nussbaum, "Beyond the Social Contract: Capabilities and Global Justice", 13.

[95] Nussbaum, "Beyond the Social Contract: Capabilities and Global Justice", 13.

[96] Caney, *Justice Beyond Borders: A Global Political Theory*, 118.

naturally evokes the notion of human rights approach to cosmopolitan justice which is discussed next.

Rights-Based Approach

The third approach to cosmopolitanism is rights-based approach to cosmopolitan justice. Some key proponents of rights-based approach to cosmopolitanism are Henry Shue, Charles Jones, David Held and Thomas Pogge. Shue[97] and Jones[98] defend the human right to subsistence. Shue argues that the right to subsistence is necessary for persons to enjoy other rights. As such it is a basic right.[99] He articulates that subsistence is one of several inherent necessities for the exercise of any right.[100] The implication is that for Shue to enjoy other civil and political rights you must first of all exercise the right to subsistence. He contends that the right to subsistence is logically a necessary component of other rights.[101] Shue argues that every human person has a basic right to minimum economic security or subsistence which encompasses "unpolluted air, unpolluted water, adequate food, adequate clothing, adequate shelter, and minimal preventive public health care."[102]

Jones proposed a similar version of right to subsistence. He argues that civic and political rights should protect important human interests. He reasoned that civil and political rights should not be indifferent to what people care about but should protect fundamental human interests.[103] These basic assumptions support the claim that persons have a right to subsistence.[104] An important human interest was identified as a person's interest in good health as well as preventing malnutrition, starvation, and disease.[105] Caney writes, "Any credible account of people's rights reflects what is important to persons—their fundamental interests."[106] Taking into consideration that subsistence is an important interest and taking into account this interpretation of rights, an inference is drawn that persons have a right to subsistence.[107]

Held proposed an expansive set of human rights in his discussion of global justice. He proposes seven types of rights that should be defended based on an ideal autonomy.[108] He identifies three types of rights which focus on economic entitle-

---

[97] Shue, *Basic Rights: Subsistence, Affluence, and U.S. Foreign Policy*, 22–27.

[98] Jones, *Global Justice: Defending Cosmopolitanism*, 50–84.

[99] Shue, *Basic Rights: Subsistence, Affluence, and U.S. Foreign Policy*, 18–20.

[100] Shue, *Basic Rights: Subsistence, Affluence, and U.S. Foreign Policy*, 26–27.

[101] Shue, *Basic Rights: Subsistence, Affluence, and U.S. Foreign Policy*, 27.

[102] Shue, *Basic Rights: Subsistence, Affluence, and U.S. Foreign Policy*, 23.

[103] Jones, *Global Justice: Defending Cosmopolitanism*, 56–57.

[104] Jones, *Global Justice: Defending Cosmopolitanism*, 59–61.

[105] Caney, *Justice Beyond Borders: A Global Political Theory*, 120.

[106] Caney, *Justice Beyond Borders: A Global Political Theory*, 120.

[107] Caney, *Justice Beyond Borders: A Global Political Theory*, 120.

[108] David Held, *Democracy and the Global Order: From the Modern State to Cosmopolitan Governance* (Cambridge: Polity Press, 1995), 192–194.

ments, including health rights, social rights and economic rights. Health rights encompass the right to good health and non-harmful environment.[109] Social rights entail the right to child support and education.[110] Economic rights involve the right to a minimum wage and the opportunity to be economically independent.[111]

Pogge also proposed a rights-based approach to global justice, which persuasively defends Articles 25 and 28 of the Universal Declarations of Human Rights.[112] Pogge defends an institutional conception of rights. He argues that a just world order is one that secures peoples enjoyment of their human rights, which includes economic rights. His defense of an institutional conception of rights disposes him to support Article 28 of the Universal Declaration of Human Rights which emphasizes institutional full realization of these rights.[113]

The pitfall of the rights-based approach to global justice is that it does not give account of the persons who have corresponding obligations to fulfill the right. Onora O'Neill forcefully articulates, "Rights are demands on others. Liberty rights demand that others not interfere with or obstruct the right-holder, rights to goods and services that others provide for the right-holder."[114] A global principle of distributive justice requires both an account of people's entitlements as well as an account of people who have obligations to provide them. A rights-based approach therefore needs to be complemented by a duty-based approach.

Duty-Based Approach

Some cosmopolitans favor a duty-based approach to global justice as against a rights-based approach.[115] O'Neill is one of the major proponents of duty- based approach to cosmopolitanism. In international politics, the focus on duties over rights is articulated in some countries' proposal for a "Universal Declaration of Human Responsibilities" to complement the "Universal Declaration of Human Rights."[116] Advocates of this proposal argue that a biased emphasis on rights has resulted in minimizing the fact that rights are complemented by corresponding duties, comprising economic and social ones.[117]

---

[109] Held, *Democracy and the Global Order: From the Modern State to Cosmopolitan Governance*, 192, 194–195.

[110] Held, *Democracy and the Global Order: From the Modern State to Cosmopolitan Governance*, 192, 195.

[111] Held, *Democracy and the Global Order: From the Modern State to Cosmopolitan Governance*, 193, 197–198.

[112] Pogge, "A Global Resources Dividend", 501.

[113] Pogge, "Cosmopolitanism and Sovereignty," 356–364.

[114] O'Neill, *Bounds of Justice*, 126.

[115] Tan, *Justice Without Borders: Cosmopolitanism, Nationalism and Patriotism*, 49.

[116] Tan, *Justice Without Borders: Cosmopolitanism, Nationalism and Patriotism*, 49.

[117] Tan, *Justice Without Borders: Cosmopolitanism, Nationalism and Patriotism*, 49.

There is a contention that a duty-based approach presents a different, and certainly superior, conceptual view on global justice.[118] In this context, Tan writes that O'Neill articulates that "rights theories are conceptually incomplete because while rights must have certain corresponding duties, not all rights correspond to an assigned duty-bearer and a clearly specified duty."[119] The implication is that every right has a corresponding obligation. Furthermore, a right is ineffective and unclaimable when there is no agent clearly assigned to bear the responsibility for the right.[120]

O'Neill defends the view that a right is effective only when a corresponding obligation is clearly specified and allocated. She argues, "When obligations are unallocated it is right that they should be met, but nobody can have an effective right –an enforceable, claimable, or waiveable right-to their being met. Such abstract rights are not effective entitlements."[121] She distinguishes between perfect obligations and imperfect obligations. Obligations are perfect when they are assigned and specified and therefore claimable and in theory enforceable.[122] Obligations are imperfect when no particular agent has been recognized, when there is substantial leeway on how an agent may fulfill the obligation, and when it is unstipulated for whom the act is to be performed.[123] O'Neill writes, "Imperfect obligations can be enforced only when they are institutionalized in ways that specify for whom the obligation is to be performed".[124]

A right without a corresponding obligation is an empty right. For example, if the claimants of a particular right such as right to food or development are unable to locate where to settle their claims, these are construed as empty manifesto rights.[125] O'Neill argues that manifesto rights are only claimed when corresponding obligations are specified and allocated. The enforcement of a right requires a corresponding duty which needs to be institutionalized. She pointed out a relationship between a meaningful right and an enforceable duty. She clearly stressed the need to institutionalize the duties of justice so as to allocate, stipulate, and enforce them.[126] She asserts that institution- building is required to specify and allocate obligations to the needy.[127]

O'Neill argues that the goal of such institutions is to regulate the actions of powerful investors who are given "excessive tax concessions" and to curb the vulnerability of poor nation states who frequently agree to terms of trade that are harmful.[128]

[118] Tan, *Justice Without Borders: Cosmopolitanism, Nationalism and Patriotism*, 49.

[119] Tan, *Justice Without Borders: Cosmopolitanism, Nationalism and Patriotism*, 50.

[120] Tan, *Justice Without Borders: Cosmopolitanism, Nationalism and Patriotism*, 50.

[121] O'Neill, *Bounds of Justice*, 126.

[122] Tan, *Justice Without Borders: Cosmopolitanism, Nationalism and Patriotism*, 50.

[123] Tan, *Justice Without Borders: Cosmopolitanism, Nationalism and Patriotism*, 50.

[124] Onora O'Neill, *Constructions of Reason: Exploration of Kant's Practical Philosophy* (New York: St Martin's Press, 1989), 191.

[125] O'Neill, *Bounds of Justice*, 126.

[126] O'Neill, *Constructions of Reason: Exploration of Kant's Practical Philosophy*, 199.

[127] O'Neill, *Bounds of Justice*, 136.

[128] O'Neill, *Bounds of Justice*, 140–141.

She favors establishing a global institutional structure which will bear the obligations of fulfilling economic rights. O'Neill clearly states "…without one or other determinate institutional structure, these supposed economic rights amount to rhetoric rather than entitlement."[129]

At the international level, no institution exists that has adequate power to coerce the states, societies and investors to agree to the Rawlsian "difference Principle" on global scale.[130] Therefore, the organizations that are focused on are networking institutions such as banks, corporations, NGO's, internet etc. which are frequently outside the bounds of the state and consequently escape the bounded justice of the state, particularly in the less developed state.[131] Jones-Pauly notes that "the solution is not to bring these institutions back into the state. Rather, the state has to deal with them as being within the bounds of a global, rather than national system of justice."[132] This entails negotiation between the bounded state and the "boundary-less"[133] non-state agents who are influencing the vulnerable in the global context.[134] While O'Neill argues for establishing of determinate institutional structure in order to fulfill economic rights of the vulnerable especially in developing countries, the current institutional arrangements and international practices are considered unjust so long as they significantly contribute to human rights violation of the poor.[135] Pogge in line with this thought argues that "… the existing medical-patent regime (trade-related aspects of intellectual property rights—TRIPS—as supplemented by bilateral agreements) is severely unjust—and its imposition a human-rights violation on account of the avoidable mortality and morbidity it foreseeably produces."[136] A review of current rules for incentivizing pharmaceutical research is discussed in the next part of this chapter.

[129] O'Neill, *Bounds of Justice*, 125.

[130] Christina Jones-Pauly, "Loosening the Bounds of Human Rights: Global Justice and the Theory of Justice," *Human Rights and Human Welfare* 1, no. 3 (July 2001): 19.

[131] Jones-Pauly, "Loosening the Bounds of Human Rights: Global Justice and the Theory of Justice," 19.

[132] Jones-Pauly, "Loosening the Bounds of Human Rights: Global Justice and the Theory of Justice," 19.

[133] O'Neill, *Bounds of Justice*, 174.

[134] Jones-Pauly, "Loosening the Bounds of Human Rights: Global Justice and the Theory of Justice," 19.

[135] Pogge, "The Health Impact Fund: How to Make New Medicines Accessible to All," 244.

[136] Pogge, "Human Rights and Global Health: A Research Program," 182.

5.2.2 *Incentivizing Pharmaceutical Research*

5.2.2.1 Global Patent System

The WTO contends that the crucial issue in authorizing patent protection for pharmaceutical products is to entrench a balance between two complementary public health goals, that of offering incentives for future inventions of new drugs, and that of guaranteeing affordable access to existing drugs.[137] But, unfortunately affordable access to essential drugs has not been realized for the global poor especially in developing countries. Data from WHO shows that in 2003, more than one-third of the population of the world continued to lack access to the drugs on the WHO Model List. The statistics was very bleak in the poorest parts of Africa and Asia that it increases to more than 50%.[138] Pogge attributes this morally troubling situation to a global institutional failure.[139]

Pogge calls attention to the huge challenge of the responsibility that wealthy countries' citizens might shoulder regarding the persistence of severe poverty and inequality in developing countries, and especially, on the correlation between their persistence and current decisions regarding the avenue for globalization.[140] Pogge articulates, "my focus is… on the present situation, on the radical inequality between the bottom half of humankind, suffering severe poverty, and those in the top seventh, whose per capital share of the global product is 180 times greater than theirs (at market exchange rates)."[141] He identified two ways of perceiving severe poverty as a moral challenge. First, is as a positive duty when we fail to accomplish our positive duty to assist persons in severe distress. Second, is as a negative duty when we fail to accomplish our stricter negative duty not to support injustice, not to promote or benefit from the unfair impoverishment of others.[142]

Pogge offers two reasons why the new global economic order is so cruel on the poor. First, is that the governments of the affluent nations have an overwhelming edge relative to bargaining power and expertise. Second, is that the representatives in international negotiations seek to advance the best interests of the people and corporations of their own country. The consideration of the needs of the global poor is excluded as part of the mandate of any of the influential parties to the negotiation. The cumulative result of such negotiations and agreements with vast power

[137] WHO and WTO, *WTO Agreements & Public Health: A Joint Study by the WHO and the WTO Secretariat* (Geneva, Switzerland: WHO and WTO, 2002), 39.

[138] WHO, *Medicines and the Idea of Essential Drugs (EDM)*, (Geneva, Switzerland: WHO, 2004), http://www.who.int/medicines/rationale.shtml. (Accesed May 16, 2014).

[139] Pogge, "Human Rights and Global Health: A Research Program," 182–209.

[140] Thomas Pogge, "Severe Poverty as a Violation of Negative Duties," *Ethics & International Affairs* 19, no.1 (March 2005): 55.

[141] Pogge, "Severe Poverty as a Violation of Negative Duties," 55.

[142] Pogge, "Eradicating Systematic Poverty: Brief for a Global Resources Dividend" 203.

differentials is obvious: a grossly unjust global order in which benefits flow largely to the affluent.[143]

The features of vast unjust global institutional order detrimental to the global poor are shown in the history of the debates regarding the design and interpretation of the TRIPS agreement, and in concerted attempts to strengthen intellectual property rights (IPRs) beyond TRIPS requirements. IPRs are further strengthened through TRIPS-plus provisions established in the current Free Trade Agreements (FTAs).[144] The distinct characteristic of many of the bilateral trade agreements currently negotiated by US is that they go beyond the multilateral standards required by the TRIPS agreement.[145] Some authors indicate that developing countries consented to the TRIPS agreement with predictable disadvantage on public health so as to obtain concessions in other aspects of economic relevance to them, such as the reducing of subsidies in agriculture in high-income countries. Economic considerations are essentially fundamental in countries in which poverty is extreme and prevalent, and critical to realizing subsistence rights of their citizens.[146]

Pogge offers rebuttal to the claim that consent to the WTO and consequently TRIPS is voluntary, for four different reasons. First is that the appeal to consent can surmount the charge of violation of rights, as long as the rights under consideration are alienable and, more precisely, can be waived by consent. However, in the context of the common notion of human rights, they cannot be thus waived.[147] Pogge argues that, "persons cannot waive their human rights to personal freedom, political participation, freedom of expression, or freedom from torture."[148] Second is that an appeal to consent obstructs the complaint of people who lack guaranteed access to the objects of some of their human rights only as long as they have themselves acceded to the government that continues their impoverishments. However, most people who are threatened by diseases or are severely deprived reside in countries that are not profoundly democratic, and consequently consent to the current global economic order by their despotic rules cannot be considered as consent by their citizens. A typical example was Nigeria's accession to the WTO in 1995 which was achieved by its ruthless military dictator Sani Abacha.[149] Third is that consent to an onerous global regime can be justified only if it was not provoked by the danger of even more burdens. Therefore, one's consent cannot validate one's enslavement

[143] Pogge, *World Poverty and Human Rights: Cosmopolitan Responsibilities and Reforms*, 26–27.

[144] Mira Johri et al., "Sharing the Benefits of Medical Innovation: Ensuring Fair Access to Essential Medicines," 2006, http://www.lawweb.usc.edu/centers/paccenter/assets/docs/Ehrenreich_Johri_2006_04_15.pdf. (Accessed March 20, 2014).

[145] Carsten Fink and Patrick Reichenmiller, *Tightening TRIPS: Intellectual Property Provisions of U.S. Free Trade Agreements* (Washington, DC: World Bank, 2005), 1–2.

[146] Mira Johri et al., "Sharing the Benefits of Medical Innovation: Ensuring Fair Access to Essential Medicines," 2006, http://www.lawweb.usc.edu/centers/paccenter/assets/docs/Ehrenreich_Johri_2006_04_15.pdf. (Accessed March 20, 2014).

[147] Pogge, "Human Rights and Global Health: A Research Program," 197–198.

[148] Pogge, "Human Rights and Global Health: A Research Program," 198.

[149] Pogge, "Human Rights and Global Health: A Research Program," 198.

when one's consent was one's only option to evade continued torture or, certainly, accidental drowning. Pogge argues that, "an appeal to consent thus blocks a complaint by the poor against the present global economic order only if, at the time of consenting, they had an alternative option that would have given them secure access to the objects of their human rights."[150] Fourth is that an appeal to consent cannot validate the severe deprivation of children who are considerably overrepresented among people experiencing serious poverty and represent about two-thirds of all deaths from causes associated with poverty estimated at about 34,000 daily.[151] Pogge argues that, "the claim that the present global economic order foreseeably and avoidably violates the human rights of children cannot be blocked by any conceivable appeal to consent."[152]

A vast body of literature indicates that IPRs as enforced in the TRIPS agreement and numerous US FTAs result in a number of ethical problems. The ethical problems highlighted by IPRs are most relevant when it involves socially valuable goods such as life-saving or essential drugs and genetically modified seeds that are granted Intellectual Property (IP) protection. The focus of the discussion in this chapter is on life-saving or essential drugs in order to explore the broader moral problems or issues precipitated by the enforcement of IPRs.[153]

Pogge presents a good synopsis of how innovation is currently incentivized within the context of TRIPS agreement and how this agreement might result in significant morally problematic results.[154] The advent of the AIDS crisis in developing countries especially in Africa has shown that the current TRIPS agreement set the critical needs of poor patients against the need of pharmaceutical companies to recover their research and development investments.[155] Producing new, safe and effective life-saving or essential drugs for the market is an exorbitant, time consuming and financially risky enterprise. This involves undertaking long clinical trials for the research and development of new drugs as well as lengthy testing and approval process. Furthermore, newly developed drugs regularly end up to be unsafe or not adequately effective, to have severe side effects, or to be unsuccessful with obtaining government approval for a specific reason, which may result in the loss of the whole investment.[156] Taken into account that pharmaceutical companies must shoulder all the costs associated with development process, it is no astonishment that such companies are hesitant to carry out research and development (R&D) of new

[150] Pogge, "Human Rights and Global Health: A Research Program," 198.

[151] Pogge, "Human Rights and Global Health: A Research Program," 199.

[152] Pogge, "Human Rights and Global Health: A Research Program," 199.

[153] Jorn Sonderholm, *Intellectual Property Rights and the TRIPS Agreement: An Overview of Ethical Problems and Some Proposed Solutions* (Washington, DC: World Bank and Human Development Network, 2010), 2–3.

[154] Pogge, "Human Rights and Global Health: A Research Program," 184.

[155] David Barnard, "In the High Court of South Africa, Case No. 4138/98: The Global Politics of Access to Low-Cost AIDS Drugs in Poor Countries," *Kennedy Institute of Ethics Journal* 12, no. 2 (June 2002): 159–174.

[156] Pogge, "Human Rights and Global Health: A Research Program," 185.

drugs unless there are clear indications of positive financial prospects of doing so. However, such positive financial prospects cannot be realized without strict enforcement of IPRs on pharmaceutical innovations.[157]

The reason for this situation is that whenever an inventor firm brings an innovation to the market, other companies would copy the innovation usually through reverse engineering, and considering that these other companies did not incur any costs relative to R&D, they would be able to charge a price for the product that is considerably lower than the one charged by the inventor firm. As a result, the market price for the product would be driven down to simply above marginal costs of production, and the inventor firm would not be able to recover its R&D costs. A macroeconomic arrangement for the buying and selling of drugs that does not grant innovators IPRs to their innovations is thus probably to result in a market failure of undersupply of pharmaceutical innovations.[158]

The solution to a market failure of undersupply of pharmaceutical innovations was enshrined in the TRIPS agreement established under the auspices of WTO in the Uruguay Round. The TRIPS agreement grants patent protection usually for 20 years to inventor firms on their inventions from the period of filing a patent application to protect them from free riding and to encourage medical innovation.[159] Furthermore, strengthening of IPRs has been continued by US through a series of bilateral FTAs that encompass additional TRIPS-plus provisions. These TRIPS-plus measures authorize patent holders to extend, or evergreen, their monopolies and they as well suppress, obstruct, and delay the production of generic drugs in various ways: through enforcement of data exclusivity, and through limitations on and political pressure against the effective utilization of compulsory licenses.[160] IPRs are construed as a socio-economic tool that establish a temporary monopoly for inventor firms and authorize such firms to charge prices for their innovations that are considerably higher than the marginal cost of production of the innovations. This enables the inventor firms to recoup their R&D costs and obtain a profit on their innovations. Therefore, in terms of increasing the financial appeal of participating in the process of producing pharmaceutical innovations, IPRs are frequently influential in rectifying the market failure of undersupply of pharmaceutical innovations.[161]

On the other hand, the introduction of IPRs for pharmaceutical innovations frequently results in another market failure that involves excluding several mutually beneficial transactions between seller and buyer. The reasonably high price of an IP protected drug drives out particular potential buyers out of the market: specifically those buyers who are able and willing to buy the product at a price fairly above its

[157] Sonderholm, *Intellectual Property Rights and the TRIPS Agreement: An Overview of Ethical Problems and Some Proposed Solutions*, 3.

[158] Pogge, "Human Rights and Global Health: A Research Program," 186.

[159] Pogge, "Human Rights and Global Health: A Research Program," 186.

[160] Thomas Pogge, "Access to Medicines," *Public Health Ethics* 1, no. 2 (July 2008): 75.

[161] Sonderholm, *Intellectual Property Rights and the TRIPS Agreement: An Overview of Ethical Problems and Some Proposed Solutions*, 4.

marginal costs of production but cannot afford to pay the profit-maximizing sale price that obtains during the period in which the product is patented.[162] This scenario is dubbed deadweight losses in economic theory, which describe the type of losses that take place when someone is able and willing to pay more fairly above the marginal cost of production for a product but unwilling or unable to pay the patent price for it.[163] The characteristic of IPRs that they drive out particular potential buyers from the market establishes what might be dubbed the "exclusion problem" or "access problem."[164] Pogge contends that the exclusion or access problem is morally disturbing especially when a group of people usually the global poor are excluded from life-saving or essential drugs and not simply from computer software, music CDs or movie discs.[165]

In the advent of the TRIPS agreement which enforces strong IPRs on all product types, the exclusion or access problem is not the only outcome. A distinct problem that also emerges is "availability problem."[166] This problem is successfully established in the context of R&D of drugs for diseases that are prevalent among people in low-income countries. Diseases which primarily result in suffering and death in low-income countries comprise malaria, leishmaniasis and Chagas' disease.[167] R&D of drugs for diseases that are prevalent among people in low-income countries is very restricted. Available data shows that less than 1% of the 1223 new drugs introduced to the international market between 1975 and 1997 were designed precisely for tropical communicable diseases.[168]

The principal reason for the prevalence of this situation among the global poor especially in developing countries is that many poor people basically do not have adequate money to pay for drugs for their sicknesses. Based on this information, for-profit pharmaceutical companies then have limited or no incentive for investing resources into the R&D of drugs for these diseases usually referred to as "neglected diseases."[169]

The availability problem stems from the fact that the incentivizing method for innovation instituted by IPRs establishes a correlation between the incentive to innovate and the price of the innovative product. In the context of TRIPS agreement,

[162] Pogge, "Human Rights and Global Health: A Research Program," 186.

[163] Michael Ravvin, "Incentivizing Access and Innovation for Essential Medicines: A Survey of the Problem and Proposed Solutions," *Public Health Ethics* 1, no. 2 (2008): 112.

[164] Ravvin, "Incentivizing Access and Innovation for Essential Medicines: A Survey of the Problem and Proposed Solutions," 116.

[165] Pogge, "Human Rights and Global Health: A Research Program," 187.

[166] Michael J. Selgelid, "A Full-Pull Program for the Provision of Pharmaceuticals: Practical Issues," *Public Health Ethics* 1, no. 2 (July 2008): 134.

[167] David B. Ridley, Henry G. Grabowski and Jeffrey L. Moe, "Developing Drugs for Developing Countries," *Health Affairs* 25, no. 2 (March 2006): 316.

[168] Patrice Trouiller et al., "Drugs for Neglected Diseases: A Failure of the Market and a Public Health Failure?", *Tropical Medicine and International Health* 6, no. 11 (November 2001): 945–951.

[169] Sonderholm, *Intellectual Property Rights and the TRIPS Agreement: An Overview of Ethical Problems and Some Proposed Solutions*, 5.

profits accrue entirely from sales, so that the greater a price a product can be marketed, the greater is the incentive to invest resources into the R&D of the product.[170] Sonderholm articulates that, "the TRIPS agreement with its strong protection of IPRs is therefore not an agreement that is conducive to the investment in R&D of products that are socially valuable to predominantly poor populations or populations that are small."[171] Socially valuable goods comprising life-saving or essential drugs are readily available abundantly for the global rich far more than they are available for the global poor.[172]

Pogge argues that the TRIPS agreement has radically limited the access to inexpensive generic copies of advanced drugs to the global poor. The lack of generic competition multiplies the prices of advanced drugs frequently 10–15 times and so effectively excludes the global poor.[173] Pogge further identified and extensively discussed the seven shortcomings of the current pharmaceutical innovation regime which is the next task of this chapter.

5.2.2.2 Disadvantages of the Present Pharmaceutical Innovation Regime

Pogge insists that the quest for a systematic solution to pharmaceutical innovation regime can begin from an analysis of the key disadvantages of the current globalized monopoly patent regime.[174] He identified seven failings of the current pharmaceutical innovation regime, including high prices, neglect of diseases concentrated among the poor, bias toward maintenance drugs, wastefulness, counterfeiting, excessive marketing and the last-mile problem.[175] A brief discussion of the seven failings of the current pharmaceutical innovation regime follows.

High Prices

A patented drug is sold close to the profit-maximizing monopoly price which is essentially determined by the demand curve of the rich. In situations where many rich or well insured people certainly need a drug, the tendency has been to significantly raise the price of the drug far above the cost of production. Pogge acknowledges that mark-ups exceed 1000% for the most part when dugs are still under

---

[170] Sonderholm, *Intellectual Property Rights and the TRIPS Agreement: An Overview of Ethical Problems and Some Proposed Solutions*, 6.

[171] Sonderholm, *Intellectual Property Rights and the TRIPS Agreement: An Overview of Ethical Problems and Some Proposed Solutions*, 6.

[172] Sonderholm, *Intellectual Property Rights and the TRIPS Agreement: An Overview of Ethical Problems and Some Proposed Solutions*, 6.

[173] Pogge, *Politics as Usual: What Lies Behind The Pro-Poor Rhetoric*, 20–21.

[174] Pogge, "The Health Impact Fund: How to Make New Medicines Accessible to All," 246.

[175] Pogge, "The Health Impact Fund: How to Make New Medicines Accessible to All," 246–247.

patent.[176] For example, Sanofi-Aventis sold its cardiovascular disease drug in Thailand for $2.20 each pill, which is about 6000% above the price the Indian generic company Emcure agreed to sale the same drug.[177] The implication is that the exorbitant prices for drugs result in significantly limiting access to just a few of the poor who can receive aid from others.[178]

Neglect of Diseases Concentrated Among the Poor

Rewarding innovators with patent-protected mark-ups results in lack of appeal and focus by pharmaceutical companies on diseases concentrated among the poor for pharmaceutical research regardless of the severity and prevalence of these diseases. This is obvious, since the demand for such a drug drops off sharply as the patent holder broadens the mark-up. The implication is that there is likely no prospect for realizing huge sales volume and a large mark-up. More so, the potential risk of driving down the price of a successful new drug to the marginal cost of production or even free of charge which results in a big loss of the innovator's investment was acknowledged as detrimental. In virtue of these concerns, pharmaceutical companies certainly target drugs for the affluent for pharmaceutical research considerably more than those of the poor.[179] The problem of neglected diseases is as well recognized as the 10/90 gap, indicating that only 10% of all pharmaceutical research is being concentrated on diseases that represent 90% of the global burden of disease.[180]

Bias Towards Maintenance Drugs

Drugs are classified into three categories, including curative, maintenance and preventive drugs. Curative drugs deal with getting rid of the diseases from the patient's body. Maintenance drugs focus on improving well-being and functioning but without eliminating the disease. Preventive drugs focus on decreasing the probability of becoming infected with the disease from the onset. The maintenance drugs are considerably the most lucrative under the current patent regime, since patients would continue to buy the drugs without being cured and do not die until after the expiration of the patent. The pharmaceutical companies gain a huge profit from such patients more than they would if they drew the same health benefit from a cure or

[176] Pogge, "The Health Impact Fund: How to Make New Medicines Accessible to All," 246.

[177] Oxfam International, "Investing for Life: Meeting Poor People's Needs for Access to Medicines through Responsible Business Practices," Briefing Paper no. 109 (November 2007): 20, http://www.oxfam.org/en/policy/bp109_investing_for_life_0711 (Accessed May 21, 2014).

[178] Pogge, "The Health Impact Fund: How to Make New Medicines Accessible to All," 246.

[179] Pogge, "The Health Impact Fund: How to Make New Medicines Accessible to All," 246.

[180] Global Forum for Health Research, *The 10/90 Report on Health Research 2003-2004* (Geneva: GFHR, 2004), http://www.globalforumhealth.org (Accessed May 21, 2014).

vaccine. Vaccines are far less profitable because they are usually bought in big quantities at discounted prices by governments.[181] Pogge argues that current regime directs pharmaceutical research in the wrong direction to the disadvantage of both the poor and the rich.[182]

Wastefulness

Within the current regime, innovators must shoulder the patents' filing cost in several national jurisdictions as well as the cost of checking these jurisdictions for likely breaches of their patents. Enormous amount of money are also expended on expensive litigation in many jurisdictions against patent holders who want to extend and prolong their patent-protected mark ups.[183] A more significant loss is incurred from the deadweight loss which stems from obstructed sales to buyers who are willing and able to pay some price between the marginal cost of production and the greater monopoly price.[184]

Counterfeiting

Big mark-ups promote the illegal production of counterfeit products that are watered down, contaminated, inactive or even lethal. Such fake products frequently jeopardize patient health. They as well significantly contribute to the emergence of drug-specific resistance, when patients consume very limited active ingredient of a watered down drug to exterminate the more resilient pathogenic agents. For example, the emergence of greatly resistant disease strains of tuberculosis presents risks to all of us.[185]

Excessive Marketing

Pharmaceutical companies usually make frantic efforts to increase their sales volume through frightening patients or by recompensing doctors especially when they keep up with very huge mark-ups. This result in fighting over market share among similar "me-too" drugs as well as incentives that persuades doctors to prescribe drugs even when they are contra-indicated or when competing drugs are expected to do better. Given big mark-ups, it is profitable to sponsor huge direct-to-consumer

[181] Pogge, "The Health Impact Fund: How to Make New Medicines Accessible to All," 247.

[182] Pogge, "The Health Impact Fund: How to Make New Medicines Accessible to All," 247.

[183] Pogge, "The Health Impact Fund: How to Make New Medicines Accessible to All," 247.

[184] Aidan Hollis, *An Efficient Reward System for Pharmaceutical Innovation*, Working Paper (Calgary: Department of Economics, University of Calgary, 2005), 8. http://www.econ.ucalgary.ca/fac-files/ah/drugprizes.pdf (Accessed May 22, 2014).

[185] Pogge, "The Health Impact Fund: How to Make New Medicines Accessible to All," 247.

advertising that induces people to take drugs they don't actually need for diseases they don't actually have, usually referred to as invented pseudo diseases.[186]

The Last-Mile Problem

Whereas the current regime offers strong incentives to sell unwanted patented drugs to people who can pay or possess insurance, it offers no incentives to guarantee that poor people benefit from drugs they need immediately. This problem is exacerbated in poor countries, which frequently do not have the infrastructure to dispense drugs as well as health care professionals to prescribe them and to guarantee their appropriate utilization. There is an understanding that the current regime offers incentives to pharmaceutical companies to discount the medical needs of the poor. A pharmaceutical company that assists poor patients to benefit from its drug under patent undercuts its own lucrativeness in three different ways: by paying for the attempt to make the drug available to them in a proficient way, by curbing a disease that its profits hinge on, and by losing rich customers who discover means of buying from inexpensive drugs intended for the poor.[187] The problems created by the current patent system led Pogge to propose new rules for reforming and incentivizing pharmaceutical research.

5.3 Global Public Good and Systematic Support

5.3.1 Reforming and Incentivizing Pharmaceutical Research

Pogge proposes two basic reform strategies for dealing with monopoly pricing issues of the current patent system, including the differential pricing strategy and the public good strategy.[188] A brief analysis of the two strategies follows.

5.3.1.1 The Differential Pricing Strategy

The differential pricing strategy usually comes in three different forms. The first form entails going back to the era before the TRIPS agreement, when IPRs, that is, patent monopolies for advanced drugs were granted and implemented in rich countries, but not in most of the poorer countries. The second form involves that inventor firms offer different prices of their patented drugs to different customers such as

[186] Pogge, "The Health Impact Fund: How to Make New Medicines Accessible to All," 247.

[187] Pogge, "The Health Impact Fund: How to Make New Medicines Accessible to All," 247.

[188] Thomas Pogge, "Pharmaceutical Innovation: Must We Exclude the Poor?" in *World Poverty and Human Rights: Cosmopolitan Responsibilities and Reforms*, Second Edition (Cambridge: Polity Press, 2008), 138.

affluent buyers and poor buyers. In this way, the firms will realize a big profit margin from sales to the more affluent customers, without giving up sales to poorer buyers at a lower margin.[189] Theoretically, pricing the product this way enables the inventor firms to obtain the best of two worlds. The firms would secure high profits on the products in markets with a high buying power without forfeiting the medium to low profits that originate from selling the product in markets with a reasonably low buying power. Furthermore, the significantly reduced price of the product in low-income countries implies that the populations of these countries would have an easier access to the product than they would have if the product was priced at high-income countries' level.[190] In this context, access problem which has been regarded as morally problematic especially relative to life-saving drugs may have been fairly alleviated, because this feature of differential pricing makes the strategy "a prima facie attractive pricing scheme for life-saving medicines."[191]

The third form of differential pricing strategy involves the rights conferred on governments as acknowledged under TRIPS rules, to issue compulsory licensing for drugs that are urgently needed in public health emergency situations.[192] For example, with the advent of the devastating impact of HIV/AIDS pandemic in Sub-Saharan Africa which has been considered a public health emergency for several countries in that region, the governments of these affected countries might authorize the manufacture and marketing of cheaper generic copies of patented HIV/AIDS drugs on the condition that the authorized generic firms pay a little license fee to the patent holders. The market entry of pharmaceutical companies manufacturing generic copies of HIV/AIDS drugs will most probably drive down the price of these drugs to simply above their marginal cost of production, and this will invariably improve access to the drugs.[193] The US has always recognized this right under 28 USC 1498, especially for cases where the licensed manufacturer is either an agency or contractor affiliated with the government, but has also been unwilling to invoke the right in the context of life-saving drugs, apparently to refrain from setting an international example disadvantageous to its pharmaceutical industry.[194] Poor countries have been encouraged in the wake of AIDS pandemic to invoke their rights to compulsory licensing in order to deal with their public health crises, but the pressure from the US and other affluent countries has barred most of them from not utilizing this alternative.[195]

[189] Pogge, "Pharmaceutical Innovation: Must We Exclude the Poor?" 238.

[190] Sonderholm, *Intellectual Property Rights and the TRIPS Agreement: An Overview of Ethical Problems and Some Proposed Solutions*, 6.

[191] Sonderholm, *Intellectual Property Rights and the TRIPS Agreement: An Overview of Ethical Problems and Some Proposed Solutions*, 6–7.

[192] Pogge, "Pharmaceutical Innovation: Must We Exclude the Poor?" 238–239.

[193] Sonderholm, *Intellectual Property Rights and the TRIPS Agreement: An Overview of Ethical Problems and Some Proposed Solutions*, 7.

[194] Pogge, "Pharmaceutical Innovation: Must We Exclude the Poor?" 239.

[195] Pogge, "Pharmaceutical Innovation: Must We Exclude the Poor?" 239.

Implementing differential pricing and compulsory licensing in the real world to deal with the problem of access to essential drugs has several limitations. Ravvin offers an overview of some of the problems that relate to these solutions to access problem. With regard to differential pricing, the principal concern is that of leakage of inexpensively sold drugs from poor countries to affluent ones through parallel trade and smuggling.[196] This view is as well stressed by other authors.[197] Pogge also discusses the risk of diversion and parallel trade.[198] He further indicates that cognizant of this risk that patent holders usually don't make efforts to defeat the second market failure through differential pricing, rather they refuse to give in to pressures to do so, and wrestle efforts to impose compulsory licensing upon them.[199] He contends that consequently, differential pricing has not gained traction and several poor buyers who would be willing and able to buy the drug at a price reasonably above the marginal cost of production are excluded from this drug because they are not willing and able to pay the much higher monopoly price.[200] Pogge argues that differential pricing strategy cannot stop the neglect of diseases that are prevalent among the poor. Differential pricing can assist in improving access to a drug at competitive market prices for the poor but only on the condition that this drug is available. However, this drug will only be available if there is adequate market demand for it, in order to make investment in its development attractive and profitable.[201] Pogge stresses that, "nearly all diseases and research avenues neglected under the current regime would continue to be neglected under a differential pricing regime."[202]

More so, an issue of social justice was also highlighted, because, the affluent people in low-income countries will have access to a specified drug at a reasonably low cost, while poor people in high-income countries will have to pay a high price for exactly the same drug. Love and Hubbard indicate that 50 million customers in India have comparable incomes to that of Europeans, and to some people it is contentious that this section of people would have access to a specified drug at a low price while uninsured, poor people in, for instance in the US would have to pay a high price for exactly the same drug.[203]

Compulsory licensing also has several limitations. First, the WTO initially just authorized national governments to issue compulsory licenses to generic producers that would manufacture products exclusively for domestic consumption. It was immediately acknowledged that this entailed that compulsory licensing cannot be

---

[196] Ravvin, "Incentivizing Access and Innovation for Essential Medicines: A Survey of the Problem and Proposed Solutions," 114.

[197] James Love and Tim Hubbard, "The Big Idea: Prizes to Stimulate R&D for New Medicines," *Chicago-Kent Law Review* 82, no. 3 (November 2007): 1519–1554.

[198] Pogge, "Pharmaceutical Innovation: Must We Exclude the Poor?" 239.

[199] Pogge, "Pharmaceutical Innovation: Must We Exclude the Poor?" 240.

[200] Pogge, "Pharmaceutical Innovation: Must We Exclude the Poor?" 240.

[201] Pogge, "Pharmaceutical Innovation: Must We Exclude the Poor?" 240.

[202] Pogge, "Pharmaceutical Innovation: Must We Exclude the Poor?" 240.

[203] Love and Hubbard, "The Big Idea: Prizes to Stimulate R&D for New Medicines," 1525.

utilized by countries that had no domestic capacity to manufacture generic drugs.[204] Most low-income countries except Brazil, India and China do not have such capacity. In 2003, the WTO General Council developed a decision that grapples with the export of pharmaceutical drugs to countries that lack domestic manufacturing capacity. In this context, countries with a domestic drug manufacturing capacity were authorized to issue a compulsory license to a domestic manufacturer which would in that case be legally allowed to export the specified drug to a low-income country that urgently need the drug to deal with a national health emergency. Some literature on compulsory licensing indicated that the 2003 WTO amendment to TRIPS has been a debacle because the judicial process involved in obtaining an export license is complicated and pervaded with practical obstacles and red tape.[205] Rimmer acknowledging this fact and based on a study of 2007 Rwanda's effort to import HIV/AIDS drugs under the WTO General Council Decision 2003 and the resulting process involving Apotex, a Canadian generic drug manufacturer, writes that it is objectionable to codify the WTO General Council Decision 2003, because it has been unsuccessful to offer a swift, efficient, and cost-effective distribution of essential medicines.[206]

Second, compulsory licensing has social costs that may counteract the short term benefits it engenders relative to improving access to life-saving medicines.[207] Paramount among these social costs include: (1) a risk of reduced direct investment in countries that turn to compulsory licensing because proprietors of patented products will look up more business-friendly legal environments, (2) a risk that the pharmaceutical company which gets a compulsory license will shadow price the original price of the patented product and so engender dead weight loss of its own in search of profits, (3) a risk that compulsory licensing will decrease the incentives of the research-driven pharmaceutical sector to innovate, (4) a risk that the governments of countries that accommodate pharmaceutical companies whose products have been bound by a compulsory license by a foreign government will hit back with trade sanctions that could gravely hurt the economy of the nation that has issued the compulsory license.[208] Bird's third point was also reiterated by Pogge as the likely long-term drawback of compulsory licensing. Pogge argues that if compulsory licenses are extensively utilized, then pharmaceutical companies are probably to be discouraged from investing in R&D of drugs that are likely to be bound by compulsory

---

[204] Jerome H. Reichman, "Compulsory Licensing of Patented Pharmaceutical Inventions: Evaluating the Options," *The Journal of Law, Medicine & Ethics* 37, no. 2 (Summer 2009): 248.

[205] Josephine Johnston and Angela A. Wasunna, "Patents, Biomedical Research, and Treatments: Examining Concerns, Canvassing Solutions," *The Hastings Center Report* 37, no. 1 (January-February 2007): 18.

[206] Matthew Rimmer, "Race Against Time: The Export of Essential Medicines to Rwanda," *Public Health Ethics* 1, no. 2 (August 2008): 89.

[207] Robert C. Bird, "Developing Nations and the Compulsory License: Maximizing Access to Essential Medicines while Minimizing Investment Side Effects," *Journal of Law, Medicine & Ethics* 37, no. 2 (Summer 2009): 209.

[208] Bird, "Developing Nations and the Compulsory License: Maximizing Access to Essential Medicines while Minimizing Investment Side Effects," 209–221.

licensing. For-profit pharmaceutical companies are as a result likely to avoid this type of R&D completely. From this perspective, compulsory licensing will be tantamount to a further obstacle to R&D of drugs for diseases that predominantly exist in developing countries.[209] Since differential pricing and compulsory licensing could not provide an attractive and effective means for easing the access problem engendered by IPRs, Pogge proposed the public good strategy as a more effective and promising strategy for improving access to essential drugs especially for the global poor in developing countries.

5.3.1.2 The Public Good Strategy

Pogge contends that the public good strategy is more promising to yield a reform plan that would circumvent the major failings of the current monopoly patent regime while simultaneously retaining most of its significant benefits.[210] The public good strategy has three critical components comprising "open access, alternative incentives, and funding."[211] A brief discussion of the three components follows.

The first component focuses on providing successful and approved new drug as a public good that all pharmaceutical companies may use free of charge. This component of the reform plan will drastically reduce the exclusion or access problem created by monopoly pricing issue. Since the new essential drugs can be freely reproduced by all pharmaceutical companies and launched in the market, the price of such drugs will most probably drop to a level simply above their marginal cost of production. Pogge argues that if this component known as open access is implemented in isolation, all economic incentive to attempt to develop new essential drugs will be destroyed.[212] Such an unpleasant situation is, nevertheless, circumvented implementing the second component of the reform plan which entails offering some alternative reward to inventors. This involves the notion that the inventor firms should be qualified to take out a multiyear patent on any essential drug they invent, and during the life of the patent, the companies should be rewarded from a centralized public fund in proportion to the impact of their invention on the global disease burden.[213]

Pogge contends that the second component of the reform plan can be identified clearly and definitely in distinct ways. These ways can be generally classified as "push" and "pull" programs. A push program chooses and finances some particular innovator such as a pharmaceutical company, maybe, or a university or a national health agency to embark on a specific research endeavor. The implication here is that, given sufficient funding, the chosen innovator will develop the required innovation, which can subsequently be made available freely for production by rival

[209] Pogge, "Pharmaceutical Innovation: Must We Exclude the Poor?" 240.
[210] Pogge, "Pharmaceutical Innovation: Must We Exclude the Poor?" 240.
[211] Pogge, "Pharmaceutical Innovation: Must We Exclude the Poor?" 241.
[212] Pogge, "Pharmaceutical Innovation: Must We Exclude the Poor?" 241.
[213] Pogge, "Human Rights and Global Health: A Research Program," 188–189.

pharmaceutical manufacturers in order to guarantee broad availability at competitive market price.[214]

On the other hand, a pull program is addressed to all potential innovators, hopeful to reward whoever is the first to accomplish a valued innovation. Pull programs possess two interconnected advantages over push programs: first, they never fund failed research endeavors and second, they produce strong financial incentives for innovators to toil towards early success. The reverse of these advantages is that, in order to evoke such a considerable research attempt, the reward must be adequately big to recompense for the risk of failure. The risk is bifold, as a research endeavor may fail either because the required drug proves evasive or because some rival innovator gets there first. Potential innovators have incentives to attempt to develop a new drug simply if the reward for success, disregarded by the likelihood of failure, is considerably greater than the anticipated cost of the research and development endeavor. From these perspectives, a pull program is akin to the current regime.[215]

Pogge contends that pull programs can be more effective than push programs because of three reasons. Push programs are more probably to fail because they obtain just one rather than several rival innovators to work on a problem. Push programs are more probably to fail because the innovator is selected on the grounds of the confidence of some outsiders in it while in pull programs the decision of each innovator to attempt is based on its own, more capable and better stimulated evaluation of its capacities. Push programs are more probably to fail because the selected innovator has much weaker incentives to toil and cost-effectively toward early success. The drawback that push programs are more probably to fail is exacerbated in reality that such failures are fully paid for, contrary to pull programs, which pay nothing for unsuccessful efforts. This reality has propensity to make push programs harder to maintain politically.[216]

Pogge further indicated that the second component of the reform plan has a number of attractive consequences. First, it will engender a strong incentive for any inventor firm to (1) sell its innovative drug inexpensively, frequently below the marginal cost of production, and (2) authorize, and even promote, other pharmaceutical companies to copy the drug.[217] Taking these steps, an inventor firm guarantees that its innovative drug would be accessible to a greater number of people in the low-income bracket, and as a result of this, the drug will have a greater impact on the global disease burden. Second, the component will establish a condition in which an inventor firm has incentives to ensure that patients are completely instructed in the proper use of its drug, including dosage and compliance. The implication is that ensuring that its product is utilized properly assists an inventor firm to circumvent the adverse situation in which its product is widely utilized but fails to make a considerable impact on the global disease burden. Third, the component will establish a condition in which an inventor firm has incentives to work hard to enhance the

[214] Pogge, "Pharmaceutical Innovation: Must We Exclude the Poor?" 241.

[215] Pogge, "Pharmaceutical Innovation: Must We Exclude the Poor?" 241–242.

[216] Pogge, "Pharmaceutical Innovation: Must We Exclude the Poor?" 242.

[217] Pogge, "Human Rights and Global Health: A Research Program," 189.

health systems of the low-income countries as well as to generate a lucrative market for pharmaceutical companies. Strong economic incentives would be established for pharmaceutical companies to attempt to develop drugs for neglected diseases such as malaria, tuberculosis and pneumonia. Taken into account that these diseases predominantly affect the global poor in the most grisly of ways, an effective drug for any of these diseases would possess an enormous impact on the global disease burden. An inventor firm that could manufacture an effective and safe drug for any of these diseases would as a result be the recipient of a reward of significant proportions.[218] This characteristic of the Health Impact Fund (HIF) is likely to ease the availability problem that exists under the current IPR driven TRIPS regime.[219]

A major task related to the second reform component involves establishing a set of principles that can guide the reward process. Pogge argues that "when two or more different medicines are alternative treatments for the same disease, then the reward corresponding to their aggregate impact must be allocated among their respective inventors on the basis of each medicine's market share and effectiveness."[220] However, Pogge recognized that things get more complicated when an essential drug is not a single product but a drug cocktail that combines a variety of drugs that have been developed and manufactured by different pharmaceutical companies.[221]

The third component of Pogge's reform plan involves developing a fair, feasible, and politically realistic allocation of the costs associated with the second component.[222] Pogge and Hollis contend that effective implementation of the reform entails that much of its cost be shouldered by high-income countries. A reasonable estimate for minimum funding level for the reform plan is about US$ 6 billion which approximately amounts to 0.01% of global income.[223] To make this increased spending for the reform plan realistic, it is pertinent to offer convincing reasons for supporting it to taxpayers and politicians of the high-income countries. Pogge opines that his plan can be substantiated by prudential considerations.[224]

First, the new incentivizing regime will result in considerably lower prices for essential drugs for consumers in high-income countries. Under the current free-market regime, consumers in these countries pay high prices for essential drugs either directly or through contributions to commercial insurance companies.[225] Second, giving the poor citizens of low-income countries no charge on the pharmaceutical research carried out for the advantage of citizens in the rich countries, the latter citizens are establishing goodwill toward themselves in the developing world

---

218 Pogge, "Human Rights and Global Health: A Research Program," 189–191.

219 Sonderholm, *Intellectual Property Rights and the TRIPS Agreement: An Overview of Ethical Problems and Some Proposed Solutions*, 26.

220 Pogge, "Human Rights and Global Health: A Research Program," 191.

221 Pogge, "Human Rights and Global Health: A Research Program," 192.

222 Pogge, "Human Rights and Global Health: A Research Program," 191.

223 Aidan Hollis and Thomas Pogge, *The Health Impact Fund: Making New Medicines Accessible for All* (New Haven, CT: Incentives for Global Health, 2008), 44.

224 Pogge, "Human Rights and Global Health: A Research Program," 192.

225 Pogge, "Human Rights and Global Health: A Research Program," 192.

by showing in a concrete way their concern for the terrible public-health problems these populations are encountering.[226] Third, the reform plan will create excellent medical-research jobs in high-income countries. Fourth, it will allow these countries to respond more effectively to public-health crises and problems in the future by securing them more briskly increasing medical knowledge combined with a stronger and more diverse arsenal of medical interventions.[227] Pogge proposes a global institution known as the Health Impact Fund (HIF) for the effective implementation of his new plan for reforming and incentivizing pharmaceutical research and innovation.

5.3.2 The Health Impact Fund

5.3.2.1 Meaning of the Health Impact Fund

Pogge describes the HIF as "a proposed pay-for-performance mechanism that would offer innovators the option—no obligation—to register any new medicine or, under certain conditions, also a traditional medicine or a new use of an existing medicine."[228] The HIF is a proposed global agency that would be primarily financed by governments of various countries. The innovator of a product would register the product at the time of marketing approval, which entails accepting responsibility to make it available, during its first 10 years on the market, wherever it is needed at the lowest possible cost of production and distribution. The innovator would additionally commit to authorizing, free of charge, generic manufacture and distribution of the product at the end of the 10-year reward period, especially if the innovator's patent has not expired on the product. In return, the registrant would receive, during those 10 years, yearly reward payments based on the assessed health impact of its product. Each reward payment would be a piece of a huge yearly pay-out with every registered product getting a share equal to its share of the assessed health impact of all HIF-registered products in a given year. If the HIF would function effectively, its annual reward pools could go up to draw an increasing share of new drugs.[229]

The HIF can be viewed as maintaining a lasting competition among innovators that spans over all countries and all diseases, with earnings linked to impact on

[226] Pogge, "Human Rights and Global Health: A Research Program," 193.

[227] Pogge, "Human Rights and Global Health: A Research Program," 193.

[228] Thomas Pogge, *The Health Impact Fund: More Justice and Efficiency in Global Health*, Development Policy Center Discussion Paper 7, Crawford School of Economic and Government (Cranberra: The Australian National University, 2011), http://devpolicy.anu.edu.au/pdf/papers/DP7_The_Health_Impact_Fund_More_justice_and_efficiency_in_global_health.pdf (Accessed May 26, 2014).

[229] Pogge, *The Health Impact Fund: More Justice and Efficiency in Global Health*, http://devpolicy.anu.edu.au/pdf/papers/DP7_The_Health_Impact_Fund_More_justice_and_efficiency_in_global_health.pdf (Accessed May 26, 2014).

health.[230] Health impact can be calculated in quality-adjusted life years (QALYs) saved. The QALY metric has been improved in the past 20 years and is already widely utilized by insurers in determining the new drugs to cover. The baseline is usually taken as the pharmaceutical arsenal before a registered drug was launched, then the HIF would measure to what extent this drug has increased the length and quality of human lives.[231] The assessment would depend on clinical and practical trials of the product, on tracking down random samples of the product to end-users made easier by serial numbers, and on statistical analysis of associations between sales data and variations in the occurrence of the target disease.[232]

The projected fund for the HIF annually would be $6 billion at the minimum, which is less than 1% of current global spending on pharmaceuticals and about 5% of current global investment in pharmaceutical research. If all countries were to participate, each would need to contribute about 0.01% of its gross national income (GNI). However, if countries representing about a third of the global product were to participate, each would need to contribute at least a fairly small 0.03% of its GNI, which is significantly reduced by enormous cost savings their governments, firms and citizens would enjoy from low-cost HIF registered drugs.[233]

Providing stable incentives entails that the HIF would need assured financing for the next 15 years in order to guarantee pharmaceutical innovators that, if they sponsor costly clinical trials now, they can receive a full decade of health impact rewards after market approval. Such a firm guarantee is as well for the advantage of the funders who would not want the incentive power of their contributions to be weakened through doubtful disregarding by potential innovators. The guarantee may be in the form of an agreement which requires each participating country to commit to the HIF a fixed part of its future gross national income (GNI). Supported by such an agreement, the HIF would naturally adjust the contributions of the different partner countries to their variable economic possessions. Furthermore, it would refrain from prolonged struggles over contribution sizes, and would guarantee each country that any additional cost it agreed to shoulder through an increase in the contribution schedule would be balanced by an equivalent increase in the contributions of all other partner countries.[234]

Kathleen Liddell presented several distinctive features of the HIF as argued by Hollis and Pogge: (1) It is committed to the sale of drugs at marginal cost production, which is the cost price. This significantly decreases the number of people priced out of the market as a result tackling the critical issue regarding cost. (2) The

[230] Thomas Pogge, "The Health Impact Fund: Enduring Innovation Incentives for Cost-Effective Health Gains," *Social Europe Journal* 5, no. 2 (June 2011): 5-9. http://www.social-europe.eu/2011/01/the-health-impact-fund-enduring-innovation-incentiv. (Accessed May 26, 2014).

[231] Pogge, "The Health Impact Fund: Enduring Innovation Incentives for Cost-Effective Health Gains," http://www.social-europe.eu/2011/01/the-health-impact-fund-enduring-innovation-incentiv. (Accessed May 26, 2014).

[232] Pogge, "The Health Impact Fund: How to Make New Medicines Accessible to All," 248.

[233] Pogge, "The Health Impact Fund: How to Make New Medicines Accessible to All," 248.

[234] Pogge, "The Health Impact Fund: How to Make New Medicines Accessible to All," 248.

size of the reward is contingent on the extent to which a new drug decreases disease, rather that the affluence of the patient, which is likely to make research on neglected diseases a more appealing proposal and, more commonly, to assess rewards in accordance with objective measure of value rather than a market-based measure. (3) The scheme is scalable. It can be extended to diagnostics, devices, mechanical inventions and Western disease research, after conducting a pilot with drugs useful in the treatment of neglected disease research. (4) It offers an incentive not just for the creation and production of new drugs, but as well private investment in health-service infrastructure. Taking into account that HIF payments are commensurate with clinical impact, pharmaceutical companies should guarantee that the correct drug gets to the correct patient, in the correct dosage, at the correct time. (5) It does not involve any considerable changes to the structure of intellectual property protection or licensing. (6) It will supposedly result in more co-operative, and then cost-efficient, relationships between patent proprietors and manufacturers of generic versions of drugs. Patent proprietors will be less probably to refuse reasonable licenses and manufacturers of generic versions of drugs less probably to violate the rights of patent proprietors by manufacturing without permission. (7) Its normative bases are both moral and prudential.[235]

The HIF has five major advantages over conventional innovation prizes, comprising advance market commitments and advance purchase commitments. First, it is a structural reform creating a lasting source of high-impact pharmaceutical innovations. Second, it is not disease-specific and consequently much less susceptible to lobbying by firms and patient groups.[236] Third, conventional prizes must elucidate the exact finish line, stating clearly at least which disease the new drug must attack, how effective and suitable it must slightly be, and how severe its side effects may be. Such precision is difficult because it presumes the very knowledge that is currently not available and whose attainment is yet to be supported. As sponsors lack this knowledge in advance, their specifications are probably to be severely suboptimal. They may be too challenging, in order that firms capitulate the effort, albeit something close to the desired drug is within their reach, or they may be inadequately challenging, in order that firms, to save time and expense, provide a drug that is just hardly good enough to win even when they could have done much better at small additional cost.[237] The HIF refrains from the problem of finish line through adaptable rewarding of any new registered drugs in proportion to its impact on the global health. Fourth, designed to avert failure and in ignorance of the real cost of innovation, specific prizes are usually much too large and therefore overpay for innovation. The HIF resolves this problem by allowing its health impact rate to

[235] Kathleen Liddell, "The Health Impact Fund: A Critique," in *Incentives for Global Public Health: Patent Law and Access to Essential Medicines*, ed. Thomas Pogge, Matthew Rimmer and Kim Rubenstein (New York: Cambridge University Press, 2010), 158–159.

[236] Pogge, "The Health Impact Fund: How to Make New Medicines Accessible to All," 248.

[237] Aidan Hollis, *Incentive Mechanisms for Innovation*, IAPR Technical Paper, (Calgary: University of Calgary, 2007), 15–16, http://www.iapr.ca/iapr/files/iapr/iapr.-tp-07005_0.pdf (Accessed May 26, 2014).

adjust itself through competition. For example, a high reward rate would rectify by bringing extra registrations, engendering an increase in the number of registered drugs, and an unappealingly low reward rate would rectify by discouraging new registrations, engendering a decrease in the number of registered drugs. Fifth, the HIF offers each registrant powerful incentives to encourage the optimal end-use of its product: to seek its extensive and effective utilization by any patients who can benefit from it.[238] The HIF provided a complete systematic solution to the major disadvantages of the current globalized patent regime which is the task of the next theme.

5.3.2.2 Solution to the Disadvantages of the Present Global Patent System

Pogge contends that all seven disadvantages of the current globalized patent regime can be significantly alleviated by supplementing the patent regime with a complementary source of incentives and rewards for developing new drugs.[239] The HIF is that mechanism Pogge argues that would provide a complete systematic solution to the seven drawbacks engendered by the current globalized patent regime. A discussion of the systematic solution by the HIF follows.

High Prices

High prices would be nonexistent for HIF registered drugs. Innovators would usually not even desire a higher price as this would decrease their health impact rewards by obstructing access to their product by most of the world's population especially the global poor. The HIF considers health benefits to the poorest of patients as equally important as the health benefits of the richest.[240]

Diseases Concentrated Among the Poor

If diseases concentrated among the poor considerably exacerbate the Global Burden of Disease (GBD), they would no longer be neglected. The more catastrophic ones among them would constitute some of the most profitable R&D opportunities for biotechnology and pharmaceutical companies. This would occur without

238 Pogge, "The Health Impact Fund: How to Make New Medicines Accessible to All," 248.

239 Pogge, "The Health Impact Fund: How to Make New Medicines Accessible to All," 247.

240 Thomas Pogge, "The Health Impact Fund: Better Pharmaceutical Innovations at Much Lower Prices," in *Incentives for Global Public Health: Patent Law and Access to Essential Medicines*, ed. Thomas Pogge, Matthew Rimmer and Kim Rubenstein (New York: Cambridge University Press, 2010), 151–152.

undercutting the profit opportunities such companies currently have by developing interventions for the diseases prevalent among the affluent.[241]

Bias Towards Maintenance Drugs

Bias towards maintenance drugs would be non-existent in the HIF-promoted R&D. The HIF measures the health impact of each registered drug related to how its utilization decreases mortality and morbidity globally—without consideration of whether it realizes this reduction through cure, symptom relief or prevention. This would help pharmaceutical companies to determine how potential research projects would maximize global public health especially about the expected global health impact of the new drug corresponding to the cost of developing it. The lucrativeness of research projects would be in order with their cost-effectiveness related to global public health.[242]

Wastefulness

Wastefulness would be drastically less high for HIF-registered products. Deadweight losses from large mark-ups would not exist. There would be very limited expensive litigation since generic competitors would be deficient of incentives to compete and innovators would lack incentives to restrain generic products because they improve the health impact reward of the innovator. Innovators may thus usually not even worry to acquire, police, and protect patents in many national jurisdictions. Registering a drug with the HIF, just involves that innovators demonstrate simply once that they have an effective and innovative product.[243]

Counterfeiting

Counterfeiting of HIF-registered products would be unappealing. With the real item extensively available close to or even below the marginal cost of production, there is little to be obtained from manufacturing and selling forged products.[244]

---

[241] Pogge, "The Health Impact Fund: Better Pharmaceutical Innovations at Much Lower Prices," 152.

[242] Pogge, "The Health Impact Fund: Better Pharmaceutical Innovations at Much Lower Prices," 152.

[243] Pogge, "The Health Impact Fund: Better Pharmaceutical Innovations at Much Lower Prices," 152.

[244] Pogge, "The Health Impact Fund: Better Pharmaceutical Innovations at Much Lower Prices," 152.

Excessive Marketing

Excessive marketing would as well be greatly decreased for HIF-registered drugs. Since each innovator is rewarded for the health impact of its addition to the medical arsenal, incentives to produce "me-too" drugs to compete with a current HIF-registered drug would be feeble. Under the current patent system, getting a patient to switch from a competitor's product to one's good product is very lucrative. However, if the latter product is HIF-registered, then the switch is not lucrative because it produces no health improvement. Moreover, innovators would have incentives to recommend an HIF-registered drug for doctors and patients simply if such marketing leads to calculable therapeutic advantages for which the innovator would afterward be rewarded.[245]

The Last-Mile Problem

The Last-mile problem would be alleviated since each HIF-registered innovator would have powerful incentives to guarantee that patients are completely instructed and appropriately provisioned so that they maximize the use of its drugs with focus on dosage, compliance etc., which will in that case, through extensive and effective utilization, have their best possible public-health impact. Instead of ignoring poor countries as unlucrative markets, pharmaceutical companies would, furthermore, have incentives to collaborate with one another and with national health ministries, international agencies and NGOs towards improving the health systems of these countries so as to increase the impact of their HIF-registered drugs there.[246] Despite Pogge's breathtaking HIF proposal for resolving the flaws of the current globalized patent system and incentives for pharmaceutical innovation in order to significantly improve affordable access to the essential drugs for the global poor especially in developing countries, critics still raised some objections to the HIF.

5.3.2.3 Criticisms of the HIF

Several objections have been raised by some authors regarding the HIF. The first major criticism focuses on its practical barriers to effective implementation.[247] First, the second component of the reform plan makes provision for the involvement of an international agency with the responsibility of assessing the impact of various

[245] Pogge, "The Health Impact Fund: Better Pharmaceutical Innovations at Much Lower Prices," 152–153.

[246] Pogge, "The Health Impact Fund: Better Pharmaceutical Innovations at Much Lower Prices," 153.

[247] Jorn Sonderholm, "A Reform Proposal in Need of Reform: A Critique of Thomas Pogge's Proposal for How to Incentivize Research and Development of Essential Drugs," *Public Health Ethics* 3, no. 2 (January 2010): 167–177.

HIF-registered drugs on the global burden of disease and implementing pay rewards to pharmaceutical companies. The participation of such an agency in the macroeconomic arrangement increases transaction costs and offers abundant opportunity for corrupt behavior with respect to the staff of the agency and those who can influence them.[248] The current system for incentivizing research and development of essential drugs also involves transaction costs. More so, this system requires both patent offices and patent courts, but, as Alex Rosenberg articulates, "a patent system's greater reliance on individuals to pursue their own interests directly, instead of through an intervening government, is generally more effective than any alternative."[249] It is projected that about 10% of the monetary resources being invested in the reform plan will be expended on administration and assessment.[250]

In respect to the issue of corruption, it is pertinent to point out the bleak empirical evidence that corrupt behavior is a critical issue prevalent among government officials in numerous developing countries in which data collection needs to be carried out.[251] Pogge has maintained that data regarding the global burden of disease and the health impact of various drugs collected in the context of the reform plan would be beneficial beyond the precise purposes of this plan.[252] The susceptibility of the assessment process to corruption would, then, undermine the value of the collected data in contrast to data generated by standard academic and governmental research programs.[253]

Second, it will be hard for the identified agency to obtain accurate and valid information regarding the assessed impact of various drugs on the global disease burden. The problem does not rest simply on establishing a reasonable metric that can be utilized to ascertain a drug's impact on the global disease burden.[254] There is a general consensus among the supporters of the HIF that the most favorable metric candidate is the QALY system which is presently being utilized by national health systems in Australia, Canada, the UK and US to assess the health impact of pharmaceuticals.[255] Additional practical concern is also highlighted with regard to applying the metric and conducting the actual field work of visiting vast, poor and frequently geographically remote populations and obtaining and an accurate summary of what

---

[248] Sonderholm, *Intellectual Property Rights and the TRIPS Agreement: An Overview of Ethical Problems and Some Proposed Solutions*, 27.

[249] Alex Rosenberg, "On the Priority of Intellectual Property Rights, Especially in Biotechnology," *Politics, Philosophy & Economics* 3, no. 1 (February 2004): 84.

[250] Hollis and Pogge, *The Health Impact Fund: Making New Medicines Accessible for All*, 31.

[251] Johann Graf Lambsdorff, *Corruption Perceptions Index (CPI)*, Transparency International 2008. http://transparency.org/news_room/latest_news/press_releases/2008/2008_09_23_cpi_20008_en. (Accessed May 27, 2014).

[252] Hollis and Pogge, *The Health Impact Fund: Making New Medicines Accessible for All*, 31.

[253] Sonderholm, *Intellectual Property Rights and the TRIPS Agreement: An Overview of Ethical Problems and Some Proposed Solutions*, 28.

[254] Sonderholm, *Intellectual Property Rights and the TRIPS Agreement: An Overview of Ethical Problems and Some Proposed Solutions*, 28.

[255] Ravvin, "Incentivizing Access and Innovation for Essential Medicines: A Survey of the Problem and Proposed Solutions," 120.

the disease burden is in the region and how various drugs are contributing to its decrease.[256]

On the other hand, Liddell raised a practical barrier issue of the HIF that focuses on inadequate commitments and contributions from participating countries. She acknowledges that the authors are mute regarding the number of partners that must participate to make this a feasible proposal. If some of the more affluent countries decline to participate, the contributions at 0.03% for each country will be inadequate. She also asserts that US$6 billion estimated as projected annual fund for the HIF appears to be a small sum, considering that the estimate of the total global spending on pharmaceuticals in 2008 was about US$735 billion. More so, she indicates that it is revealing that the total operating budget for the United Nations' initiative is estimated at US$4.19 billion, and even at this rate numerous countries are late in their payments. From this perspective, she inferred that the HIF proposal is an enormous sum of money, and may not likely be fulfilled by foreign aid budgets.[257]

Another concern raised by Liddell relates to the issue of the relationship between the HIF and the patent system. She acknowledges that it is not clear whether it is essential to attach the HIF system to patent protection, because it establishes eligibility criteria of uncertain assistance and the territorial nature of patent protection introduces a lot of difficult policy questions for a system with global ambitions.[258]

On the other hand, Liddell also notes that there is an unnerving feeling that the HIF proposal plays right into the hands of the pharmaceutical companies. She argues that "it fuels their search for profits, offering them yet another optional method to increase their existing profit margins at the expense of the public purse, when they are already amongst the very wealthiest industries."[259] The explanation is that there is a general false assumption that anything that falls short of an appealing profit stands little opportunity of being endorsed by the politically influential pharmaceutical companies and their governments.[260]

Furthermore, Liddell identifies another issue which recognizes that there is inadequate empirical evidence to support the crucial premises in the HIF proposal or to demonstrate that the benefits of the HIF validate such a substantial policy enterprise. She indicates that this offers an uncomfortable contradiction in the sense that "the HIF is a proposal that seeks to organize the cost and direction of scientific research on the basis of proven utility, yet the regulatory tools enlisted to achieve this lack an equivalent evidence base."[261]

Sonderholm on the other hand, raised concerns regarding the prudential appeal of the HIF.[262] He cites an example with two distinct drugs with varying cost of

---

256 Sonderholm, *Intellectual Property Rights and the TRIPS Agreement: An Overview of Ethical Problems and Some Proposed Solutions*, 28.

257 Liddell, "The Health Impact Fund: A Critique," 161–162.

258 Liddell, "The Health Impact Fund: A Critique," 179.

259 Liddell, "The Health Impact Fund: A Critique," 179.

260 Liddell, "The Health Impact Fund: A Critique," 179.

261 Liddell, "The Health Impact Fund: A Critique," 180.

262 Sonderholm, "A Reform Proposal in Need of Reform: A Critique of Thomas Pogge's Proposal for How to Incentivize Research and Development of Essential Drugs," 167–177.

production and impact on the global disease burden. The first drug decreases the symptoms of diarrhea in infants and averts the symptoms for 4 weeks. It is effective in 40% of cases and sales in the form of two pills at the cost of production of $2. The second drug also decreases the symptoms of diarrhea in infants and averts the symptoms for 4 weeks. It is effective in 90% of cases and sales in the form of a powder that requires to be dissolved in 25 centiliter of clean water. It should also be stored at refrigerator temperature. The cost of production is one quarter of the first drug.[263]

In this situation it is expected that the manufacturer of the first drug would receive a greater reward than the manufacturer of the second drug. This is based on the analysis that the first drug is probably to have a greater impact on the global disease burden than the second drug. This results from the fact that the effectiveness of the first drug does not require something which is frequently lacking in developing countries and which the second drug requires so as to be effective, including clean drinking water and cooled storage capacity. Therefore, specific infrastructure features of the regions in which infants with diarrhea generally live contribute in a very substantial way to the fairly small reward that the manufacturer of the second drug would receive under Pogge's reform plan.[264]

The implication is that pharmaceutical companies that are motivated by the profit would recognize that the economic prospects of developing high-tech essential drugs intended for the medical needs of the populations in developing countries are inadequate.[265] A high-tech drug means a drug that needs clean drinking water, electricity and/or educated health professionals to be effective. On the other hand, a low-tech drug does not need any of these things to be effective.[266] From this perspective, pharmaceutical companies would inevitably reorganize some of their R&D efforts towards low-tech drugs. There would also inevitably be a rise of new pharmaceutical companies that would exclusively focus on the development of low-tech essential drugs that tackle the medical needs of the populations in developing countries.[267]

Sonderholm indicates that Pogge would embrace these developments, but, however, he challenges the prudential reasons for citizens of high-income countries to support a reform plan that engender these developments.[268] It is pertinent to note that the rise of this new opportunity of drug development will probably generate

---

[263] Sonderholm, *Intellectual Property Rights and the TRIPS Agreement: An Overview of Ethical Problems and Some Proposed Solutions*, 29.

[264] Sonderholm, *Intellectual Property Rights and the TRIPS Agreement: An Overview of Ethical Problems and Some Proposed Solutions*, 30.

[265] Sonderholm, *Intellectual Property Rights and the TRIPS Agreement: An Overview of Ethical Problems and Some Proposed Solutions*, 30.

[266] Sonderholm, *Intellectual Property Rights and the TRIPS Agreement: An Overview of Ethical Problems and Some Proposed Solutions*, 30.

[267] Sonderholm, Intellectual Property Rights and the TRIPS Agreement: An Overview of Ethical Problems and Some Proposed Solutions, 30.

[268] Sonderholm, "A Reform Proposal in Need of Reform: A Critique of Thomas Pogge's Proposal for How to Incentivize Research and Development of Essential Drugs," 167–177.

new jobs, but the funding of these jobs would come from the fund that pays for the second component of the reform plan. This implies that the huge resources that are required to fund these new jobs in developed countries would come from these developed countries.[269]

The final issue was raised by Michael Selgelid which emphasized a problem of causal attribution for the reimbursement process critical to the HIF.[270] This is the problem that focuses on assessing the extent to which any decrease in the global disease burden, or the burden of any particular disease, is the outcome of one intervention in contrast with another.[271] The causal attribution problem is considered a serious danger to the feasibility of health impact measurement by either QALYs or Disability-Adjusted-Life Years (DALYs). Successively, it is also a danger to the feasibility of the HIF per se. This is as result of the fact that the critical reimbursement process of the HIF is precisely premised on the notion that pharmaceutical companies are rewarded proportionate to the effect of their products on the size of the global disease burden. The import of the problem is additionally highlighted by the fact that numerous successful medical interventions are ones that include a few different active ingredients.[272]

5.4 Conclusion

Concluding remarks on this chapter which discusses international responsibility and the health impact fund focuses on a recapitulation of Pogge's contention for a stronger interpretation of global responsibilities for providing affordable access to drugs for participants and host populations in developing countries. This dominant approach invoked by Pogge in arguing for post-trial access to drugs is called cosmopolitanism. He forcefully argues that our critical duty of justice to take effective action on the issue of affordable access to essential drugs is grounded on human rights which extend universally to all individuals.

In contrast to Rawls, Pogge challenges the thesis of explanatory nationalism which appeals to domestic factors as engendering persistent severe poverty and global inequality. He insists that global factors perpetuate severe poverty and global inequality and calls for moral responsibility on the part of affluent countries to redress this situation. He contends that unjust global institutional order imposed on the global poor perpetuates harm and consequently violates their negative rights. He

[269] Sonderholm, *Intellectual Property Rights and the TRIPS Agreement: An Overview of Ethical Problems and Some Proposed Solutions*, 30–31.

[270] Selgelid, "A Full-Pull Program for the Provision of Pharmaceuticals: Practical Issues," 140.

[271] Sonderholm, *Intellectual Property Rights and the TRIPS Agreement: An Overview of Ethical Problems and Some Proposed Solutions*, 31.

[272] Sonderholm, *Intellectual Property Rights and the TRIPS Agreement: An Overview of Ethical Problems and Some Proposed Solutions*, 32.

argues then, that citizens of affluent countries have not merely positive duties to assist, but also more stringent negative duties not to harm the global poor.

Pogge contends that increasing massive mortality and morbidity rates prevalent in developing countries can be addressed through guaranteeing improved access to essential drugs. To accomplish this objective of significantly reducing the global disease burden especially among the global poor in developing countries, Pogge proposes a concrete, feasible and politically realistic plan for reforming current globalized patent rules for incentivizing pharmaceutical research.

In discussing Pogge's cosmopolitan approach to the issue of post-trial access to drugs in developing countries, meaning of cosmopolitanism and four different approaches to cosmopolitanism were explicated as prelude. The current rules for incentivizing pharmaceutical research were discussed. In this context, the current global patent system was argued as an institutional failure, because the TRIPS agreement and the TRIPS-plus provisions significantly impede affordable access to essential drugs for the global poor in developing countries. The TRIPS agreement and TRIPS-plus provisions created access problem and essentially excluded the global poor from life-saving or essential drugs.

Pogge's proposed new rules for reforming and incentivizing pharmaceutical research were discussed in response to the problems engendered by the current patent system. He proposed two basic reform strategies for dealing with the monopoly pricing issues and for improving access to essential drugs, including the differential pricing strategy and the public good strategy. Differential pricing and compulsory licensing were argued as falling short in resolving the drawbacks from the current patent system and in providing an attractive and effective means for improving affordable access to essential drugs. The public good strategy was defended as more promising in generating a reform plan that would prevent the chief failings of the current monopoly patent regime, while at the same time, preserving most of its important benefits.

The HIF was discussed as a global institution for the effective implementation of Pogge's new plan for reforming and incentivizing research and innovation. The HIF's significant structural reform of the current global patent system was discussed. Furthermore, the HIF's systematic solution to seven disadvantages of the current global patent scheme was discussed. Some key objections by some authors were also discussed. Just as the weakness of the statist approach of Rawls is its exaggerated emphasis upon each individual nation, similarly the weakness of the cosmopolitan approach of Pogge is an exaggerated emphasis upon international relations. The strengths of each approach, avoiding the main weaknesses in each, can be combined to establish a paradigm of global health justice to provide affordable access to drugs in developing nations, as discussed in the next chapter.

Chapter 6
Global Health Responsibility

6.1 Introduction

The World Health Organization (WHO) views health in the twenty-first century as "a shared responsibility, involving equitable access to essential care and collective defense against transnational threats."[1] Lawrence O. Gostin et al. articulate that these two critical aspects of global health identified by the WHO "require global leadership, sustainable resources, collaboration, and mutual support among states, businesses, philanthropy, and civil society."[2] The Global Strategy for Women's and Children's Health emphasizes that, "Global partnership and the sufficient and effective provision of aid and financing are essential."[3] In essence, improved global health governance is a desideratum for global health.[4]

A logical and consistent system of global health governance can be rooted on the common interests of states and their partners. All states possess self-interests in promoting global health governance as a collective protection from transnational health, controlling infectious diseases where they arise and preventing global spread

---

[1] WHO, "About WHO", (Geneva: World Health Organization, 2010), http://www.who.int/about/en/ (Accessed June 8, 2014).

[2] Lawrence O. Gostin et al., *The Joint Action and Learning Initiative on National and Global Responsibilities for Health*, World Health Report, Background Paper, No. 53 (Geneva: World Health Organization, 2010), http://www.who.int/healthsystems/topics/financing/healthreport/JALI_No.53.pdf (Accessed June 8, 2014).

[3] United Nations, *Global Strategy for Women's and Children's Health*, (New York: United Nations, 2010), http://www.who.int/pmnch/topics/maternal/20100914_gwwch_en.pdf (Accessed June 8, 2014).

[4] Gostin et al., *The Joint Action and Learning Initiative on National and Global Responsibilities for Health*, World Health Report, Background Paper, No. 53 (Geneva: World Health Organization, 2010), http://www.who.int/healthsystems/topics/financing/healthreport/JALI_No.53.pdf (Accessed June 8, 2014).

E.C. Obi, *Post-Trial Access to Drugs in Developing Nations*, Advancing Global Bioethics 7, DOI 10.1007/978-3-319-60028-4_6

of health hazards.[5] More so, states possess self-interests in guaranteeing equitable access to essential services which involve "health systems, including cost-effective drugs and vaccines, and other human health needs (e.g., safe water, nutrition, sanitation, vector control, and tobacco reduction) to all people."[6]

The responsibility for guaranteeing the right to health for all rests not exclusively with states and their obligations to their own people, but as well with the global community.[7] About half of the Millennium Development Goals (MDGs) tackle basic human needs, indicating an understanding that all states possess an interest in guaranteeing that crucial needs are fulfilled for all human beings everywhere.[8]

Nevertheless, more than a decade following the endorsement of the MDGs, notwithstanding some progress, such as the decrease in child mortality[9] and broadening of AIDS treatment[10] the global community has not attained essential improvements in global health or considerably reduced health inequalities.[11] The devastating impact of HIV/AIDS is still significantly evident in developing countries especially in Sub-Saharan Africa, where millions are not yet able to afford anti-retroviral drugs for treating HIV/AIDS. The global community has fallen short of effectively achieving basic human needs. There are profound structural reasons for the lack of substantial progress, such as "the absence of leadership, fragmentation and lack of coordination of multiple actors, persisting inadequate levels of domestic and international health spending, and foreign aid and programs that do not match national priorities."[12]

---

[5] Gostin et al., *The Joint Action and Learning Initiative on National and Global Responsibilities for Health*, World Health Report, Background Paper, No. 53 (Geneva: World Health Organization, 2010), http://www.who.int/healthsystems/topics/financing/healthreport/JALI_No.53.pdf (Accessed June 8, 2014).

[6] Gostin et al., *The Joint Action and Learning Initiative on National and Global Responsibilities for Health*, World Health Report, Background Paper, No. 53 (Geneva: World Health Organization, 2010), http://www.who.int/healthsystems/topics/financing/healthreport/JALI_No.53.pdf (Accessed June 8, 2014).

[7] Gostin et al., *The Joint Action and Learning Initiative on National and Global Responsibilities for Health,* World Health Report, Background Paper, No. 53 (Geneva: World Health Organization, 2010), http://www.who.int/healthsystems/topics/financing/healthreport/JALI_No.53.pdf (Accessed June 8, 2014).

[8] United Nations, *Millennium Development Goals*, (New York: United Nations, 2010), http://www.un.org/millenniumgoals (Accessed June 10, 2014).

[9] Julie Knoll Rajaratnam et al., "Neonatal, Postneonatal, Childhood, and Under- 5 Mortality for 187 Countries, 1970–2010: A Systematic Analysis of Progress Towards Millennium Development Goal 4," *Lancet* 375, no. 9730 (June 2010): 1988–2008.

[10] World Health Organization, *More than Five Million People Receiving HIV Treatment*, Media Release, (Geneva: World Health Organization, July 19, 2010), http://www.who.int/mediacentre/news/releases/2010/hiv_treatment_20100719/en/index.html (Accessed June 10, 2014).

[11] Gostin et al., *The Joint Action and Learning Initiative on National and Global Responsibilities for Health,* World Health Report, Background Paper, No. 53 (Geneva: World Health Organization, 2010), http://www.who.int/healthsystems/topics/financing/healthreport/JALI_No.53.pdf (Accessed June 8, 2014).

[12] Gostin et al., *The Joint Action and Learning Initiative on National and Global Responsibilities for Health,* World Health Report, Background Paper, No. 53 (Geneva: World Health Organization, 2010), http://www.who.int/healthsystems/topics/financing/healthreport/JALI_No.53.pdf (Accessed June 8, 2014).

Rawls's and Pogge's approaches for global health justice as discussed in chapters four and five respectively are insufficient. Rawls's statist version of relational justice emphasizes the national responsibility of each country to fulfill the right to health of all its citizens and effectively excludes the global responsibility of individual nations. Pogge's cosmopolitan approach focuses too much on global responsibility without a sufficient attention to the national responsibility of individual societies.[13]

The paradigm proposed in this chapter combines these two approaches by adopting their strengths and avoiding their weaknesses. The paradigm refers to a sliding scale of national and global responsibilities about the right to health in general and affordable access to drugs. This combined approach considers "global responsibility as supplementing, not replacing national responsibility for health."[14] The implication is that the primary responsibility for realizing the right to health is a national responsibility. However, when poor countries exhaust their domestic resources and are still not able to fulfill the right to health of their citizens, rich countries can step in to exercise their global responsibility as a secondary responsibility for the realization of the right to health.[15]

The first part of the chapter deals with responsibility for health inequalities focusing on sliding scale of national and global responsibilities. The second part deals with challenges of global health as well as the global responsibility and health inequalities between nations. The third part focuses on the global capacity to redress health inequalities which consists of the health development paradigm, the medical relief paradigm and the new global health paradigm. The new global health paradigm evolved from the global AIDS response and is at the intersection of health development paradigm and medical relief paradigm.

The fourth part deals with responsibility for post-trial access to drugs which focuses on human rights and global health justice. The fifth part of the chapter addresses access to post-trial drugs in developing nations focusing on ethical issues in Global Health Fund and as well as responsibilities of Global Health Fund which result in the effective realization of the core content of the right to health including affordable access to essential drugs for all. A more detailed explanation of a paradigm for global health justice will begin with sliding scale of national and global responsibilities.

---

[13] Norman Daniels, *Just Health: Meeting Health Needs Fairly* (Cambridge: Cambridge University Press, 2008), 345–355.

[14] Gorik Ooms and Rachel Hammonds, "Taking Up Daniels' Challenge: The Case for Global Health Justice," *Health and Human Rights* 12, no.1 (2010): 29.

[15] Ooms and Hammonds, "Taking Up Daniels' Challenge: The Case for Global Health Justice," 40–41.

6.2 Responsibility for Health Inequalities

6.2.1 Sliding Scale of Responsibilities

6.2.1.1 Perspectives in Global Health Justice

In his landmark work, Just Health: Meeting Health Needs Fairly published in 2008, Daniels develops a complicated theory of justice with a concluding challenge regarding International health inequalities and Global Justice.[16] He made a compelling case for national health justice which involves a case for obligations of mutual assistance to decrease health inequalities at the national level. However, Daniels doubts whether he can extend his theory to the global level to as well make a convincing case for obligations of mutual assistance beyond state borders. He concedes that arguments of relational justice which serve as the foundation of his theory of health justice cannot simply be extended to the global level. Nevertheless, in his call for action, he articulates, "Despite the lack of closure on these matters, the account developed here provides an integrated theory that helps us see the path to pursue in promoting population health and distributing it fairly, globally as well as domestically."[17]

This chapter in response to Daniels' challenge for action pursues the path to global health justice by developing a paradigm for global health justice which emphasizes a sliding scale of national and global Responsibilities. The paradigm combines approaches of Rawls and Pogge to global health justice by adopting their strengths and avoiding their weaknesses. Rawls' strength is his acknowledgement of the national responsibility of each country to fulfill the right to health of all its citizens. Pogge's strength is his recognition of the global responsibility of individual nations. The paradigm also builds on Daniels's several arguments but goes further to develop a thesis that emphasizes the global responsibility for health.[18] The implication is that global responsibility is viewed as augmenting, not substituting, the national responsibility. Other areas of agreements and disagreements with Daniels in developing the paradigm will be highlighted in this chapter.

Daniels thoroughly investigates two approaches of attempting to resolve the impasse between the statist and cosmopolitan perspectives. One approach which is cosmopolitan involves a minimalist strategy that emphasizes an international obligation of justice to avoid harming people by engendering deficits in fulfilling their human rights.[19] The minimalist nature of this perspective stems from the fact that people may concur on negative duties not to harm even if they are not in agreement

[16] Daniels, *Just Health: Meeting Health Needs Fairly*, 333–355.

[17] Daniels, *Just Health: Meeting Health Needs Fairly*, 355.

[18] Gorik Ooms, *From the Global AIDS Response towards Global Health?* A Discussion Paper for the Helene De Beir Foundation and the International Civil Society Support Group, (January 2009), http://www.hdbf.org/?pageid=345&lang=en (Accessed June 13, 2014).

[19] Thomas Pogge, "Severe Poverty as a Violation of Negative Duties," *Ethics and International Affairs* 19, no. 1 (March 2005): 55–83.

regarding positive duties to aid. Daniels contends that this approach deals with some international health issues better than others, but its limitations were clearly exposed in relation to sources of international health inequalities, some of which are not tackled by negative duties.[20] Daniels as well explores a more encouraging relational justice approach that requires that we determine a more intermediary conception of justice suitable for developing international institutions and rule-making bodies, leaving it open simply how fundamental issues of equality would be in such a context.[21] Such an approach if appropriately developed could tackle broader sources of international health inequalities.[22] Therefore, the broader sources of inequality can only be tackled by more robust accounts of global justice.[23]

Daniels argues that Pogge's minimalist strategy articulated in his harms to health argument has severe limitations. There is no clarity for identifying the baseline for measuring harm. Determining when there is a deficit in a human right to health is also not clear. There is no clear way of specifying what to utilize as a baseline in measuring a deficit in the right to health. Pogge's minimalist strategy was applied to the brain drain of health personnel from low-income countries to the Organization for Economic Cooperation and Development (OECD) countries.[24] Shortcomings were also identified in applying the minimalist strategy to some international health issues. For example, the issue of international property rights and the incentives they generate goes beyond the problem of access to existing drugs, such as the anti-retroviral cocktails that have been the emphasis in recent years. Patent holders on anti-retroviral drugs championed a fight to limit access to generic copies of their drugs. The consequence was direct harm to people who could have benefited from anti-retroviral drugs and died in alternative. However, these generics that apparently save other lives would not have surfaced without the incentives engendered by the current patent system. Furthermore, large multinational pharmaceutical companies have been faulted for a research and development bias against drugs needed in developing countries. They have responded to current incentives by focusing on "blockbuster" drugs for more affluent markets, encompassing many "me too" drugs that slightly enhance the effectiveness of older drugs or decrease their side effects marginally.[25]

Daniels contends that despite the issue of vagueness, Pogge's proposal cannot be defended by appealing exclusively to the "no-harm" principle. He articulates that, "the proposed incentive fund would better help to realize human rights to health as Pogge argues, but "not optimally helping" is not the same as "harming," and so the justification has shifted."[26] There may be cogent reasons for an account of global

[20] Daniels, *Just Health: Meeting Health Needs Fairly*, 337.

[21] Joshua Cohen and Charles Sabel, "Extra Rempublicam Nulla Justitia? *Philosophy and Public Affairs* 34, no. 2 (March 2006): 147–175.

[22] Daniels, *Just Health: Meeting Health Needs Fairly*, 337.

[23] Daniels, *Just Health: Meeting Health Needs Fairly*, 342.

[24] Daniels, *Just Health: Meeting Health Needs Fairly,* 337–340.

[25] Daniels, *Just Health: Meeting Health Needs Fairly*, 339.

[26] Daniels, *Just Health: Meeting Health Needs Fairly*, 340.

justice to take into consideration the interests of people impacted by existing property right protections more meticulously than those agreements currently do, but that leads us into more disputed area than the minimalist strategy.[27]

International harming is complicated in several ways. The harms are usually not deliberate and occasionally benefits were debatably intended. Harms are frequently mixed with benefits. At any rate, great caution must be employed to explicate the baseline in measuring harm. Daniels argues that, "such a complex story about motivations, intentions, and effects might seem to weaken the straightforward appeal of the minimalist strategy, but the complexity does not undermine the view that we have obligations of justice to avoid harming health."[28]

Gorik Ooms and Rachel Hammonds indicate that Daniels presents the differences between global and national responsibilities in health justice as characterized by "an innate tension, with the path to a successful integration being one that requires careful, constant negotiation between dangerous but opposing alternatives."[29] The proposed paradigm discussed in this chapter on the other hand, envisages the space between these alternatives which implies the space to devise and establish global health justice. Ooms and Hammonds figuratively compare this scenario to the narrow strait between Scylla and Charybdis, the two great sea monsters in Greek mythology that Odysseus had to hold equally distant to guarantee the security of his journey through the strait, even when the equally distant, the adjacency of each endangered his ships and its sailors.[30] The implication in the context of the two extreme poles of the debate on global health justice refers to the pull of the cosmopolitan intuition of Pogge and strongly statist versions of relational justice of Rawls which carry severe risks. Daniels in response to this situation advocates for an intermediary ground that resists the pull of two opposing alternatives. He contends that investigation should concentrate on a middle ground between strongly statist claims which indicate that egalitarian requirements of social justice are exclusively the realm of the nation-state and its well-defined basic structure as articulated by Rawls and Thomas Nagel and strongly cosmopolitan claims that principles of justice apply globally to individuals, despite the relations in which they stand or the institutional structures that provide framework for them to interact.[31] He calls for some explanations of what it would imply for these intermediary institutions to make decisions or enforce practices that tackle gross global health inequalities as issues of justice.[32]

The paradigm of global justice proposed in this chapter also acknowledges the risks of both claims highlighted by Daniels. The paradigm also supports Daniels' call to resist the pull of Pogge's cosmopolitan intuition which focuses too much on global responsibility, without a sufficient attention to the preeminence of national

[27] Daniels, *Just Health: Meeting Health Needs Fairly*, 340.

[28] Daniels, *Just Health: Meeting Health Needs Fairly*, 340.

[29] Ooms and Hammonds, "Taking Up Daniels' Challenge: The Case for Global Health Justice," 29–30.

[30] Ooms and Hammonds, "Taking Up Daniels' Challenge: The Case for Global Health Justice," 30.

[31] Daniels, *Just Health: Meeting Health Needs Fairly,* 346.

[32] Daniels, *Just Health: Meeting Health Needs Fairly*, 346.

responsibility that could easily and significantly undermine the latter.[33] Daniels acknowledges the lack of institutions from the cosmopolitan perspective that can provide just outcomes in a consistent and sustained fashion for individuals. He thinks that the cosmopolitan theory inherently does not offer any meaningful clue to how a commitment to justice can be maintained by global institutions. It also does not make provision for any difference in justice concerns that may be suitable to institutions of different types.[34] He considers justice as a "stable product of institutions structured in certain ways."[35] There is a consensus that the global institution essential to regulate the relationship between national and global responsibility is deficient, and that this lack should offer adequate motivation to establish such an institution.[36]

Similarly, it is critical to resist what Daniels regards as "strongly statist versions of relational justice."[37] There is a contention that beyond the state that a moral order exists, but it is restricted to more basic humanitarian obligations to help those grappling with serious risks and possessing pressing needs. More so, it must as well not infringe on some basic human rights, and the agreements must be complied with.[38] There are no obligations of justice to "distribute health fairly, or to protect equality of opportunity, or to assist other societies to become as well off as they can be in satisfying rights to health or education or political participation."[39] Actually, if States were seen exclusively as institutions that could regulate health justice, the lack of a "global state" would absolve states from all responsibility for the consequences of their behaviors outside their borders.[40]

Daniels rejects the idea of anchoring the global health justice on international human rights law. He writes, "Recasting the problem as one of human rights, specifically a human right to health and health care, does not help us answer these questions about international justice for two reasons."[41] First, the international legal obligation to obtain a human right to health for a population rests primarily with each state for its own population. In spite of the fact that international human rights agreements and proclamations postulate international obligations to assist other states in fulfilling human rights,[42] the international obligations cannot constitute the

[33] Daniels, *Just Health: Meeting Health Needs Fairly*, 348.

[34] Daniels, *Just Health: Meeting Health Needs Fairly*, 348.

[35] Daniels, *Just Health: Meeting Health Needs Fairly*, 348.

[36] Ooms and Hammonds, "Taking Up Daniels' Challenge: The Case for Global Health Justice," 30.

[37] Daniels, *Just Health: Meeting Health Needs Fairly*, 349.

[38] Daniels, *Just Health: Meeting Health Needs Fairly*, 349.

[39] Daniels, *Just Health: Meeting Health Needs Fairly*, 349.

[40] Ooms and Hammonds, "Taking Up Daniels' Challenge: The Case for Global Health Justice," 30.

[41] Daniels, *Just Health: Meeting Health Needs Fairly*, 335.

[42] Committee on Economic, Social and Cultural Rights, *General Comment No. 14: The Right to the Highest Attainable Standard of Health*, UN Doc. No. E/C.12/2000/4, (New York: United Nations, 2000), http://www.unhchr.ch/tbs/doc.nsf/%28symbol%29/E.C.12.2000.4.En. (Accessed June 19, 2014).

principal elements of the human right to health and health care.[43] Second, health inequalities may persist, even when a right to health is obtained to the extent possible in different states.[44] In contrast, the new paradigm proposed in this chapter grounds the global health justice on international human rights law. In this context, the international human rights law provides a theoretical framework for national and global responsibility for health.

Establishing a global institution to govern the relationship between national and global responsibility for health forms a critical part of the paradigm. Ooms and Hammonds consider the Global Fund to fight AIDS, Tuberculosis and Malaria (the Global Fund) that already is in existence as a prototype.[45] The structure of the Global Fund is currently explained more by its practical action than by any theoretical foundation of global health justice established in advance. Envisaging the Global Fund as a paradigm effective for realizing the global health justice based on a middle ground can as well offer a theoretical basis to fortify the work of the Global Fund itself.[46] A critical component of the paradigm for global health proposed in this chapter is the focus on government programs to resolve the problem of affordable access to anti-retroviral drugs in developing countries which is discussed next.

6.2.1.2 Government Programs and Access to Drugs in Developing Countries

The paradigm proposed in this chapter advocates for affluent developed countries to fund government programs for resolving the problem of affordable access to drugs in developing countries. There is a contention that this view contrasts with the perspective argued by article 34 of the recent version of the Declaration of Helsinki, which stipulates, "In advance of a clinical trial, sponsors, researchers and host country governments should make provisions for post-trial access for all participants who still need an intervention identified as beneficial in the trial. This information must also be disclosed to participants during the informed consent process."[47] A discussion regarding the justification of the emphasis on government programs to address the problem of affordable access to drugs for participants and host populations in developing countries is therefore imperative here.

The affordable access to essential drugs for participants and host populations of clinical trials in developing countries falls within the scope of right to health. Realizing the right to health belongs to the state. The state still has the primary responsibility to fulfill the progressive realization of the right to health of its citizens

[43] Daniels, *Just Health: Meeting Health Needs Fairly*, 335.

[44] Daniels, *Just Health: Meeting Health Needs Fairly*, 335.

[45] Ooms and Hammonds, "Taking Up Daniels' Challenge: The Case for Global Health Justice," 30.

[46] Ooms and Hammonds, "Taking Up Daniels' Challenge: The Case for Global Health Justice," 30.

[47] World Medical Association, *Declaration of Helsinki: Ethical Principles for Medical Research Involving Human Subjects*, (Edinburgh, Scotland: World Medical Association, 2013), 7, Article 34, (Accessed June 20, 2014).

including affordable access to essential drugs even after the initiation and execution of clinical trials by multinational pharmaceutical companies. The State would not abdicate its responsibility to multinational pharmaceutical companies but can collaborate with pharmaceutical companies at the end of clinical trials to enhance affordable access to essential drugs for participants and host populations in developing countries. Daniels writes, "Primary responsibility for realizing rights to health and health care in a population should rest with each state."[48] Despite the fact that the primary responsibility for population health is shouldered by each state, this does not imply that the state has sole responsibility.[49] Other actors such as multinational pharmaceutical companies, international agencies or institutions and governments of various countries especially affluent countries could also contribute significantly in improving the population health, but they simply supplement the efforts of the state in fulfilling the rights to health of its citizens. Ashcroft argues that realizing the requirement of post-trial provisions of successful products to participants and host populations would inevitably involve a more pronounced "collaboration between researchers, funders, hosts, and health systems, if this part of the Declaration is to be more than simply aspirational."[50] Pharmaceutical companies can assist in providing access to drugs to participants and host populations in developing countries, but this does not mean that the state is relieved of its primary responsibility to provide access to essential drugs for its citizens. Improving the population health of each state falls to state's ability to establish and implement good health policy.[51] Ruger also articulates that, "Individual nation-states have primary and prior obligations to deal with health inequalities."[52] The implication is that states have obligations to maintain health equity at the state level.[53]

On the other hand, Grady discusses broadened possible strategies for guaranteeing continued post-trial access of successful products to participants and host populations. The first strategy emphasizes that various stakeholders in a clinical research including investigators, sponsors, communities, national health systems, and international organizations should shoulder responsibility for guaranteeing continued post-trial access of beneficial treatment to participants and host populations. Possible strategies of dealing with continued post-trial access should be discussed and negotiated before starting a clinical research. Researchers and sponsors are instructed not to ignore this critical issue. Similarly, it is pertinent to note that researchers and sponsors cannot be burdened with the exclusive responsibility of treating people who should be obtaining treatment through the regular health care

---

[48] Daniels, *Just Health: Meeting Health Needs Fairly*, 344.

[49] Daniels, *Just Health: Meeting Health Needs Fairly*, 344.

[50] Richard Ashcroft, "After the Trial is Over: What are the Sponsor's Obligations? *Policy Briefs*, May 1, 2005. http://www.scidev.net/en/policy-briefs/after-the-trial-is-over-what-are-the-sponsor-s-obl... (Accessed June 21, 2014).

[51] Daniels, *Just Health: Meeting Health Needs Fairly*, 344.

[52] Jennifer P. Ruger, "Ethics and Governance of Global Health Inequalities," *Journal of Epidemiology and Community Health* 60, no. 11 (November 2006): 1001.

[53] Ruger, "Ethics and Governance of Global Health Inequalities," 1001.

infrastructure.[54] Grady articulates that, "expecting researchers and sponsors to fill that gap is not only an unrealistic expectation but would also act as a powerful negative disincentive."[55] The clinical research for possible prevention of HIV in Cambodia was cited as an example in which the research was stopped because the Women's Network for Unity, a Cambodian sex workers union insisted that participants would receive health care for thirty years at the end of the clinical trial.[56]

Stakeholders involved in research and health care delivery should continue to collaborate to devise creative strategies to provide continued treatment for participants and host populations at the end of the clinical research. A typical example of a creative strategy with multiple stakeholders is "the HIV Netherlands, Australia, Thailand Research Collaboration (HIV-NAT) co-payment and sliding scale drug fund program."[57] HIV-NAT is a non-government, non-profit organization that involves three collaborators including the Thai Red Cross AIDS Research Center in Thailand, the National Center in HIV Epidemiology and Clinical Research in Sydney, and the International Antiviral Therapy Evaluation Center in Amsterdam.[58] The HIV-NAT drug fund program at first assess all patients who applied and thereafter reassess them yearly by experienced social workers to ascertain their ability to pay, and the case is at that moment reviewed by the drug fund committee who determines an amount to be subsidized. The committee is entrusted with the responsibility of exploring possible ways to decrease costs without endangering the patient as well as supervising the bulk purchase of drugs to get low prices. The support provided to patients might be in the form of cash or drugs or a combination of the two.[59]

The second strategy stresses that the problem of post-trial access to beneficial treatment for participants should be examined in the context of other considerations for ethical research. Providing treatment to a few individuals during or at the end of a research does not completely remove or tackle concerns regarding exploitation. Negotiating fair benefits in the context of research to minimize exploitation of participants and host populations is a critical part of collaborative research.[60]

The third strategy emphasizes that the world health community must continue to be dedicated to discovering strategies to encourage improved access to required

---

[54] Christine Grady, "The Challenge of Assuring Continued Post-Trial Access to Beneficial Treatment," *Yale Journal of Health Policy, Law and Ethics* 5, no. 1 (Winter 2005): 433–434.

[55] Grady, "The Challenge of Assuring Continued Post-Trial Access to Beneficial Treatment," 434.

[56] Grady, "The Challenge of Assuring Continued Post-Trial Access to Beneficial Treatment," 432.

[57] Grady, "The Challenge of Assuring Continued Post-Trial Access to Beneficial Treatment," 434.

[58] Jintanat Ananworanich et. al, "Creation of A Drug Fund for Post-Clinical Access to Antiretrovirals," *The Lancet* 364, no. 9428 (July 2004): 101–102.

[59] Ananworanich et. al, "Creation of A Drug Fund for Post-Clinical Access to Antiretrovirals," 101.

[60] Ezekiel J. Emmanuel et al., "Addressing Exploitation: Reasonable Availability Versus Fair Benefits," in *Exploitation and Developing Countries: The Ethics of Clinical Research*, ed. Jennifer S. Hawkins and Ezekiel J. Emmanuel (Princeton, New Jersey: Princeton University Press, 2008), 286–310.

health care and treatment globally. This involves the energy and creativity of policymakers, scientists, clinical providers, politicians, and communities.[61]

The fourth strategy stresses that sponsors and researchers should shoulder responsibility for certain short-term solutions when suitable. For instance, it would be essential to offer medications to participants while waiting for the approval of the tested drug, or to establish or promote patient assistance programs for costly treatments. Continued focus on reducing the costs of treatments for people who need them is as well demanded.[62]

Justifications that would put the burden or responsibility on government programs such as beneficence and justice entails that the governments would have similar responsibilities to people not enrolled in the clinical research. They would not usually justify merely moving resources from non-participants to participants. Conversely, justifications that would entail an obligation owed exclusively to research participants such as compensation for harm or the research-participant relationship tend to be those that oblige simply the researchers and their sponsors.[63]

Justifications given for providing post-trial anti-retroviral drugs need to coincide with mechanisms proposed for providing it. Joseph Millium offered several justifications for providing post-trial antiretroviral drugs to participants and host populations of clinical research. Six different justifications offered for providing post-trial antiretroviral drugs for participants and host populations include harm to participants, fiduciary relationship, reciprocity, duty of rescue, imperfect duty of beneficence and global justice.[64] Justifications focused on obligation owed only to participants such as harm to participants, fiduciary relationship and reciprocity place obligation on researchers and their sponsors. On the other hand, justifications focused on obligation owed to people in urgent need, people in need and people in unjust situations such as duty of rescue, imperfect duty of beneficence and global justice respectively place obligation on all those who can help[65] including researchers, sponsors, host country governments, international governmental or non-governmental aid agencies and governments of affluent developed countries.

Duties of beneficence are general duties that might be shouldered by everyone. Moral agents have general duties simply as a consequence of their agency, and general duties are owed to moral patients simply as a consequence of their moral status.[66] Millum argues that, "… unlike, say, duties arising from the researcher-participant relationship, a duty of beneficence to supply (antiretroviral) ART could fall to governments or international bodies who are entirely independent of the research

[61] Grady, "The Challenge of Assuring Continued Post-Trial Access to Beneficial Treatment," 435.

[62] Grady, "The Challenge of Assuring Continued Post-Trial Access to Beneficial Treatment," 435.

[63] Joseph Millum, "Post-Trial Access to Antiretrovirals: Who Owes What to Whom?" Bioethics 25, no. 3 (2011): 153.

[64] Millum, "Post-Trial Access to Antiretrovirals: Who Owes What to Whom?" 148–153.

[65] Millum, "Post-Trial Access to Antiretrovirals: Who Owes What to Whom?" 153.

[66] Millum, "Post-Trial Access to Antiretrovirals: Who Owes What to Whom?" 152.

enterprise."[67] The implication is that all the mechanisms that have been proposed for providing ART to research participants would be valid strategies to fulfill a duty of beneficence.[68]

Similar to duties of beneficence, the duty to promote global justice is probably to be general which implies that everyone is responsible and not simply those people who are actively involved in the research and in the interaction with the global poor. Therefore, "there is no reason to think that researchers working in the developing world have any greater duty to promote justice than people who are not, where those people could also make a difference."[69] The implication is that similar to beneficence, that all the proposed mechanisms for providing ART could be valid strategies to accomplish a duty to promote justice. Duties of beneficence and duties to correct injustice are based on the unfortunate condition of the beneficiaries; they are not contingent on the beneficiaries engaging in clinical trials.[70]

The Committee on Economic, Social and Cultural Rights allude to resources within the state as well as "those available through international cooperation and assistance"[71] in fulfilling the core obligations of the right to health. From this perspective, it is evident that developed countries have obligations of international assistance and cooperation for access to essential drugs[72] including anti-retroviral drugs to participants and host populations in developing countries. Another component of the proposed paradigm is the special obligations of researchers and their sponsors to participants and host populations and whether such obligations can be transferred to other benefits other than just providing successful drugs is discussed next. We will also engage and respond to some authors who argue that pharmaceutical companies do not have any moral obligations to research participants and host communities after the end of clinical trials.

6.2.1.3 Obligations of Researchers and Sponsors to Participants and Host Populations

There is a consensus that researchers and sponsors have obligations to participants and host populations at the end of a clinical trial. However, there is a contentious debate regarding what exactly is owed to participants and host populations, whether it is the successful product or intervention or whether it is other benefits negotiated by the stakeholders. Furthermore, there is a controversy regarding whether

---

[67] Millum, "Post-Trial Access to Antiretrovirals: Who Owes What to Whom?" 152.

[68] Millum, "Post-Trial Access to Antiretrovirals: Who Owes What to Whom?" 152.

[69] Millum, "Post-Trial Access to Antiretrovirals: Who Owes What to Whom?" 152.

[70] Millum, "Post-Trial Access to Antiretrovirals: Who Owes What to Whom?" 152.

[71] Committee on Economic, Social and Cultural Rights, *General Comment No. 3: The Nature of States Parties' Obligations*, UN Doc. No. E/C.12/1991/23, (New York: United Nations, 1990), Para. 13.

[72] Emily A. Mok, "International Assistance and Cooperation for Access to Essential Medicines," *Health and Human Rights* 12, no. 1 (2010): 76.

providing a successful intervention adequately and fairly compensates the participants and host populations of the clinical research. Another aspect of debate deals with some authors who contend that pharmaceutical companies do not have any moral obligations to research participants and host communities after the conclusion of clinical trials.

International ethical guidelines and comprehensive reports on international research address the post-trial obligations of sponsors and researchers to participants and host populations. All the guidelines and reports indicate the common notion that research must be responsive to the health needs and priorities of the population where the research is executed and should likely benefit that population.[73]

Previous versions of the Declaration of Helsinki (DoH) did not mention a requirement regarding making successful products available to participants or to host populations after the conclusion of a trial. However, the 2000 revision of DoH tackles the issue in paragraphs 19 and 30. Paragraph 19 indicates that, "Medical research is only justified if there is a reasonable likelihood that the populations in which the research is carried out stand to benefit from the results of the research."[74] The rendering of this paragraph presents some limitations such as difficulty to ascertain the criteria for the likelihood of benefit and the degree of likelihood necessary. DoH as well tackles in a strong requirement the issue of benefits that accrue to the participants in paragraph 30: "At the conclusion of the study, every patient entered into the study should be assured of access to the best proven prophylactic, diagnostic and therapeutic methods identified by the study."[75] The strong requirement articulated by DoH coincides with Guidance Point 2 of the UNAIDS Guidance Document for preventive vaccine research, about what should be made reasonably available to research participants: "Any HIV preventive vaccine demonstrated to be safe and effective… should be made available as soon as possible to all participants in the trials in which it was tested as well as to other populations at high risk of HIV infection …plans should be developed at the initial stages of HIV vaccine development to ensure such availability"[76] Unlike DoH, UNAIDS Guidance document stresses the importance of extending the benefits to others in the community or country at the end of successful trials. It also emphasizes the discussion regarding making a successful vaccine available before the commencement of the trials.[77]

[73] Ruth Macklin, *Double Standards in Medical Research in Developing Countries* (Cambridge: Cambridge University Press, 2004), 82.

[74] World Medical Association, "Declaration of Helsinki: Ethical Principles for Medical Research Involving Human Subjects," *Bulletin of the World Health Organization* 79, no. 4 (2001): 374.

[75] World Medical Association, "Declaration of Helsinki: Ethical Principles for Medical Research Involving Human Subjects," 374.

[76] Joint United Nations Programme on HIV/AIDS (UNAIDS), *Ethical Considerations in HIV Preventive Vaccine Research: UNAIDS Guidance Document*, (Geneva: UNAIDS, 2000), 13.

[77] Joint United Nations Programme on HIV/AIDS (UNAIDS), *Ethical Considerations in HIV Preventive Vaccine Research: UNAIDS Guidance Document,* 14.

The Council for International Organizations of Medical Sciences (CIOMS) in its 2002 version emphasizes two important aspects of international research including the research being responsive to the health needs and priorities of the community in which it is conducted and making any successful product reasonably available to the population or community that hosted the trial.[78]

The US National Bioethics Advisory Commission (NBAC) report and the Nuffield Council report tackle two significant points concerning availability of successful products at the conclusion of a trial: "availability to the research participants themselves (the only point addressed in the Declaration of Helsinki), and availability of successful products to others in the country or community."[79] The NBAC recommends that research proposals must incorporate an explanation how effective new interventions would be made available to some or all the populations of the countries that are hosting the research, simultaneously with research participants themselves at the conclusion of the research.[80] The Nuffield Council report stresses that researchers must commit before beginning a trial, to guarantee post-trial access to effective interventions to participants and host populations at the end of the trial. The research proposals are also required to incorporate an explanation of how new proven interventions would be made available to both research participants and the host populations. The report also acknowledges that post-trial access to effective interventions would be contingent on several factors, such as the result of the research, the cost of providing the intervention and overseeing its provision, threat engendered by the disease and the obligation of making a successful intervention available is primarily shouldered by national government.[81] The Universal Declaration on Bioethics and Human Rights signed by 191 countries articulated in Article 15 that "benefits resulting from any scientific research and its application should be shared with society as a whole and within the international community, in particular with developing countries."[82] However, the benefits can take several forms, in agreement with the principles of the Declaration, but not essentially continuity of treatment.[83]

A vast number of available literature endorsed post-trial obligations of researchers and sponsors to participants and host populations after the conclusion of a clinical trial. A survey conducted in developing countries, with researchers in the HIV/AIDS area, endorsed that the participant population of the studies should benefit

---

[78] Council for International Organizations of Medical Sciences, *International Ethical Guidelines for Biomedical Research Involving Human Subjects*, (Geneva: CIOMS, 2002), 51.

[79] Macklin, *Double Standards in Medical Research in Developing Countries,* 83.

[80] National Bioethics Advisory Commission (NBAC), *Ethical and Policy Issues in International Research: Clinical Trials in Developing Countries* (Bethesda, MD: NBAC, 2001), 74.

[81] Nuffield Council on Bioethics, *The Ethics of Research Related to Healthcare in Developing Countries*, (London, UK: Nuffield Council on Bioethics, 2002), 120–125.

[82] Hans Galjaard, "Sharing Benefits" in *The UNESCO Universal Declaration on Bioethics and Human Rights: Background, Principles and Application*, ed. Henk A.M.J. ten Have and Michele S. Jean (Paris, France: UNESCO Publishing, 2009), 231.

[83] Galjaard, "Sharing Benefits" in *The UNESCO Universal Declaration on Bioethics and Human Rights: Background, Principles and Application*, 231.

from the study, and about more than half of the researchers from U.S. and developing countries surveyed endorsed that interventions proven effective should be provided to the host population at the conclusion of the study for two to five years.[84]

A qualitative study, conducted through focal groups in Kenya, with 89 research participants comprising potential patients for HIV/AIDS studies, researchers and administrators concluded that it would be unreasonable to stop providing therapy after HIV/AIDS studies to patients, except in cases where it is completely justified.[85] Zhiyong Zong also discussed the issue of post-trial provision of beneficial experimental interventions especially in developing countries citing international guidelines and recommendations that addressed the subject. Zong endorses planning in advance and establishing a collaborative partnership among pertinent parties such as sponsors, researchers, local healthcare system, the Research Ethics Committee and participants as a viable strategy for addressing the issues concerning post-trial provision.[86]

A study published in 2009 by Seema Shah, Stacey Elmer and Christine Grady discusses planning for post-trial access to antiretroviral treatment for research participants in developing countries with focus on the implementation process. The study investigated whether the National Institutes of Health (NIH) guidelines have been implemented in ART trials funded by NIH in developing countries. The 18 studies identified in the database of the Division of AIDS (DAIDS) included plans for post-trial access for participants. More than 70% that is, about 13 of 18 trials had specific mechanisms for realizing post-trial access, but none of them ensured long-term access. Half of the trials incorporated explanations of post-trial access that involved collaboration with outside sources or national access programs, established by the governments of the countries hosting the trials. None of the studies advocated for priority access for trial participants in connection to other patients in the country. The authors contend that the strength and form of the NIH guidelines support researchers to explore alternatives and collaboration to expedite access to antiretroviral treatment. Similarly, the flexibility of the guidelines expedites and promotes the learning of practical difficulties, a more effective strategy than establishing stringent requirements that researchers may be unable to fulfill.[87]

Three competing paradigms have been proposed regarding providing benefits to participants and host populations at the conclusion of a trial including reasonable

---

[84] Nancy Kass and Adnan A. Hyder, "Attitudes and Experiences of U.S. and Developing Country Investigators Regarding U.S. Human Subjects Regulations," in *Ethical and Policy Issues in International Research: Clinical Trials in Developing Countries*, Vol. II, Commissioned Papers and Staff Analysis, (Bethesda, MD: National Bioethics Advisory Commission, 2001), B39–40.

[85] D.N. Shaffer et al., "Equitable Treatment for HIV/AIDS Clinical Trials Participants: A Focus Group Study of Patients, Clinician Researchers and Administrators in Western Kenya," *Journal of Medical Ethics* 32, no. 1 (January 2006): 55–60.

[86] Zhiyong Zong, "Should Post-Trial Provision of Beneficial Experimental Interventions be Mandatory in Developing Countries," *Journal of Medical Ethics* 34, no. 3 (2008): 188–192.

[87] Seema Shah, Stacey Elmer and Christine Grady, "Planning for Posttrial Access to Antiretroviral Treatment for Research Participants in Developing Countries," *American Journal of Public Health* 99, no. 9 (September 2009): 1556–1562.

availability, fair benefits framework and human development approach. The concept of reasonable availability is ambiguous. CIOMS recognizing this ambiguity of reasonable availability indicates that "the issue of reasonable availability is complex and will need to be determined on a case- by-case basis and then enumerates countless "relevant considerations."[88] Four primary issues require stipulation: "(1) the nature of the commitment; (2) who is responsible for fulfilling the requirement; (3) what constitutes making something reasonably available; and (4) who must have access."[89] Each of these issues has attracted a variety of answers. Therefore, regardless of the consensus on reasonable availability requirement, there is a considerable controversy on how it should be stipulated and essentially implemented.[90]

Nevertheless, proponents of reasonable availability consider it as a requirement of ethical research in developing countries that is critical to avert exploitation of communities.[91] CIOMS articulates, "If there is good reason to believe that a product developed or knowledge generated by research is unlikely to be reasonably available to…the population of a proposed host country or community after the conclusion of the research, it is unethical to conduct the research in that country or community."[92]

Reasonable availability requirement has been criticized for several reasons. First, it grapples with a mistaken conception of exploitation, because, while reasonable availability concentrates on a type of benefit such as a proven intervention, exploitation concentrates on a fair level of benefits. The emphasis is not on what people obtain but how much they obtain. Second, reasonable availability struggles with a narrow conception of benefits, because it considers only access to a drug, vaccine or intervention as a benefit, and disregards others benefits such as training, infrastructure, or health services. Third, reasonable availability deals with excessively broad group of beneficiaries. It requires post-trial access for host community or country. On the other hand, tackling exploitation requires benefits just for those shouldering risks or burdens of research, without any justification to bestow benefits on a whole country that does not shoulder a burden or risk of research. Fourth, reasonable availability grapples with the issue that no single trial is conclusive. It requires access to a drug, vaccine, or intervention after a single trial. However, it frequently requires many confirmatory trials to justify the safety and efficacy of an intervention. Fifth, reasonable availability that entails providing one drug can be "golden handcuff" because if research demonstrates that another intervention is more effective, the

[88] Council for International Organizations of Medical Sciences, *International Ethical Guidelines for Biomedical Research Involving Human Subjects,* 53.

[89] Ezekiel J. Emmanuel, "Benefits to Host Countries," in *The Oxford Textbook of Clinical Research Ethics*, ed. Ezekiel J. Emmanuel et al., (New York: Oxford University Press, 2008), 721.

[90] Ruth Macklin, "After Helsinki: Unresolved Issues in International Research," *Kennedy Institute of Ethics Journal* 11, no. 1 (March 2001): 17–36.

[91] Emmanuel et al., "Addressing Exploitation: Reasonable Availability Versus Fair Benefits," 290.

[92] Council for International Organizations of Medical Sciences, *International Ethical Guidelines for Biomedical Research Involving Human Subjects*, 53.

community may be assured the old drug not the newer and more effective one.[93] Sixth, it is not within the scope of the authority of researchers and numerous sponsors to ensure reasonable availability. Clinical researchers as well as some sponsors in developed countries, such as the NIH and the Medical Research Council of the United Kingdom (MRC), do not regulate drug approval processes in their own countries, let alone other countries. In the same way, they do not regulate budgets for health ministries or foreign or development aid in order to put into practice research results such as provision of drugs or vaccines, and may be, by law, prohibited from providing successful interventions at the conclusion of trials as in the case of NIH. Seventh, the requirement of reasonable availability implies that the population is deprived of making its own autonomous decisions regarding what benefits merit the risks of a research trial.[94] Due to numerous shortcomings of reasonable availability requirement, the fair benefits framework was proposed by the participants of the 2001 Conference on Ethical Aspects of Research in Developing Countries at Malawi.[95]

The fair benefits requirement was highlighted by the DoH paragraph 33 with its reference to "access to interventions identified as beneficial in the study or to other appropriate care or benefits"[96] at the conclusion of the trial. The UNESCO Universal Declaration on Bioethics and Human Rights also alludes to sharing various forms of benefits from scientific research with the entire society especially in developing countries. It provides a comprehensive list of benefits such as: (a) special and continuous aid to, and recognition of, the individuals and groups that have participated in the research; (b) affordable access to quality health care; (c) provision of new therapeutic interventions originating from scientific research; (d) assistance for health services; (e) access to knowledge generated from science and technology; (f) capacity-building facilities for research goals; (g) other types of benefits in consonant with the principles established in this declaration.[97]

The fair benefits framework enunciates two basic assumptions. First, the solution to avoiding exploitation is to guarantee that people who shoulder the risks and burdens of research obtain fair benefits through the conduct and/ or results of research. Second, all forms of benefits that accrue from research, not only access to a tested drug, vaccine or intervention, must be examined in ascertaining the fair benefits.[98] The population at risk for exploitation is the pertinent group to obtain benefits; this encompasses the participants in the clinical research and any members of the community who might as well shoulder burdens and risks for conducting the research.[99]

---

[93] Emmanuel, "Benefits to Host Countries," 723.

[94] Emmanuel et al., "Addressing Exploitation: Reasonable Availability Versus Fair Benefits," 297.

[95] Emmanuel et al., "Addressing Exploitation: Reasonable Availability Versus Fair Benefits," 307–310.

[96] World Medical Association, *Declaration of Helsinki: Ethical Principles for Medical Research Involving Human Subjects*, (Edinburgh, Scotland: World Medical Association, 2008), 5.

[97] Galjaard, "Sharing Benefits" 231.

[98] Emmanuel, "Benefits to Host Countries," 724–725.

[99] Emmanuel, "Benefits to Host Countries," 725.

Providing benefits just to research participants would broaden health inequalities in the resource-poor host country and consequently highlight issues regarding causing injustice.[100] Therefore, providing benefits to the host country should be executed in a fashion that improves rather than exacerbates health inequalities.[101]

The fair benefits framework maintains that "there should be a comprehensive delineation of tangible benefits to the research participants and the population from both conduct and the results of research."[102] Some of the benefits consist of: (a) Improved health of research subjects; (b) Posttrial access to medications for research subjects; (c) Health services and public health measures accessible to the population; (d) Employment and economic activity; (e) Availability of interventions at the conclusion of research; (f) Improvements to the health care infrastructure, training of health care and research professionals and research capacity; (g) Long-term research collaboration; (h) Sharing of financial rewards, including intellectual property rights.[103] Such benefits guarantee that community where the research is conducted will obtain benefits in return for engaging in the research. Building infrastructure has been identified as a good way, researchers can help offer sustainable improvements that will assist to shrink health inequalities between rich and poor nations.[104]

The form and amount of such collateral or secondary benefits to participants and communities should be negotiated among the sponsors, researchers, and host-country partners before the research is initiated. Overcoming disparities in negotiating power requires that agreements should be made public in order that other communities and countries will know what benefits might be realized.[105]

The fair benefits framework would seem to be effective at averting exploitation at the level of the individual research participants and the level of the host population, community, or country, and it provides more understandable guidance than DoH or CIOMS guidelines.[106] Nevertheless, some authors contend that someone could give a justification for any research with human participants as long as other benefits, not connected to the research itself, can be utilized to justify unneeded research. Most people agree that research on male pattern baldness and cosmetic surgery, currently, should not be considered to be highly significant, but it might be justified in the context of fair benefits approach.[107]

---

[100] Bernard Lo, Nancy Padian and Mark Barnes, "The Obligation to Provide Antiretroviral Treatment in HIV Prevention Trials," *AIDS* 21, no. 10 (June 2007): 1229–1231.

[101] Bernard Lo, *Ethical Issues in Clinical Research: A Practical Guide* (Philadelphia: Lippincott Williams & Wilkins, 2010), 203.

[102] Emmanuel, "Benefits to Host Countries," 725.

[103] Emmanuel, "Benefits to Host Countries," 725.

[104] Lo, *Ethical Issues in Clinical Research: A Practical Guide*, 203.

[105] Lo, *Ethical Issues in Clinical Research: A Practical Guide*, 203.

[106] Adil E. Shamoo and David B. Resnik, *Responsible Conduct of Research*, Second Edition, (New York: Oxford University Press, 2009), 335.

[107] Adil E. Shamoo, "Debating Moral Issues in Developing Countries," *Applied Clinical Trials* 14, no. 6 (June 2005): 86–96.

Some authors argue that fair benefits framework does not go far enough to tackle issues of exploitation or benefit for the host population.[108] Alex John London points out the shortcoming of the fair benefits framework on the basis that its idea of exploitation and justice are too restricted. He contends that to comprehend exploitation and justice one must examine beyond specific transactions or relationships and examine the larger social, economic, cultural, and political context. For London, the fair benefits framework is contingent on a fair agreement between researchers/sponsors and a host population, community, or country. The agreement is fair if both parties give their assent and benefits are fairly distributed. He contends that the issue with this notion is that it disregards the broader context in which the agreement is established, such as extreme poverty, famine and disease in the host country, or the history of relationship between two countries which may involve racism, slavery, theft, or exploitation. He argues that an agreement cannot be really fair without one tackling this broader context. Accomplishing this goal requires that researchers and sponsors do more than merely providing fair benefits. They must engage in rectifying past injustices and support social, economic, and political development in the host country. This perspective was dubbed the human development approach by London.[109] In a nutshell, he contends that "a better approach will reframe the question of justice in international research in a way that makes explicit the links between medical research, the social determinants of health, and global justice."[110]

Several criticisms were leveled against the human development approach. Shamoo acknowledges London's human development approach being admirable in several ways, but argues that it is as well too idealistic and unrealistic. It overly requires researchers and sponsors to do a lot more for the host countries than they can probably be expected to do. The human development approach if implemented would make biomedical research in developing countries exorbitant and complex. Sponsors would opt to refrain from research in developing countries to circumvent paying the exorbitant costs of nation building.[111] In as much as promoting economic, social, and political development in developing countries is a valuable goal, it is a responsibility best entrusted to the United Nations, the International Monetary Fund, the World Bank, and other organizations whose primary goal is development. The primary goal of research is research. Researchers and sponsors should offer meaningful and fair benefits to the population hosting the research, but they need not overextend themselves in what they do.[112]

On the other hand, Emmanuel criticized the human development approach in various areas. He highlights the abstract nature and ambiguity of the human

---

[108] Shamoo and Resnik, *Responsible Conduct of Research*, 335.

[109] Alex John London, "Justice and the Human Development Approach to International Research," *Hastings Center Report* 35, no. 1 (January-February 2005): 24–37.

[110] London, "Justice and the Human Development Approach to International Research," 24.

[111] Mary Terrell White, "A Right to Benefit from International research: A New Approach to Capacity Building in Less-Developed Countries," *Accountability in Research: Policies and Quality Assurance* 14, no. 2 (2007): 73–92.

[112] Shamoo and Resnik, *Responsible Conduct of Research*, 335.

development that makes it hard to be sure what it requires. The human development approach misconstrues the problem that guaranteeing benefits to host countries is intended to tackle. Most people agree that global injustice and exploitation are critical ethical issues in the world. Nevertheless, the aim of identifying clearly the extent of the obligation to provide benefits to developing countries that participate in biomedical research is to reduce the possibility of exploitation by developed country researchers and sponsors. Such benefits are not intended to tackle fundamental background global injustice. The human development approach is therefore pointing to a different problem from that being tackled by the reasonable availability requirement or fair benefits framework. There is a detach between the ethical challenges presented by conducting clinical trials in developing countries and the issues the human development approach considers itself to be tackling.[113] Furthermore, the human development approach appears most pertinent in helping to identify clearly what research questions are being pursued in developing countries, rather than the benefits that result from particular research protocols.[114]

Another aspect of the debate deals with engaging and responding to some authors who argue that big pharmaceutical companies do not have any moral obligations to research participants and host communities after the conclusion of the trial. Some authors argue that it is justified if the participants and host communities do not receive any treatment at all prior to the research because participation in the trial is better than nothing.[115] They contend that research participants were no worse off than if they were not in the trial,[116] and were not being deprived of any benefit as it provided opportunity for the host populations to access health care they wouldn't have ordinarily had.[117] Resnik contends that " provision of AZT, unavailable to the general population of the host country, is supererogatory requirement."[118] He argues that ethical obligations of researchers are dependent on socio-economic context of the host populations.[119] In contrast, Angell argues that, "ethical standards governing research should vary with the political and economic conditions of the region."[120]

[113] Emmanuel, "Benefits to Host Countries," 726–727.

[114] Emmanuel, "Benefits to Host Countries," 727.

[115] Reidar K. Lie, Ethical Issues in Clinical Trial Collaborations with Developing Countries – with Special Reference to Preventive HIV Vaccine Trials with Secondary Endpoints, (Bergen, Norway), https://enterics.tghn.org/site_media/media/articles/trialprotocoltool.../Ethicals.pdf (Accessed March 25, 2017).

[116] Marcia Angell, "Investigators' Responsibilities for Human Subjects in Developing Countries," The New England Journal of Medicine 342, no. 13 (March 30, 2000): 967–969.

[117] Paquita De Zulueta, "Randomised Placebo-Controlled Trials and HIV-Infected Pregnant Women in Developing Countries. Ethical Imperialism or Unethical Exploitation," Bioethics 15, no. 4 (August 2001): 294.

[118] Zulueta, "Randomised Placebo-Controlled Trials and HIV-Infected Pregnant Women in Developing Countries. Ethical Imperialism or Unethical Exploitation," 300.

[119] David B. Resnik, "The Ethics of HIV Research in Developing Nations," Bioethics 12, no. 4 (October 1998): 286–306.

[120] Angell, "Investigators' Responsibilities for Human Subjects in Developing Countries," The New England Journal of Medicine, 967–969.

She argues that our ethical standards should not be determined by where the research is conducted.[121]

When big pharmaceutical companies fail to facilitate post-trial access of successful drugs, it is a failure of their obligations to protect the best interests of their research participants and host populations and to promote distributive justice. Furthermore, local economic conditions cannot determine and justify solutions for post-trial benefits to participants and host populations. A global notion of justice, which emphasizes international equity contradicts any international research guidelines that permits local conditions to define the scope of obligations of big pharmaceutical companies to research participants and host populations. Benatar and Singer argue that, "Research ethics must be deeply rooted in the context of global health, and that …it must ultimately be concerned with reducing inequalities in global health and achieving justice in health research and health care."[122] Globalization reinforced by a market ethic does not serve the interests of the worst off but rather the opposite.[123] Zulueta argues that, " a global ethics, underpinned by an ethic of distributive justice, and framed in terms of human rights, has the potential to redress the balance and promote greater equity."[124]

A framework for conducting international clinical trials in developing countries that would foster justice and fairness to facilitate post-trial access to research participants and host populations is imperative. Big pharmaceutical companies can take several steps to facilitate post-trial access to drugs – lower prices, donate some drugs, or negotiate in advance to make successful drugs available after the end of the trial.[125] There is increasing support not to justify the view that a research that simply contributes to scientific knowledge is relevant to conditions in developing countries.[126] Macklin articulates that, "When research involves testing a product whose effectiveness can ameliorate disease conditions but will never be made available in such countries, it cannot be considered relevant to those countries."[127] As clinical trials carried out in low-income countries would likely continue, there is a pressing need to consider post-trial access interventions for participants and host populations. The issue of post-trial access to successful interventions may not be as urgent in nations where participants routinely access healthcare and medicines through public schemes, but it is essentially critical in places where this is nonexistent. The need to discuss concrete strategies to guarantee post-trial access to

[121] Angell, "Investigators' Responsibilities for Human Subjects in Developing Countries," The New England Journal of Medicine, 967–969.

[122] Solomon R. Benatar and Peter A. Singer, "A New Look at International Research Ethics," BMJ 321, (September 2000): 824–826.

[123] Zulueta, "Randomised Placebo-Controlled Trials and HIV-Infected Pregnant Women in Developing Countries. Ethical Imperialism or Unethical Exploitation," 309.

[124] Zulueta, "Randomised Placebo-Controlled Trials and HIV-Infected Pregnant Women in Developing Countries. Ethical Imperialism or Unethical Exploitation," 309.

[125] Macklin, *Double Standards in Medical Research in Developing Countries,* 118.

[126] Macklin, *Double Standards in Medical Research in Developing Countries,* 118.

[127] Macklin, *Double Standards in Medical Research in Developing Countries,* 118.

beneficial interventions is critical.[128] Another critical issue addressed in the context of a paradigm for global justice is the global responsibility for growing health inequalities discussed in the next part of this chapter.

6.2.2 Global Health Inequalities

6.2.2.1 Challenges of Global Health

The increasing global health inequalities resulting in poor health outcomes especially among the world's global poor in developing countries created an urgent need for "fair and effective global governance for health – the organization of national and global norms, institutions, and processes that collectively shape the health of the world's population."[129] The relevance of global governance for health extends beyond the health arena. It entails rectifying the presently unfair and harmful health effects of international regimes such as international trade, intellectual property and finance, and establishing secure, active, democratic political institutions.[130]

The Joint Action and Learning Initiative on National and Global Responsibilities for Health (JALI) was established by a coalition of civil society and academics, with a shared vision of the right to health.[131] JALI attempts to establish a post-Millennium Development Goal (MDG) framework for global health, one entrenched in the right to health and intended for obtaining universal health coverage for all people.[132] JALI establishes an international agreement regarding solutions to four critical challenges of global health: (I) explaining essential health services and goods; (II) elucidating governments' obligations to their country's populations; (III) investigating the responsibilities of all governments towards the global poor; and (IV) introducing a global structure to enhance health as a matter of social justice.[133] A brief discussion of the four critical challenges of global health follows.

---

[128] Kori Cook, Jeremy Snyder and John Calvert, "Attitudes Toward Post-Trial Access to Medical Interventions: A Review of Academic Literature, Legislation, and International Guidelines", *Developing World Bioethics* 16, no. 2 (August 2016): 78.

[129] Lawrence O. Gostin et al., "The Joint Action and Learning Initiative: Towards a Global Agreement on National and Global Responsibilities for Health," *PLoS Medicine* 8, no. 5 (May 2011): e1001031.

[130] Gostin et al., "The Joint Action and Learning Initiative: Towards a Global Agreement on National and Global Responsibilities for Health," e1001031.

[131] Gostin et al., "The Joint Action and Learning Initiative: Towards a Global Agreement on National and Global Responsibilities for Health," e1001031.

[132] Gostin et al., "The Joint Action and Learning Initiative: Towards a Global Agreement on National and Global Responsibilities for Health," e1001031.

[133] Lawrence O. Gostin et al., "National and Global Responsibilities for Health," *Bulletin of the World Health Organization* 88, (2010): 719.

I. Essential Services and Goods Ensured to Every Person under the Human Right to Health

The first crucial challenge for JALI is to ascertain essential health services and goods that every person has a right to anticipate. Gostin et al. writes "without articulating these, it is impossible to define each state's obligation to its own inhabitants, as well as the duties of high-income countries towards low- and middle-income countries."[134] The World Health Organization (WHO) has highly prioritized the place of universal health coverage on the global health agenda,[135] expounding three dimensions of coverage: "(1) the proportion of the population served; (2) the level of services; and (3) the proportion of health costs covered by prepaid pooled funds."[136] Universal coverage is defined "as access to key promotive, preventive, curative and rehabilitative health interventions for all at an affordable cost."[137]

The human right to health which is an international treaty obligation offers crucial understanding about how states should work towards realizing universal coverage.[138] Core obligations provide criteria to evaluate progress towards universal coverage, for example "non-discrimination, equitable distribution of health facilities, and essential services for all, including those addressing underlying determinants of health."[139]

States are required by the core principle of equality to emphasize covering 100% of their populations. Even though 100% coverage of all health services may not be feasible right away, full coverage of essential health interventions should be an initial standard or criterion towards universal coverage.[140] The right to health framework works against a restricted definition of essential services. The essential services should entail WHO's building blocks for health services such as "services, workforce, information and financing and governance; essential vaccines, medicines and technologies; and fundamental human needs (e.g.,

[134] Gostin et al., "National and Global Responsibilities for Health," 719.

[135] World Health Assembly, *Primary Health Care, Including Health System Strengthening: WHA62.12* (Geneva: WHO, 2010), http://apps.who.int/gb/ebwha/pdf_files/WHA62-RECI/WHA62_RECI-en-P2.pdf (Accessed July 1, 2014).

[136] World Health Organization, *World Health Report: Health Systems Financing: the Path to Universal Coverage* (Geneva: WHO, 2010), http://www.who.int/whr/2010/en/index.html (Accessed July 1, 2014).

[137] World Health Assembly, *Social Health Insurance: Sustainable Health Financing*, Universal Coverage and Social Health Insurance, Report by Secretariat, A58/20 (Geneva: WHO, 2005), http://apps.who.int/gb/ebwha/pdf_files/WHA58/A58_20-en.pdf. (Accessed July 1, 2014).

[138] Gostin et al., "The Joint Action and Learning Initiative: Towards a Global Agreement on National and Global Responsibilities for Health," e1001031.

[139] Committee on Economic, Social and Cultural Rights, *General Comment No. 14: The Right to the Highest Attainable Standard of Health,* UN Doc. No. E/C.12/2000/4, (Geneva : United Nations, 2000), http://www.unhchr.ch/tbs/doc.nsf/%28symbol%29/E.C.12.2000.4.En. (Accessed June 19, 2014).

[140] Gostin et al., "The Joint Action and Learning Initiative: Towards a Global Agreement on National and Global Responsibilities for Health," e1001031.

sanitation, nutritious food, potable water, safe housing, vector abatement, tobacco control, and healthy environments)."[141]

The provision of each of these essential services should signify just one critical step towards attaining the highest attainable standard of health. States, including affluent ones will need to continue to work towards achieving universal coverage.[142] The right to health requires these essential services to be universally available, acceptable, accessible and of good quality.[143]

II. States' Responsibilities for the Health of their Populations

States possess the primary responsibility to fund and guarantee all the essential goods and services within the context of the right to health.[144] The estimate of the WHO as the minimum annual cost for providing all the essential goods and services under the right to health is US$ 40 per person,[145] with the exclusion of basic survival needs. Nevertheless, the obligation of states should not be restricted just to their populations, but also to the global community to control health threats that jeopardize other countries and region. Gostin et al. argue that in most cases, "state obligations should extend to fostering a functioning interdependent global community, in which everyone recognizes that our mutual survival is a matter of common concern."[146]

There is no consensus on the level of health sector funding sufficient to fulfill the population's needs. African heads of state pledged to allocate at least 15% of national budgets to the health sector in 2001, in Abuja, Nigeria.[147] More so, about 32 African countries established an aspirational target of public sector budget allocations for sanitation and hygiene programs to attain at least 0.5% of gross domestic product.[148] In 2007, the average per capita of government health

---

[141] Gostin et al., "The Joint Action and Learning Initiative: Towards a Global Agreement on National and Global Responsibilities for Health," e1001031.

[142] Gostin et al., "The Joint Action and Learning Initiative: Towards a Global Agreement on National and Global Responsibilities for Health," e1001031.

[143] Committee on Economic, Social and Cultural Rights, *General Comment No. 14: The Right to the Highest Attainable Standard of Health,* UN Doc. No. E/C.12/2000/4, (Geneva: United Nations, 2000), http://www.unhchr.ch/tbs/doc.nsf/%28symbol%29/E.C.12.2000.4.En. (Accessed June 19, 2014).

[144] Gostin et al., "National and Global Responsibilities for Health," 719.

[145] Guy Carrin, David Evans, and Ke Ku, "Designing Health Financing Policy towards Universal Coverage," *Bulletin of the World Health Organization* 85, no. 9 (September 2007): 652.

[146] Gostin et al., *The Joint Action and Learning Initiative on National and Global Responsibilities for Health,* World Health Report, Background Paper, No. 53 (Geneva: World Health Organization, 2010), http://www.who.int/healthsystems/topics/financing/healthreport/JALI_No.53.pdf (Accessed June 8, 2014).

[147] Organization of African Unity, *Abuja Declaration on HIV/AIDS, Tuberculosis and Other Related Diseases*, (Abuja, Nigeria: OAU, 2001), http://www.un.org/ga/aids/pdf/abujadeclaration.pdf (Accessed July 2, 2014).

[148] Second African Conference on Sanitation and Hygiene, *The eThekwini Declaration*, (Durban, South Africa: AfricaSan + 5 Conference, 2008), 1–8, http://www.wsp.org/wsp/sites/wsp.org/files/publications/eThekwiniAfricaSan.pdf. (Accessed July 2, 2014)

investment in Africa is US$ 34, corresponding to a mean of 9.6% budget allocation, which is compared with US$ 1374 and 17.1% in the Americas.[149] This encompasses 15 African countries that invest as small as US$ 2–10 per capita, which cannot start to fulfill the population's health needs.[150] Furthermore, numerous low- and middle-income countries decrease domestic health spending for every dollar they obtain in foreign health assistance.[151]

Additionally, States have a responsibility to "govern well – honestly, transparently and accountably – with the full participation of civil society."[152] However, data shows that health systems among low-income countries are among the ones that are most badly governed.[153]

III. Responsibilities of All Countries to Guarantee the Health of the World's Population.

Resource-limited countries lack capacity to guarantee all of their populations even the essential health goods and services, let alone a fuller realization of the right to health. Countries well-placed to assist are required to do so in the context of the principles of international law and global social justice.[154] The Committee on Social, Economic and Cultural Rights has affirmed that cooperation towards fulfilling the right to health is "an obligation of all states," especially those "in a position to assist others."[155] All countries have mutual responsibilities towards guaranteeing the health of the most disadvantaged population of the world.[156]

Formulating global health funding as "aid" is basically faulty because it presumes an intrinsically unequal benefactor – dependent relationship. Essentially, global collaboration obliges a collective response to shared risks and basic rights, where all states possess mutual responsibilities.[157] Charitable giving gen-

[149] World Health Organization, *World Health Statistics 2010*, (Geneva: WHO, 2010), http://www.int/whosis/whostat/EN_WHS10_Part2pdf (Accessed July 2, 2014)

[150] African Civil Society, Letter to July 2010 African Union Summit on Upholding African Health and Social Development Commitments, EQUINET (Regional Network on Equity in Health in Southern Africa): 2010. http://www.equinetafrica.org/newsletter/index.php?id=7411 (Accessed July 2, 2014).

[151] Chunling Lu et al., "Public Financing of Health in Developing Countries: A Cross-National Systematic Analysis," *The Lancet* 375, no. 9723 (April 2010): 1375–1387.

[152] Gostin et al., "National and Global Responsibilities for Health," 719.

[153] Transparency International, *Global Corruption Report 2006: Corruption and Health* (London: Pluto Press, 2006), http://www.transparency.org/publications/gcr/gcr_2006 (Accessed July 2, 2014).

[154] Gostin et al., "The Joint Action and Learning Initiative: Towards a Global Agreement on National and Global Responsibilities for Health," e1001031.

[155] Committee on Economic, Social and Cultural Rights, *General Comment No. 14: The Right to the Highest Attainable Standard of Health*, UN Doc. No. E/C.12/2000/4, (Geneva: United Nations, 2000), http://www.unhchr.ch/tbs/doc.nsf/%28symbol%29/E.C.12.2000.4.En. (Accessed June 19, 2014).

[156] Gostin et al., "The Joint Action and Learning Initiative: Towards a Global Agreement on National and Global Responsibilities for Health," e1001031.

[157] Gostin et al., "National and Global Responsibilities for Health," 719.

erally signifies that "the donor decides how much to give, and for what and to whom."[158] Therefore, "aid" is "not predictable, scalable or sustainable."[159] It undercuts ownership of and responsibility for health programs of the host country.[160]

Apart from development assistance, coordination and coherence is desperately needed across sectors, as global health can be enhanced or harmed through state and international policies and rules that regulate areas such as international trade, intellectual property, health worker migration, international financing, and debt relief. These responsibilities include the use of state authority and influence over global institutions such as the World Bank, International Monetary Fund, and World Trade Organization.[161] Furthermore, high-income countries have not come nearer to realizing their pledge made in 1970 to spend 0.7% of their gross national product annually on Official Development Assistance (ODA).[162] After four decades, their average contribution is at 0.31%. Exploring innovative strategies to guarantee sufficient and lasting funding with agreed-upon priorities, will be critical in guaranteeing that poor countries obtain the capacity to realize the right to health.[163]

IV. Global Health Governance Required to Guarantee that All States Live up to their Mutual Responsibilities

Global health governance is necessary because states will not embrace international norms without true global partnerships, fair burden sharing and efficient programs that enhance health outcomes.[164] Gostin et al. writes "translating a shared understanding of national and global responsibilities into new realities requires effective and democratic governance for health."[165] Despite the Paris Declaration on Aid Effectiveness, global health grapples with grand challenges of poor leadership, poor coordination and underfunded priorities, and a deficiency in transparency, accountability, and enforcement.[166] More so, political, legal and economic challenges obstruct effective governance. Health ministries of various countries frequently lack fundamental knowledge of, and control over, foreign-supported programs. Gostin and Mok contend that we need a

[158] Gostin et al., "National and Global Responsibilities for Health," 719.

[159] Gostin et al., "National and Global Responsibilities for Health," 719.

[160] Gostin et al., "National and Global Responsibilities for Health," 719.

[161] Gostin et al., "The Joint Action and Learning Initiative: Towards a Global Agreement on National and Global Responsibilities for Health," e1001031.

[162] United Nations Millennium Project, *The 0.7% Target: An In-depth Look* (New York: United Nations Millennium Project, 2005), http://www.unmillenniumproject.org/press/07.htm (Accessed July 2, 2014).

[163] Gostin et al., "National and Global Responsibilities for Health," 719.

[164] Gostin et al., "National and Global Responsibilities for Health," 719.

[165] Gostin et al., "The Joint Action and Learning Initiative: Towards a Global Agreement on National and Global Responsibilities for Health," e1001031.

[166] Lawrence O. Gostin and Emily A.Mok, "Grand Challenges in Global Health Governance," *British Medical Bulletin* 90, no. 1 (2009): 7–18.

system of governance that promotes effective partnerships and coordinates initiatives to establish collaborations and circumvent destructive competition.[167]

More importantly, global health governance should strengthen the leadership and normative function of WHO which, as a United Nations agency, must have the legitimacy, authority and resources to assist all countries in ensuring the right to health.[168] Furthermore, state policies such as agricultural subsidies, intellectual property, and foreign affairs can effectively impact health in resource-poor countries. Consequently, states should endorse a "health-in-all-policies" approach where all ministries deal with the health effects of their policies and programs.[169] Effective governance must encompass active participation of the citizen to guarantee "transparency, collaboration, and accountability while maximizing creativity, and resource mobilization by states, international organizations, businesses, and civil society."[170]

The global health structure should make provision to hold stakeholders accountable, with clear criteria for success, monitoring progress and enforcement all of which have been deficient.[171] Lack of adequately exact obligations and compliance mechanisms in the context of the right to health impedes accountability, although encouraging signs of better approaches abound.[172] Human rights organizations and UN special rapporteurs are increasing the clarity of state obligations in the context of the right to health, which is imperative for meaningful accountability, just as are constitutional court decisions affirmed in Argentina, India, and South Africa.[173] Establishing the global responsibility for growing health inequalities among nations is discussed next.

6.2.2.2 Global Responsibility and Health Inequalities Between Nations

Ooms and Hammonds argue that there is global responsibility for global health and there are obligations of justice to help fulfill the right to health in other countries, because of the increasing wealth inequality between nations which has significant direct impact on health inequity.[174] Health inequity is explained as the "unjust

[167] Gostin and Mok, "Grand Challenges in Global Health Governance," 7–18.

[168] Gostin et al., "National and Global Responsibilities for Health," 719.

[169] Gostin et al., "The Joint Action and Learning Initiative: Towards a Global Agreement on National and Global Responsibilities for Health," e1001031.

[170] Gostin et al., "The Joint Action and Learning Initiative: Towards a Global Agreement on National and Global Responsibilities for Health," e1001031.

[171] Gostin et al., "The Joint Action and Learning Initiative: Towards a Global Agreement on National and Global Responsibilities for Health," e1001031.

[172] Gostin et al., "The Joint Action and Learning Initiative: Towards a Global Agreement on National and Global Responsibilities for Health," e1001031.

[173] Hans V. Hogerzeil et. al, "Is Access to Essential Medicines as Part of the Fulfillment of the Right to Health Enforceable through the Courts? *The Lancet* 368, no. 9532 (July 2006): 305–311. http://www.who.int/medicines/news/Lancet_EssMedHumanRight.pdf. (Accessed July 2, 2014)

[174] Ooms and Hammonds, "Taking Up Daniels' Challenge: The Case for Global Health Justice," 30.

distribution of the socially controllable factors affecting population health and its distribution,"[175] The concept of "health-related goods" introduced by John Arras and Elizabeth Fenton has been used as short hand of Daniels' "socially controllable factors affecting population health and its distribution"[176] in the discussion of growing health inequalities between nations.

There is a direct correlation between wealth inequalities and health inequalities between nations. Health-related goods are associated with costs in money. Health-related goods such as health care, prevention, water, sanitation, and nutrition involve a substantial spending for any nation. The implication is that what governments can spend on the distribution of health-related goods is exclusively determined by their revenue, which is invariably affected by their wealth.[177]

Studies on the evolution of global wealth inequalities conducted by Branko Milanovic showed that wealth inequalities between countries, articulated as an inter-country Gini coefficient, are progressively growing.[178] A Gini coefficient of zero utilized for inter-country wealth distribution would signify that all countries have precisely the same average Gross Domestic Product (GDP) per capita, which denotes complete inter-country equality. Additionally, Gini coefficient of one used for inter-country wealth distribution would signify that one single country would have the whole GDP of the world's economy, which denotes maximum inequality. Milanovic examines the inter-country Gini coefficients between countries from 1980–2000 and indicates how wealth inequality between countries is actually increasingly moving toward maximum inequality and away from maximum equality.[179]

The increasing inter-country inequality is attributed to several factors. The long-lasting effects of slavery and colonization[180] which constitute a significant factor in the inter-country inequality should not be minimized. Nunn attributes Africa's poor economic performance and underdevelopment in the second half of the twentieth century to its history of extraction, characterized by the events of slave trades and colonialism.[181] Nunn argues that there is a strong negative correlation between the number of slaves exported from a country and current economic performance. He argues further that evidence indicates that slave trades had an adverse effect on

[175] Daniels, *Just Health: Meeting Health Needs Fairly*, 101.

[176] John D. Arras and Elizabeth M. Fenton, "Bioethics and Human Rights: Access to Health-Related Goods," *Hastings Center Report* 39, no. 5 (2009): 27–38.

[177] Ooms and Hammonds, "Taking Up Daniels' Challenge: The Case for Global Health Justice," 30.

[178] Branko Milanovic, *Worlds Apart: Measuring International and Global Inequality* (Princeton: Princeton University Press, 2005), 39–40.

[179] Milanovic, *Worlds Apart: Measuring International and Global Inequality*, 39, 181.

[180] Nathan Nunn, "The Long Term Effects of Africa's Slave Trades," *Quarterly Journal of Economics* 123, no. 1 (2008): 139–176.

[181] Nunn, "The Long Term Effects of Africa's Slave Trades," 139

economic development of various countries in Africa.[182] Another important factor is that rich countries use their economic and political power to negotiate unequal or unfair trade agreement.[183] Stiglitz argues that the WTO rules governing international trade are extremely unfair because they have been intended to benefit the developed countries, to a certain extent at the expense of the developing countries.[184] Another factor worth noting is the shift of financial resources from poor to rich countries that are traced to illegal or at least illicit activities and causes, that is, "illicit financial flows" that significantly contribute to increasing inter-country inequality and can obfuscate international assistance from rich to poor countries.[185] The report findings show that developing countries lost an estimated $858.6 billion-$1.06 trillion in illicit financial outflows in 2006.[186]

Robert Merton's "Matthew Effect"[187] can as well be utilized in explaining the increasing inter-country inequality. This alludes to a verse in the biblical Gospel of Matthew – "For to all those who have, more will be given, and they will have an abundance; but from those who have nothing, even what they have will be taken away,"[188] – Merton highlights the improper allocation of credit for contributions in science. He explicates how scientists with significant reputation in their field are more likely to be acknowledged and awarded for their scientific work than scientists who have not created any impression, even when both equally contribute to a scientific advancement.[189]

Gunnar Myrdal built on the same biblical quotation to explicate his theory of "circular and cumulative causation," which estimates growing inequalities within and between countries engaging in a free market.[190] Myrdal's theory contends "that the play of the forces of the market normally tends to increase, rather than decrease, the inequalities between regions."[191] Centers of strong economic growth draw capital and skills, and can fund an efficient logistical infrastructure, as a result growing

---

[182] Nunn, "The Long Term Effects of Africa's Slave Trades," 139.

[183] Joseph E. Stiglitz, "The Future of Global Governance," in *The Washington Consensus Reconsidered: Towards a New Global Governance* ed. Narcis Serra and Joseph E. Stiglitz (New York: Oxford University Press, 2008), 312–313.

[184] Stiglitz, "The Future of Global Governance," 309.

[185] Dev Kar and Devon Cartwright-Smith, *Illicit Financial Flows from Developing Countries: 2002–2006* (Washington, D.C.: Center for International Policy, 2009), http://www.gfip.org/storage/gfip/economist%20-%20final%20version%201-2-09.pdf. (Accessed July 4, 2014).

[186] Dev Kar and Devon Cartwright-Smith, *Illicit Financial Flows from Developing Countries: 2002-2006* (Washington, D.C.: Center for International Policy, 2009), http://www.gfip.org/storage/gfip/economist%20-%20final%20version%201-2-09.pdf. (Accessed July 4, 2014).

[187] Robert K. Merton, "The Matthew Effect in Science: The Reward and Communication Systems of Science are Considered," *Science* 159, no. 3810 (January 1968): 56–63.

[188] Matthew 25:29, as translated in the *New Revised Standard Version.*

[189] Merton, "The Matthew Effect in Science: The Reward and Communication Systems of Science are Considered," 56–63.

[190] Gunnar Myrdal, *Rich Lands and Poor: The Road to World Prosperity* (New York: Harper and Row, 1957), 12.

[191] Myrdal, *Rich Lands and Poor: The Road to World Prosperity*, 26.

even faster. In their direct outskirts, they may engender "spread effects," that is, benefits for regions that are within the direct outskirts of economic growth centers. On the other hand, further from the center, the existence of these "economic growth centers" engenders "backwash effects," as far away regions experience the flight of their capital and skills toward economic growth centers. Within affluent countries, spread effects can be stronger than backwash effects, and "state policies have been initiated which are directed toward greater regional equality: the market forces which result in backwash effects have been offset, while those resulting in spread effects have been supported."[192] Conversely, poor countries are inclined to mainly experience the backwash effects from economic growth centers, since such growth centers are frequently situated in other countries. Myrdal indicates that if from one perspective the explication of the current and ever-increasing global inequalities is the cumulative propensity intrinsic in the unimpeded play of the market forces in circumstances where the effectiveness of the spread effects is feeble, from another perspective the explication is the lack of a world state which could intervene in the interest of equality of opportunity.[193]

It is pertinent to note that the influence of apparent global misconduct such as colonization and slavery, unfair trade rules, and illicit financial flows should not be underrated.[194] Pogge posited various ways in which affluent countries are contributing to both the continued severe poverty of poor countries and severe poverty within poor countries. He contends that obligations of global justice are primarily obligations of rectification to recompense for failure to accomplish the negative duty of doing no harm.[195] Ooms and Hammonds assert the crucial importance of rectifying the apparent harm that is being done by affluent countries to the world's poor. However, even if it were feasible to rectify or recompense for all past and prior apparent misconduct, even if an equal opportunity could be created for global free trade, global free trade would nonetheless engender some winners and some losers.[196] They further argue that "…if winners are allowed to invest their present gains in future comparative advantages without global corrective measures, the Gini coefficient for inter-country wealth inequality will continue to grow toward one and away from zero."[197] The global-level Matthew Effect is apparently a less type of harm and may then not require rectification on the basis of the negative duty of doing no harm. However, it requires rectification from the point of view that it

[192] Myrdal, *Rich Lands and Poor: The Road to World Prosperity*, 39–40.

[193] Myrdal, *Rich Lands and Poor: The Road to World Prosperity*, 63–64.

[194] Ooms and Hammonds, "Taking Up Daniels' Challenge: The Case for Global Health Justice," 31.

[195] Thomas Pogge, "Severe Poverty as a Human Rights Violation," in *Freedom from Poverty as a Human Right: Who Owes What to the Very Poor?* ed. Thomas Pogge (New York: Oxford University Press, 2007), 30–53.

[196] Ooms and Hammonds, "Taking Up Daniels' Challenge: The Case for Global Health Justice," 31.

[197] Ooms and Hammonds, "Taking Up Daniels' Challenge: The Case for Global Health Justice," 31–32.

decreases the capacity of some countries to distribute health-related goods. From this perspective, Daniels argues that obligations of mutual assistance at the national level are imperative. He emphasizes the special moral significance of health "because protecting normal functioning helps to protect the range of exercisable opportunities open to people and because various theories of justice support the idea that we have an obligation to protect opportunity and thus health."[198] Obligations of mutual assistance beyond borders are desiderata, to defend equal opportunity globally and taking the Matthew Effect into consideration at the global level.[199]

Another option for addressing inter-country wealth inequalities which undercuts the ability of poor countries to buy health-related goods for its populations was proposed by Robert Archer. Archer describes the great inequalities that occurred between rich and poor over a century ago in the industrialized nations. As a result of this difficult and challenging situation, "many governments in richer countries came to realize, or were pressured to accept, that extreme social and economic inequities were unsustainable."[200]

To surmount these inequalities, Archer insists that "systems of universal health care, social security, unemployment insurance and public housing were put in place."[201] These social protection schemes fundamentally function by collecting financial resources in conformity with participants' means and directly redistributing them "in the form of health-related goods, education-related goods, or other social rights related goods, in accordance with participants' needs."[202] The just distribution of these goods was not ensured by the primary distribution of wealth, ensuing from free markets. It was then imperative to introduce a secondary system for redistribution of wealth, through redistribution that included either money transfers or social rights-related goods transfers, such as funding of health care services for individuals who need it by the government.[203] Some societies rather than establishing a secondary redistribution to correct the primary, attempted to change the primary distribution of resources, and espoused communism. These attempts seemingly resulted in establishing other types of injustice.[204] Ooms and Hammonds argued for an analogous system of secondary redistribution of wealth at the global level which could favorably counteract the Matthew Effect, and authorize a less unfair

---

198 Daniels, *Just Health: Meeting Health Needs Fairly*, 140.

199 Ooms and Hammonds, "Taking Up Daniels' Challenge: The Case for Global Health Justice," 32.

200 Robert Archer, *Duties Sans Frontiers: Human Rights and Global Social Justice* (Versoix, Switzerland: International Council on Human Rights Policy, 2003), 1. http://www.ichrp.org/files/reports/43/108_report_en.pdf. (Accessed July 4, 2014).

201 Archer, *Duties Sans Frontiers: Human Rights and Global Social Justice*, 1.

202 Ooms and Hammonds, "Taking Up Daniels' Challenge: The Case for Global Health Justice," 32.

203 Ooms and Hammonds, "Taking Up Daniels' Challenge: The Case for Global Health Justice," 32.

204 Ooms and Hammonds, "Taking Up Daniels' Challenge: The Case for Global Health Justice," 43.

distribution of health-related goods to occur.[205] They further articulate "it is precisely because a secondary redistribution of wealth system fails to occur at the global level, as Myrdal notes, that we argue for the need to recognize and support global responsibility for growing health inequalities."[206] Essentially, the global free market should not be abrogated but rather attuned to tackle and rectify the inequalities on the global level that impede the ability of poor countries to buy and distribute health-related goods for their citizens.[207] Critical to the solution of redressing growing global health inequalities is to establish the global capacity for the needed intervention which is discussed in the next part of this chapter.

6.2.3 Global Capacity and Health Inequalities

There are two global elements drawn from the World Health Organization mandate, "a globally shared responsibility for the health of all people, and global threats posed by infectious diseases."[208] These two global elements seen as critical components of an emerging global health paradigm are considered not mutually exclusive. The fact that infectious diseases do not respect national borders significantly contributes to a consciousness of global responsibility for the health of all people. More so, the risk of uncontrolled pandemic increasing rapidly from low-income to middle- and high-income motivates the more affluent to assist poor people because the more affluent do not want to contract diseases of the poor people.[209] This explains most likely why Official Development Assistance (ODA) for health appears to concentrate excessively on infectious diseases.[210]

Nevertheless, a global responsibility for the health of all people should go beyond a readily disposition to address the global threats presented by infectious diseases and guarantee that there is equal focus and solidarity for non-infectious diseases.[211]

---

[205] Ooms and Hammonds, "Taking Up Daniels' Challenge: The Case for Global Health Justice," 32.

[206] Ooms and Hammonds, "Taking Up Daniels' Challenge: The Case for Global Health Justice," 32.

[207] Ooms and Hammonds, "Taking Up Daniels' Challenge: The Case for Global Health Justice," 32.

[208] Ooms, *From the Global AIDS Response towards Global Health?* A Discussion Paper for the Helene De Beir Foundation and the International Civil Society Support Group, (January 2009), 5. http://www.hdbf.org/?pageid=345&lang=en (Accessed June 13, 2014).

[209] Ooms, *From the Global AIDS Response towards Global Health?* A Discussion Paper for the Helene De Beir Foundation and the International Civil Society Support Group, (January 2009), 5. http://www.hdbf.org/?pageid=345&lang=en (Accessed June 13, 2014).

[210] David Stuckler et al., "WHO's Budgetary Allocations and Burden of Disease: A Comparative Analysis," *Lancet* 372, no. 9649 (November 2008): 1563–1569.

[211] Ooms, *From the Global AIDS Response towards Global Health?* A Discussion Paper for the Helene De Beir Foundation and the International Civil Society Support Group, (January 2009), 5. http://www.hdbf.org/?pageid=345&lang=en (Accessed June 13, 2014).

Ooms argues that "there is some trans-national solidarity to promote the health of all people, but it is limited and it is most often intended to be temporary: the objective is to help other countries assume their responsibilities towards their inhabitants, within a foreseeable future."[212] He acknowledges basic differences between the way people residing in high-income countries exercise solidarity for health within their countries' borders, and the way those same people exercise solidarity in health beyond the borders of the countries they reside in.[213]

First, there is an enormous difference in quantity. It is not unusual for people in high-income countries to expend higher than 10% of their Gross Domestic Product (GDP) on health. For example, in 2005, people in Germany spent equal to 10.7% of their GDP on health.[214]

The Second significant difference refers to the intention of guiding the exercise of solidarity. Solidarity in health, within the borders of a country does not have an objective to be temporary, its objective entails continuing reciprocal solidarity. Concluding the solidarity is not the intention guiding the act of solidarity in this context. The intention is to help the beneficiary to recover soon and possibly become prolific in such as a way as to contribute to continuing reciprocal solidarity. On the other hand, the intention guiding solidarity changes basically beyond the borders of a country. The intention of trans-national solidarity does not entail that the beneficiaries would turn into contributors to a continuing reciprocal solidarity mechanism between countries. The intention is that all beneficiaries would turn out to be self-sufficient within a foreseeable future, and consequently trans-national solidarity could terminate, as solidarity within countries would be adequate.[215] Ooms points out that "for some reason, we can endorse the metaphor of a "single global market," but not the metaphors of a "single global hospital" and a "single global school."[216]

The consequences of dealing with trans-national health solidarity as a temporary issue and invariably a rejection of a Global Health paradigm can be far-reaching especially for the 1.3 billion people residing in low-income countries. In low-income countries, which had the total GDP of US$810 billion for 1.3 billion people, or US$600 per person per year, domestic public health expenditure of US$18 per

---

[212] Ooms, *From the Global AIDS Response towards Global Health?* A Discussion Paper for the Helene De Beir Foundation and the International Civil Society Support Group, (January 2009), 5. http://www.hdbf.org/?pageid=345&lang=en (Accessed June 13, 2014).

[213] Ooms, *From the Global AIDS Response towards Global Health?* A Discussion Paper for the Helene De Beir Foundation and the International Civil Society Support Group, (January 2009), 5. http://www.hdbf.org/?pageid=345&lang=en (Accessed June 13, 2014).

[214] World Health Organization Website, National Health Accounts, Country Information, (2008), http://www.who.int.nha/country/nha_ratios_and_percapita_levels_2001-2005.xls (Accessed July 8, 2014).

[215] Ooms, *From the Global AIDS Response towards Global Health?* A Discussion Paper for the Helene De Beir Foundation and the International Civil Society Support Group, (January 2009), 6. http://www.hdbf.org/?pageid=345&lang=en (Accessed June 13, 2014).

[216] Ooms, *From the Global AIDS Response towards Global Health?* A Discussion Paper for the Helene De Beir Foundation and the International Civil Society Support Group, (January 2009), 6. http://www.hdbf.org/?pageid=345&lang=en (Accessed June 13, 2014).

person per year is certainly a severe challenge. It involves government revenue except for grants of 20% of GDP, and 15 % of government revenue for public health expenditure. The identified two targets are fairly challenging.[217] Provision of universal access to primary health care at US$18 per person per year is considerably a huge challenge, and thus considered a "mission impossible."[218] Data from the Commission on Macroeconomics and Health (CMH) shows that governments are required to spend at least US$40 per person per year on basic health, and this does not completely include comprehensive primary health care.[219]

The domestic public health expenditure can be complemented by ODA for health, but providing ODA for health within the context of the present development paradigm which implies without a global health paradigm, constitutes a severe problem, primarily because ODA has been in the past unreliable. The unreliability of ODA for health is actually expected as ODA is meant to be temporary and thus not reliable in the long run.[220]

One critical consequence that is identified regarding the unreliability of ODA for health in the long run is that occasionally the fund is being poorly utilized, which may mean not being used where it is greatly needed or could offer the largest benefit, or even not used at all. Moreover, the International Monetary Fund (IMF) and the World Bank deter low-income countries' governments from growing the levels of recurring health expenditure, including health professionals' salaries, utilizing ODA for health since ODA for health is ultimately unreliable.[221] Information from the Independent Evaluation Office of the IMF, indicates that more ODA is rechanneled to grow the foreign exchange reserves of low-income countries, than is utilized for the objective for which it was meant, which is to grow public expenditure.[222]

---

[217] Ooms, *From the Global AIDS Response towards Global Health?* A Discussion Paper for the Helene De Beir Foundation and the International Civil Society Support Group, (January 2009), 7. http://www.hdbf.org/?pageid=345&lang=en (Accessed June 13, 2014).

[218] Ooms, *From the Global AIDS Response towards Global Health?* A Discussion Paper for the Helene De Beir Foundation and the International Civil Society Support Group, (January 2009), 7. http://www.hdbf.org/?pageid=345&lang=en (Accessed June 13, 2014).

[219] Commission on Macroeconomics and Health, *Investing in Health for Economic Development* (Geneva: World Health Organization, 2001), http://www.cid.harvard.edu/archive/cmh/cmhreport.pdf. (Accessed July 8, 2014).

[220] Ooms, *From the Global AIDS Response towards Global Health?* A Discussion Paper for the Helene De Beir Foundation and the International Civil Society Support Group, (January 2009), 7. http://www.hdbf.org/?pageid=345&lang=en (Accessed June 13, 2014).

[221] Ooms, *From the Global AIDS Response towards Global Health?* A Discussion Paper for the Helene De Beir Foundation and the International Civil Society Support Group, (January 2009), 7. http://www.hdbf.org/?pageid=345&lang=en (Accessed June 13, 2014).

[222] Independent Evaluation Office of the International Monetary Fund, *The IMF and Aid to Sub-Saharan Africa* (Washington D.C.: Independent Evaluation Office of the International Monetary Fund, 2007), http://www.imf.org/external/np/ieo/2007/ssa/eng/pdf/report.pdf. (Accessed July 8, 2014)

Furthermore, Nancy Birdsall explains Overseas Development Assistance (ODA) as guilty of "seven deadly sins."[223] She identifies seven deadly sins associated with ODA as (i) "impatience"- in relation to institution building and possessing a restricted commitment to longer-term support; (ii) "envy" – focuses on lack of success to effectively coordinate and at other times to conspire with one another and not essentially in the involved developing countries' interests; (iii) "ignorance" and a lack of success to effectively assess development interventions; (iv) "pride" – refers particularly to a lack of success to exit when suitable; (v) "sloth" – refers to carelessness with concepts and their application, and notably feigning that participation is the same with developing country ownership; (vi) "greed" – distinguished by unreliable and insufficient "stingy transfers" ; and (vii) "foolishness" – distinguished by insufficient obligations to funding global and regional public goods.[224]

Despite the inherent impediments associated with the ODA for health, the global AIDS response which is a synthesis of medical relief and health development paradigms creates a promising global health paradigm. Hammonds and Ooms argue that "the approach adopted by AIDS activists – and their ability to remain outside the development paradigm, often termed "AIDS exceptionalism" – is at the root of their success."[225] A brief discussion of the three paradigms, including the health development paradigm, the medical relief paradigm and the new global health paradigm for redressing the health inequalities follows.

6.2.3.1 The Health Development Paradigm

The health development paradigm aims for sustainability which is usually defined within the context of self-sufficiency.[226] The Office of Sustainable Development of the Bureau for Africa of the United States Agency for International Development provides the definition of sustainability as "the ability of host country entities (community, public and/or private) to assume responsibility for programs and/or outcomes without adversely affecting the ability to maintain or continue program objectives or outcomes."[227] It describes financial sustainability as "having enough reliable funding", meaning funding "generated from a country's own resources."[228]

---

[223] Nancy Birdsall, "Seven Deadly Sins: Reflections on Donor Failings," in *Reinventing Foreign Aid* ed. William Easterly (Cambridge, MA: MIT Press, 2008), 515–551.

[224] Birdsall, "Seven Deadly Sins: Reflections on Donor Failings," 515–551.

[225] Rachel Hammonds and Gorik Ooms, "Global Solidarity for Health," *Global Future*, no. 2 (2009): 6.

[226] Ooms, *From the Global AIDS Response towards Global Health?* A Discussion Paper for the Helene De Beir Foundation and the International Civil Society Support Group, (January 2009), 11. http://www.hdbf.org/?pageid=345&lang=en (Accessed June 13, 2014).

[227] Subhi Mehdi, *Health and Family Planning Indicators: Measuring Sustainability* (Washington D.C.: US Agency for International Development, Office of Sustainable Development, Bureau of Africa, 1999), 1. http://sara.aed.org/publications/cross_cutting/indicators2.pdf. (Accessed July 8, 2014).

[228] Mehdi, *Health and Family Planning Indicators: Measuring Sustainability* (Washington D.C.: US Agency for International Development, Office of Sustainable Development, Bureau of Africa, 1999), 8. http://sara.aed.org/publications/cross_cutting/indicators2.pdf. (Accessed July 8, 2014).

The International development organizations, such as the World Bank, are motivated by global solidarity but their emphasis is on sustainable interventions that result in self-sufficiency. They focus on guaranteeing an exit plan. Development practitioners concentrate on enhancing health or education for all within a country for a restricted time and view the long-term sustainability of an intervention a critical factor in establishing the goal, design, implementation and evaluation of a project.[229]

In the area of international assistance for health, Enrico Pavignani and Sandro Colombo remark: "Sustainability is continuously invoked as a key criterion to assess any aid-induced activity or initiative. Sometimes, the concept is given the weight of a decisive argument. Thus, to declare something "unsustainable" may sound as equivalent of "worthless" or even "harmful", in this way overruling any other consideration."[230] This is one of the limitations of health development paradigm because some effective health interventions are crippled because they cannot be sustained by domestic resources. Sustainability in the sense of self-sufficiency is further limited because it cannot be used as a criterion to assess medical relief but rather as criterion to assess health development. It resonates with the United Nations Office for the Coordination of Humanitarian Affairs (OCHA) definition mentioned earlier which indicates: "If the appropriate response is within the capacity of the affected states, communities or individuals, then health development is the right answer, and the response should be sustainable."[231] The problem is even worse because most of the states, communities and individuals affected by severe emergency health crises cannot effectively handle such situations. For example, providing a life-long treatment for people living with AIDS is prohibitively costly that it is unaffordable for developing nations. The health development paradigm offers no effective solution to many of such cases of health care interventions because of its unnecessary emphasis on the traditional concept of sustainability.

Akin to this perspective is the observation of Pablo Gottret and Georges Schieber that "Sustainability has generally been described in terms of self-sufficiency."[232] The implication is that from this context, international assistance would not promote any distribution of health-related goods that could not be sustained by the beneficiary country. Therefore, international assistance would merely endorse distribution of health-related goods endeavors that the beneficiary country could sustain on its own, and if it could sustain these endeavors on its own, the international assistance then would not be required. This explains why the international assistance

[229] Hammonds and Ooms, "Global Solidarity for Health," 6.

[230] Enrico Pavignani and Sandro Colombo, *Analysing Disrupted Health Sectors* (Geneva: World Health Organization, 2009), 176. http://www.who.int/hac/techguidance/tools/disrupted_sectors/adhsm_en.pdf. (Accessed July 9, 2014).

[231] Ooms, "The Right to Health and the Sustainability of Healthcare: Why a New Global Health Aid Paradigm is Needed," 14.

[232] Pablo Gottret and Georges Schieber, *Health Financing Revisited: A Practitioner's Guide* (Washington, D.C.: The World Bank, 2006), 138. http://siteresources.worldbank.org/INTHSD/Resources/topics/Health-Financing/HFRFull.pdf. (Accessed July 9, 2014).

has not made a significant impact in some areas of global health especially in "maternal and child health."[233]

More so, if US$40 per person per year is needed for equitably providing a basic set of required health services, in that case, sustainability in the sense of financial self-sufficiency might not be practical. First, let us suppose that developing countries can raise government revenue equivalent to 20% of the GDP and assign 15% of government revenue to health expenditure, both of which are fairly ambitious. Second, if a comparison of these assumptions was made with current levels of government revenue and allocation to health expenditure, just countries with a GDP of US$1,333 per person can attain government revenue of US$266 per person per year and government health expenditure of US$40 per person per year. It is quite evident that low-income countries cannot attain this, based on the categorization of the World Bank in which their Gross National Income (GNI) is just US$935 or less.[234] Ooms also observes that "All trendy development approaches point out that sustainable health care – narrowly defined as independent from international aid – is illusionary in the world's poorest countries."[235] Therefore, public health budgets in low-income countries must be grown and critically need more guaranteed national and international financial commitments.[236]

Sustainability in the sense of self-sufficiency is then not compatible with an equitable provision of a basic set of health services for the 1.3 billion people residing in low-income countries. It makes it hard to associate medical relief with health development.[237] Ooms contends that "In the field of health care, the issue of sustainability creates a dichotomy between medical relief and health development, because relief is unaffected by the condition of self-reliance."[238] The dichotomy creates turf battles between the proponents of medical relief and the proponents of health development.[239] Essentially, sustainability as the criterion is not actually taken into account when ascertaining the appropriateness of the medical relief response, due to the fact that the crisis is supposed to be temporary and consequently there is an assumption that long-term response would not be required. It is not completely accurate that sustainability in the sense of self-sufficiency is appealed to as a criterion for all

[233] Hammonds and Ooms, "Global Solidarity for Health," 6.

[234] World Bank, Website, Data, Country Classification, 2008. http://go.worldbank.org/K2CKM78CCO (Accessed July 9, 2014).

[235] Gorik Ooms, "Health Development Versus Medical Relief: The Illusion Versus the Irrelevance of Sustainability," *PLos Medicine* 3, no. 8 (August 2006): e345.

[236] Ooms, "Health Development Versus Medical Relief: The Illusion Versus the Irrelevance of Sustainability," e345.

[237] Ooms, *From the Global AIDS Response towards Global Health?* A Discussion Paper for the Helene De Beir Foundation and the International Civil Society Support Group, (January 2009), 11. http://www.hdbf.org/?pageid=345&lang=en (Accessed June 13, 2014).

[238] Ooms, "Health Development Versus Medical Relief: The Illusion Versus the Irrelevance of Sustainability," e345.

[239] Ooms, "Health Development Versus Medical Relief: The Illusion Versus the Irrelevance of Sustainability," e345.

aid-induced activities. Definitely, it is not needed in humanitarian or medical relief interventions.[240]

6.2.3.2 The Medical Relief Paradigm

The medical relief paradigm was initially established to respond to severe health crises. In the event of a natural disaster or war that significantly devastates a population, emergency humanitarian organizations such as the International Red Cross and Red Crescent Movement or Medecins Sans Frontieres (MSF) usually respond. They typically appeal also to global solidarity in requesting for funds and their commended work responds to the pressing desire that the majority of people feel to "do something" to assist those they perceive suffering.[241] Until lately, humanitarian organizations were controlled by the notion of "temporary disruption of a pre-existing equilibrium", or as humanitarian aid opponents explain, to "help populations get back to where they were before disaster struck."[242] From this perspective, the issue of sustainability is not taken into consideration when ascertaining the scope of the medical relief response. For example, one can utilize helicopters to rescue people following severe incidents of floods in countries that would not be able to fund a helicopter fleet themselves; since the incidents of floods are exceptional, the response is then not supposed to be sustainable.[243]

Nevertheless, numerous humanitarian crises continue for decades and in such cases some type of sustainability is then needed. In such cases, the sustainability of medical relief depends on sustained international aid as an alternative type of sustainability different from self-sufficiency.[244]

The practitioners of medical relief paradigm have been seriously challenged by chronic health crises especially the epidemics of AIDS, tuberculosis and malaria, but as well as recurring occurrences of malnutrition, or still prevalent lack of access to the most basic level of health care.[245] The World Disasters Report 2008 of the International Federation of Red Cross and Red Crescents Societies (IFRC) concen-

[240] Ooms and Hammonds, "Taking Up Daniels' Challenge: The Case for Global Health Justice," 33.

[241] Hammonds and Ooms, "Global Solidarity for Health," 6.

[242] Pippa Trench et al., *Beyond Any Drought*, (London: Sahel Working Group, 2007), http://www.careinternational.org.uk/download.php?id=617 (Accessed July 10, 2014).

[243] Ooms, *From the Global AIDS Response towards Global Health?* A Discussion Paper for the Helene De Beir Foundation and the International Civil Society Support Group, (January 2009), 12. http://www.hdbf.org/?pageid=345&lang=en (Accessed June 13, 2014).

[244] Ooms, *From the Global AIDS Response towards Global Health?* A Discussion Paper for the Helene De Beir Foundation and the International Civil Society Support Group, (January 2009), 12. http://www.hdbf.org/?pageid=345&lang=en (Accessed June 13, 2014).

[245] Ooms and Hammonds, "Taking Up Daniels' Challenge: The Case for Global Health Justice," 33.

trates on the AIDS epidemic, as "a disaster in many ways."[246] Data shows that more people are currently dying in Mozambique as a result of these chronic crises than during its 20 years of war, and average life expectancy has decreased from 40 to 27 years.[247] In some areas of the Democratic of Congo not ravaged by conflict, mortality goes beyond emergency thresholds.[248] Taking into account these prevalent situations in developing countries, the goal to "help populations get back to where they were before disaster struck" is futile.[249]

The Office for the Coordination of Humanitarian Affairs (OCHA) describes "humanitarian crises" or "emergencies" as "any situation in which there is an exceptional and widespread threat to life, health or basic subsistence, that is beyond the capacity of individual and the community."[250] If the lack of capacity of the affected individuals, communities or countries to deal with a situation is what changes a development problem to a humanitarian crisis, therefore 1.3 billion people residing in low-income countries are experiencing a permanent humanitarian crisis as neither they nor their communities are capable to offer what it takes to deal with the situation.[251]

This widespread state of emergency is not merely rhetoric because it has resulted in humanitarian actors and providers of medical relief such as the IFRC and Medecins Sans Frontieres (MSF) broadening their meanings of humanitarian crises or disasters to intervene in the fights against AIDS, tuberculosis and malaria.[252] More so, there is a contention that if the President's Emergency Plan for AIDS Relief (PEPFAR), established by the United States, includes "relief" in its name, it is not simply because "PEPFAR sounds better than PEPFADA – which could have been the acronym of a President's Emergency Plan For AIDS Development

[246] International Federation of Red Cross and Red Crescent Societies, *World Disasters Report 2008 – Focus on HIV and AIDS* (Geneva: International Federation of Red Cross and Red Crescent Societies, 2008), http://www.ifrc.org/Docs/pubs/disasters/wdr2008/WDR2008-full.pdf. (Accessed July 10, 2014).

[247] Karen A. Stanecki, *The AIDS Pandemic in the 21st Century* (Washington D.C.: US Census Bureau, 2004), http://www.census.gov/prod/2004pubs/wp02-2.pdf. (Accessed July 10, 2014).

[248] Alain Kassa et al., *Access to Health Care, Mortality, and Violence, in Democratic Republic of the Congo* (Brussels: Medecins Sans Frontieres, 2005), http://www.doctorswithoutborders.org/publications/reports/2005/drc_healthcare_11-2005.pdf. (Accessed July 10, 2014).

[249] Ooms, *From the Global AIDS Response towards Global Health?* A Discussion Paper for the Helene De Beir Foundation and the International Civil Society Support Group, (January 2009), 12. http://www.hdbf.org/?pageid=345&lang=en (Accessed June 13, 2014).

[250] United Nations Office for the Coordination of Humanitarian Affairs, *Humanitarian Strategic Framework for Southern Africa 2005* (Johannesburg: United Nations Office for the Coordination of Humanitarian Affairs, 2005), http://www.sahims.net/doclibrary/Sahims_Documents/050613_och1.pdf. (Accessed July 10, 2014).

[251] Ooms, *From the Global AIDS Response towards Global Health?* A Discussion Paper for the Helene De Beir Foundation and the International Civil Society Support Group, (January 2009), 12. http://www.hdbf.org/?pageid=345&lang=en (Accessed June 13, 2014).

[252] Ooms, *From the Global AIDS Response towards Global Health?* A Discussion Paper for the Helene De Beir Foundation and the International Civil Society Support Group, (January 2009), 12. http://www.hdbf.org/?pageid=345&lang=en (Accessed June 13, 2014).

Assistance – but because PEPFAR was conceived as a medical relief programme: not aiming for self-reliance within a foreseeable future, but an emergency response to a crisis."[253]

It is pertinent to note that some people are not contented with this situation. Firstly, the more expanded meaning of humanitarian crises establishes a field of intervention that is too huge for humanitarian organizations. Consequently, criteria for intervention unavoidably encompass some arbitrary choices.[254] Alan Whiteside and Amy Whalley condemn humanitarian actors for their lack of success to "provide clear guidelines as to when an event is severe enough to be declared an emergency" and "recognize change in the nature of disasters", and consequently for not addressing the real humanitarian crises.[255] It might be possible that basically, humanitarian organizations were not designed to respond to chronic crises because their dependence on expatriate implementers and parallel management systems and their need to remain independent from governments which coincidentally are all strategies intended for acute crises in armed conflicts especially, severely restricted their potential as a mobilizer to respond to chronic health crises and to improve primary health care for all.[256] In fact, the two health paradigms namely the health development paradigm and the medical relief paradigm were not sufficiently designed to effectively redress global health inequalities.[257] Ooms and Hammonds articulating their limitations write, "One is too focused on domestic self-reliance; the other has to remain independent from the governments of the countries in which it operates."[258] It was then imperative to introduce another health paradigm known as Global Health paradigm which is designed to effectively redress global health inequalities.

[253] Ooms, *From the Global AIDS Response towards Global Health?* A Discussion Paper for the Helene De Beir Foundation and the International Civil Society Support Group, (January 2009), 12. http://www.hdbf.org/?pageid=345&lang=en (Accessed June 13, 2014).

[254] Ooms, *From the Global AIDS Response towards Global Health?* A Discussion Paper for the Helene De Beir Foundation and the International Civil Society Support Group, (January 2009), 12. http://www.hdbf.org/?pageid=345&lang=en (Accessed June 13, 2014).

[255] Alan Whiteside and Amy Whalley, *Reviewing "Emergencies" for Swaziland: Shifting the Paradigm in a New Era* (Durban, South Africa: HEARD, 2007), http://data.unaids.org/pub/Report/2007/swaziland%20emergency%20report_final%20pdf_en.pdf. (Accessed July 10, 2014).

[256] Ooms, *From the Global AIDS Response towards Global Health?* A Discussion Paper for the Helene De Beir Foundation and the International Civil Society Support Group, (January 2009), 13. http://www.hdbf.org/?pageid=345&lang=en (Accessed June 13, 2014).

[257] Ooms and Hammonds, "Taking Up Daniels' Challenge: The Case for Global Health Justice," 33.

[258] Ooms and Hammonds, "Taking Up Daniels' Challenge: The Case for Global Health Justice," 33.

6.2.3.3 Global AIDS Response – A New Global Health Paradigm

The global AIDS response started with a medical relief paradigm because of necessity, not only because the HIV/AIDS pandemic produced a crisis situation in developing nations with high prevalence, but also because the health development paradigm could not contain the costs of AIDS treatment.[259] The global AIDS response constitutes a new global health paradigm which is at the intersection of health development paradigm and medical relief paradigm.[260]

Pavignani and Colombo remark: "Sustainability tends to be employed as an all-encompassing term, but it seems useful to distinguish between technical sustainability, which relates to the capacity to carry out certain functions, and financial sustainability, which results from resource availability, fiscal capacity and the relative priority of health care provision."[261] Exploring a clear distinction between technical sustainability and financial sustainability is critical to establishing the foundation of a Global Health paradigm. The Global Health paradigm would aim for operational sustainability in the conventional sense of self-sufficiency at the national level borrowed from the health development paradigm. It would also give authorization for unlimited reliance on international financial support like the medical relief paradigm.[262] Ooms argues that "In doing so it would recognize a globally shared responsibility for the health of all people and respond to the need for a new approach to providing basic health care to people in middle-income and low-income countries."[263]

Actually, the Global Fund is already implementing this: it has discarded financial sustainability in the conventional sense as a criterion for support, but unlike PEPFAR it does oblige technical or operational sustainability of the interventions it endorses. Therefore, the Global Fund did not discard financial sustainability entirely; rather it coined a new concept of sustainability, sustainability at the international level, depending on sustained international solidarity as well as on domestic resources. This implies that when countries utilize their grants from the Global Fund judiciously and effectively, they can depend on continued support from the Global

---

[259] Ooms and Hammonds, "Taking Up Daniels' Challenge: The Case for Global Health Justice," 33.

[260] Ooms and Hammonds, "Taking Up Daniels' Challenge: The Case for Global Health Justice," 33.

[261] Enrico Pavignani and Sandro Colombo, *Analysing Disrupted Health Sectors* (Geneva: World Health Organization, 2009), 176. http://www.who.int/hac/techguidance/tools/disrupted_sectors/adhsm_en.pdf. (Accessed July 9, 2014).

[262] Ooms, *From the Global AIDS Response towards Global Health?* A Discussion Paper for the Helene De Beir Foundation and the International Civil Society Support Group, (January 2009), 14. http://www.hdbf.org/?pageid=345&lang=en (Accessed June 13, 2014).

[263] Ooms, *From the Global AIDS Response towards Global Health?* A Discussion Paper for the Helene De Beir Foundation and the International Civil Society Support Group, (January 2009), 14. http://www.hdbf.org/?pageid=345&lang=en (Accessed June 13, 2014).

Fund.[264] Michel Kazatchkine, the executive director of the Global Fund, highlighted the invented new concept of sustainability in his closing speech at the XVII International AIDS Conference: "The Global Fund has helped to change the development paradigm by introducing a new concept of sustainability. One that is not based solely on achieving domestic self-reliance but on sustained international support as well."[265]

The Global Fund did not coin this new concept of sustainability without prior concerted efforts from the global community. The United Nations General Assembly Special Session on HIV/AIDS of June 2001 resulted in a declaration in which member states pledged to "make every effort to provide progressively and in a sustainable manner, the highest attainable standard of treatment for HIV/AIDS."[266] Furthermore, the follow-up assembly of June 2006 also resulted in a declaration, in which member states committed themselves "to supporting and strengthening existing financial mechanisms, including the Global Fund to Fight AIDS, Tuberculosis and Malaria, as well as relevant United Nations organizations, through the provision of funds in a sustained manner."[267] The WHO's 2008 report recognizes that the sudden increase in external funds aimed at health through bilateral channels or through the new generation's global financing instruments has improved the energy or enthusiasm of the health sector.[268] However, the report highlights immediately that "these additional funds need to be progressively re-channeled in ways that help build institutional capacity towards a longer-term goal of self-sustaining, universal coverage."[269]

The Global Health paradigm with its emphasis on the new concept of sustainability within which the Global Fund functions was established out of necessity rather than theory. Establishing a theoretical framework will be necessary to ground the practice of the global fund and other associated practices in realizing global health justice.[270]

---

[264] Ooms, *From the Global AIDS Response towards Global Health?* A Discussion Paper for the Helene De Beir Foundation and the International Civil Society Support Group, (January 2009), 14. http://www.hdbf.org/?pageid=345&lang=en (Accessed June 13, 2014).

[265] Michel Kazatchkine, *Closing Speech at the XVII International AIDS Conference in Mexico* (Geneva: Global Fund to Fight AIDS, Tuberculosis and Malaria, 2008), http://www.theglobalfund.org/en/pressreleases/?pr=pr_080811 (Accessed July 10, 2014).

[266] United Nations General Assembly, *Declaration of Commitment on HIV/AIDS*, A/RES/S-26/2, (Geneva: Joint UN Programme on HIV/AIDS, 2001), 22. http://www.un.org/ga/aids/docs/aress262.pdf. (Accessed July 10, 2014).

[267] United Nations General Assembly, *Political Declaration on HIV/AIDS*, A/RES/60/262, (2006), Paragraph 41. http://data.unaids.org/pub/Report/2006/20060615_HLM_PoliticalDeclaration_ARES60262_en.pdf. (Accessed July 10, 2014).

[268] Wim Van Lerberghe et al., *The World Health Report 2008: Primary Health Care – Now More Than Ever* (Geneva: World Health Organization, 2008), 106. http://www.who.int/whr/2008/whr08_en.pdf. (Accessed July 10, 2014).

[269] Lerberghe et al., *The World Health Report 2008: Primary Health Care – Now More Than Eve*r, 106.

[270] Ooms and Hammonds, "Taking Up Daniels' Challenge: The Case for Global Health Justice," 34.

6.3 Responsibility for Affordable Access to Drugs

6.3.1 Human Rights and Global Health Justice

The Global Health paradigm is argued as grounded on international human rights law. Ooms and Hammonds articulate that "Ethics and value lie at the heart of the formal framework of international human rights law."[271] This part of the chapter discusses the scope to which the formal framework of international law offers a basis for a different aid paradigm, particularly the Global Health paradigm, in which international assistance becomes an obligation, responding to an entitlement, and hence unlimited rather than temporary, under specific conditions.[272]

The 1948 Universal Declaration of Human Rights[273] is the basis of the modern human rights movement, and despite the fact that it is not a legally binding document in itself, later international human rights treaties that are established on the values originated from the Universal Declaration of Human Rights engender legally binding obligations on governments.[274] Sofia Gruskin and Daniel Tarantola explicate that "in practical terms, international human rights law is about defining what governments can do to us, cannot do to us and should do for us."[275]

The two crucial treaties ensuing from the Universal Declaration of Human Rights, the International Covenant on Civil and Political Rights[276] and the International Covenant on Economic, Social and Cultural Rights[277] include legally binding obligations for the states that approve them.[278] The International Covenant

[271] Gorik Ooms and Rachel Hammonds, "Correcting Globalisation in Health: Transactional Entitlements Versus the Ethical Imperative of Reducing Aid-Dependency," *Public Health Ethics* 1, no. 2 (January 2008): 157.

[272] Ooms, *From the Global AIDS Response towards Global Health?* A Discussion Paper for the Helene De Beir Foundation and the International Civil Society Support Group, (January 2009), 26. http://www.hdbf.org/?pageid=345&lang=en (Accessed June 13, 2014).

[273] United Nations, *Universal Declaration of Human Rights*, G.A. RES. 217A (III), UN Doc. A/180, (1948), http://www.un.org/Overview/rights.html. (Accessed July 12, 2014).

[274] Ooms, *From the Global AIDS Response towards Global Health?* A Discussion Paper for the Helene De Beir Foundation and the International Civil Society Support Group, (January 2009), 26. http://www.hdbf.org/?pageid=345&lang=en (Accessed June 13, 2014).

[275] Sofia Gruskin and Daniel Tarantola, *Health and Human Rights*, Working Paper, (Boston: Francois-Xavier Bagnoud Center for Health and Human Rights, Harvard School of Public Health, 2000), 4. http://www.hsph.harvard.edu/fxbcenter/FXBC_WP10-Gruskin_and_Tarantola.pdf. (Accessed July 12, 2014).

[276] United Nations, *International Covenant on Civil and Political Rights*, G.A. Res. 2200A (XXI), UN Doc. A/6316 (1966). http://www2.ohchr.org/english/law/pdf/ccpr.pdf. (Accessed July 12, 2014).

[277] United Nations, *International Covenant on Economic, Social and Cultural Rights*. G.A. Res. 2200A (XXI), UN Doc. A/6316 (1966). http://www.2.ohchr.org/english/law/pdf/cescr.pdf. (Accessed July 12, 2014).

[278] Ooms, *From the Global AIDS Response towards Global Health?* A Discussion Paper for the Helene De Beir Foundation and the International Civil Society Support Group, (January 2009), 26. http://www.hdbf.org/?pageid=345&lang=en (Accessed June 13, 2014).

on Economic, Social and Cultural Rights is the most pertinent and the focus for this discussion because it explicates states obligations with respect to the right to health. Nevertheless, it is pertinent to note that human rights cannot be achieved separately, which basically implies the right to health cannot be realized without improving the right to education and respect for civil and political rights.[279] A more detailed discussion of international human rights law as a foundation for global health justice begins with the definition and scope of the right to health.

6.3.1.1 Meaning and Scope of the Right to Health

Clearly, the right to health does not imply "the right of everyone to be healthy or to be provided with health."[280] Asbjorn Eide argues that "No state and no institution can guarantee our health, but more or less optimal conditions for the enjoyment of good health can be created, and this is what the rights to health is all about."[281] States can neither guarantee good health nor provide defense against every likely cause of human disease. Genetic factors, individual vulnerability to disease and the acceptance of unhealthy or perilous lifestyle may play critical roles with regard to an individual's health.[282] Eide writes "The right to health is therefore a right to have optimal conditions for as many as possible to live a long and healthy life."[283]

Furthermore, article 12 of the International Covenant on Economic, Social and Cultural Rights defines the right to health as "the right to the highest attainable standard of physical and mental health"[284] and the associated obligations of the state encompass the provisions of medical services and the underlying preconditions necessary to health, comprising of things such as clean water, sanitation, hospitals, clinics, trained medical and professional officials and essentials drugs.[285] This basic

[279] Ooms, *From the Global AIDS Response towards Global Health?* A Discussion Paper for the Helene De Beir Foundation and the International Civil Society Support Group, (January 2009), 26. http://www.hdbf.org/?pageid=345&lang=en (Accessed June 13, 2014).

[280] Asbjorn Eide, *The Health of the World's Poor – A Human Rights Challenge* (Oslo: Norwegian Directorate of Health, 2011), 29. http://www.helsedirektoratet.no/english/publications/the-health-of-the-worlds-poor-a-human-rights-challenge/Publikasjoner/the-health-of-the-worlds-ppor-a-human-rights-challenge.pdf. (Accessed July 13, 2014).

[281] Eide, *The Health of the World's Poor – A Human Rights Challenge*, 29.

[282] Committee on Economic, Social and Cultural Rights, *General Comment No. 14: The Right to the Highest Attainable Standard of Health,* UN Doc. No. E/C.12/2000/4, (Geneva : United Nations, 2000), http://www.unhchr.ch/tbs/doc.nsf/%28symbol%29/E.C.12.2000.4.En. (Accessed June 19, 2014).

[283] Eide, *The Health of the World's Poor – A Human Rights Challenge*, 29.

[284] Committee on Economic, Social and Cultural Rights, *General Comment No. 14: The Right to the Highest Attainable Standard of Health,* UN Doc. No. E/C.12/2000/4, (Geneva : United Nations, 2000), http://www.unhchr.ch/tbs/doc.nsf/%28symbol%29/E.C.12.2000.4.En. (Accessed June 19, 2014).

[285] Committee on Economic, Social and Cultural Rights, *General Comment No. 14: The Right to the Highest Attainable Standard of Health,* UN Doc. No. E/C.12/2000/4, (Geneva : United Nations, 2000), http://www.unhchr.ch/tbs/doc.nsf/%28symbol%29/E.C.12.2000.4.En. (Accessed June 19, 2014).

definition of the right to health has been endorsed and broadened in later international conventions, comprising the 1989 Convention on the Rights of the Child, as well as other national and international legislation.[286] Additionally, numerous United Nations Committees have engaged in an active role in more explication of essential elements of the right to health in their General Comments and in assessments of States' compliance with obligations in the context of the International Covenant on Economic, Social and Cultural Rights as well as other more current treaties. Therefore, there is a dynamic development of the understanding of the right to health in international law; "it is not just frozen in the bare bones definition from the mid-1960s."[287]

The 1989 Convention on the Rights of the Child, which has been approved by all States with the exception of the United States and Somalia, can be seen as a sign of global intentions on the development of understanding of the right to health and the obligations it requires.[288] Article 24 of the Convention on the Rights of the Child offers more direction in understanding what the right to health implies as well as creating norms for governments concerning the right to health of children.[289] A typical example of how these legal documents help form government policy stems from the UN Committee on the Rights of the Child which has elucidated Article 24 of the Convention on the Rights of the Child as obliging governments to take some particular actions to guarantee the right to health of children. First, a government must offer reliable data on the health of children to the Committee on the Rights of the Child. Second, a government must demonstrate that it is taking necessary steps to guarantee that it sufficiently funds the health of children. Third, a state must take necessary steps to guarantee that the health of all children is respected. Individual government conformity to these actions and other obligations is examined by the Committee on the Rights of the child, when governments turn in their periodic reports.[290]

A further significant development took place in 2000 when the Committee on Economic, Social and Cultural Rights released a General Comment 14 on the right to health, tackling the scope of the right to health and the significance of international

---

[286] Ooms, *From the Global AIDS Response towards Global Health?* A Discussion Paper for the Helene De Beir Foundation and the International Civil Society Support Group, (January 2009), 27. http://www.hdbf.org/?pageid=345&lang=en (Accessed June 13, 2014).

[287] Ooms, *From the Global AIDS Response towards Global Health?* A Discussion Paper for the Helene De Beir Foundation and the International Civil Society Support Group, (January 2009), 27. http://www.hdbf.org/?pageid=345&lang=en (Accessed June 13, 2014).

[288] Ooms, *From the Global AIDS Response towards Global Health?* A Discussion Paper for the Helene De Beir Foundation and the International Civil Society Support Group, (January 2009), 27. http://www.hdbf.org/?pageid=345&lang=en (Accessed June 13, 2014).

[289] United Nations, *Convention on Rights of the Child*, United Nations Human Rights, (1989), http://www.ohchr.org/en/professionalinterest/pages/crc.aspx. (Accessed July 13, 2014)

[290] UN Human Rights Committee (HRC), *Covenant on Civil and Political Rights (CCPR) General Comment No. 17: Article 24 Rights of the Child*, (April 7, 1989), http://www.refworld.org/docid/45139b464.html (Accessed July 13, 2014)

cooperation in realizing the right to health.[291] The general comment on the right to health elucidates the scope of national and international obligations which was not clearly addressed by the language of article 2(1) of the ICESCR.

The scope of the right to health was addressed with the introduction of the concept of the progressive realization. A critical element of economic and social rights is that they can simply be realized in a progressive fashion, in due course and not immediately, as it relates to numerous civil and political rights.[292] With respect to the right to health the Committee on Economic, Social and Cultural Rights remarks: "The concept of progressive realization constitutes a recognition of the fact that full realization of all economic, social and cultural rights will generally not be able to be achieved in a short period of time."[293] The principle of progressive realization is "critical for resource-poor countries that are responsible for striving towards human rights goals to the maximum extent possible."[294]

Furthermore, the concept of progressive realization should not be misunderstood as validating incessant delays in the achievement of economic, social, and cultural rights, while expecting for economic growth and adequate domestic resources to become accessible.[295] It is not to be seen as "an escape hatch (for) recalcitrant states."[296] Such an elucidation would divest economic, social, and cultural rights of any meaningful value, particularly for the deprived and vulnerable.[297] Hence, the Committee on Economic, Social and Cultural Rights comments that States parties have "an obligation to move as expeditiously and effectively as possible."[298]

---

[291] Committee on Economic, Social and Cultural Rights, *General Comment No. 14: The Right to the Highest Attainable Standard of Health*, UN Doc. No. E/C.12/2000/4, (Geneva : United Nations, 2000), http://www.unhchr.ch/tbs/doc.nsf/%28symbol%29/E.C.12.2000.4.En. (Accessed June 19, 2014).

[292] United Nations, *International Covenant on Economic, Social and Cultural Rights*. G.A. Res. 2200A (XXI), UN Doc. A/6316 (1966), Art. 2, para.1. http://www.2.ohchr.org/english/law/pdf/cescr.pdf. (Accessed July 12, 2014).

[293] Committee on Economic, Social and Cultural Rights, *General Comment No. 3, The Nature of States Parties' Obligations*, UN Doc. No. E/C.12/1991/23 (1990), Art.2, para.1.9. http://www.unhchr.ch/tbs/doc.nsf/(symbol)/CESCR+General+comment+3.En. (Accessed July 13, 2014).

[294] Gruskin and Tarantola, *Health and Human Rights*, Working Paper, (Boston: Francois-Xavier Bagnoud Center for Health and Human Rights, Harvard School of Public Health, 2000), 4. http://www.hsph.harvard.edu/fxbcenter/FXBC_WP10-Gruskin_and_Tarantola.pdf. (Accessed July 12, 2014).

[295] Ooms, *From the Global AIDS Response towards Global Health?* A Discussion Paper for the Helene De Beir Foundation and the International Civil Society Support Group, (January 2009), 28. http://www.hdbf.org/?pageid=345&lang=en (Accessed June 13, 2014).

[296] Scott Leckie, "Another Step Towards Indivisibility: Identifying the Key Features of Violations of Economic, Social and Cultural Rights," *Human Rights Quarterly* 20, no.1 (February 1998): 94.

[297] Ooms, *From the Global AIDS Response towards Global Health?* A Discussion Paper for the Helene De Beir Foundation and the International Civil Society Support Group, (January 2009), 28. http://www.hdbf.org/?pageid=345&lang=en (Accessed June 13, 2014).

[298] Committee on Economic, Social and Cultural Rights, *General Comment No. 3, The Nature of States Parties' Obligations*, UN Doc. No. E/C.12/1991/23 (1990), Art.2, para.1.9. http://www.unhchr.ch/tbs/doc.nsf/(symbol)/CESCR+General+comment+3.En. (Accessed July 13, 2014).

Pertinent to note is that progressive realization as well applies to resource-rich countries.[299]

To refute the notion that "progressive realization" may entail "no immediate obligations," the Committee on Economic, Social and Cultural Rights stresses a variety of principles that explicate the nature of obligations of States parties: the principle of non-retrogression which entails that a State should not take steps backwards, the obligations to provide international assistance and the principle of core obligations.[300] The focus here is on the principle of core obligations and the obligation to provide international assistance because they engender obligations of global health justice as well as offer useful framework for grounding a Global Health paradigm.[301]

6.3.1.2 Core Obligations and the Obligation to Provide Assistance

The Committee on Economic, Social and Cultural Rights explained the core content of the right to health within the context of its explanation of the core obligations that stem from the right to health.[302] Esin Orucu in 1986 provided a detailed explanation of the notion of the "core content" of a human right: the essential substance of a right, its reason for being, in the absence of which it would be devoid of meaning.[303] The Maastricht Guidelines prepared in 1997 by international legal experts further broadened on this idea with reference to requirements for the fulfillment of a minimum core obligation.[304] The concept of "core content" was approved by the Committee in General Comment No. 3. It elucidated that there are limits to the compromises that states can make with respect to achieving economic, social and cultural rights by appealing to the clearly acknowledged unfeasibility of achieving

---

[299] United Nations General Assembly, *Declaration on the Right to Development*, G.A. Res. 41/128, UN GAOR (1986), Art. 3. http://www.un.org/documents/ga/res/41/a41r128.htm (Accessed July 13, 2014).

[300] Committee on Economic, Social and Cultural Rights, *General Comment No. 3, The Nature of States Parties' Obligations*, UN Doc. No. E/C.12/1991/23 (1990), Art.2, para.1.9. http://www.unhchr.ch/tbs/doc.nsf/(symbol)/CESCR+General+comment+3.En. (Accessed July 13, 2014).

[301] Ooms, *From the Global AIDS Response towards Global Health?* A Discussion Paper for the Helene De Beir Foundation and the International Civil Society Support Group, (January 2009), 28. http://www.hdbf.org/?pageid=345&lang=en (Accessed June 13, 2014).

[302] Committee on Economic, Social and Cultural Rights, *General Comment No. 14: The Right to the Highest Attainable Standard of Health*, UN Doc. No. E/C.12/2000/4, (New York: United Nations, 2000), http://www.unhchr.ch/tbs/doc.nsf/%28symbol%29/E.C.12.2000.4.En. (Accessed June 19, 2014).

[303] Esin Orucu, "The Core of Rights and Freedoms: The Limits of Limits," in *Human Rights: From Rhetoric to Reality* ed. Tom Campbell et. al (New York: Blackwell, 1986), cited in Ooms and Hammonds, "Correcting Globalisation in Health: Transactional Entitlements Versus the Ethical Imperative of Reducing Aid-Dependency," 158.

[304] Urban Morgan Institute for Human Rights, "The Maastricht Guidelines on Violations of Economic, Social and Cultural Rights," *Human Rights Quarterly* 20, no. 3 (August 1998): 691–704.

all of them entirely and at once. More so, there is a minimum threshold, a minimum essential level or core content, which must be achieved without further delay.[305] The Committee further explained that neither resource constraints nor progressive realization can justify non-conformity to the core obligations remarking that the burden lies with the State to show that it has utilized all available resources to fulfill its core obligations, which are non-derogable.[306]

Brigit C.A. Toebes explaining the core content of the right to health categorized the elements that contribute to the health status of persons separating them into two sub-groups: elements of healthcare and the underlying preconditions for health.[307] In 2000, the Committee's General Comment No. 14 on the right to health acknowledged the existence of a minimum essential level of the right to health.[308]

The Committee on Economic, Social and Cultural Rights explains the minimum essential level of the right to health indirectly, through the explanation of the core obligations of States parties concerning the right to health. Core obligations encompass obligations to ensure access to essential health services and support for preconditions of health. The core obligations also entail the obligation to provide essential drugs, as defined by the WHO. The fundamental significance of non-discrimination is stressed all through, since it is the obligation for a state to focus particularly on vulnerable or marginalized groups.[309]

Basing on their experience, most health practitioners in developing countries think that this definition of the minimum essential level of the right to health is a distant dream. Public health expenditure was less than US$10 per person per year in 2004, in about 37 of the world's low-income countries.[310] The implication is that low-income countries are just very poor to provide a basic package of health services, including AIDS treatment which is estimated to cost US$40.00 per person per

[305] Committee on Economic, Social and Cultural Rights, *General Comment No. 3, The Nature of States Parties' Obligations*, UN Doc. No. E/C.12/1991/23 (1990), Art.2, para.1.9. http://www.unhchr.ch/tbs/doc.nsf/(symbol)/CESCR+General+comment+3.En. (Accessed July 13, 2014).

[306] Committee on Economic, Social and Cultural Rights, *General Comment No. 14: The Right to the Highest Attainable Standard of Health*, UN Doc. No. E/C.12/2000/4, (Geneva : United Nations, 2000), http://www.unhchr.ch/tbs/doc.nsf/%28symbol%29/E.C.12.2000.4.En. (Accessed June 19, 2014).

[307] Brigit C.A. Toebes, *The Right to Health as a Human Right in International Law* (Amsterdam: Hart/ Intersentia, 1999), cited in Ooms and Hammonds, "Correcting Globalisation in Health: Transactional Entitlements Versus the Ethical Imperative of Reducing Aid-Dependency," 158.

[308] Committee on Economic, Social and Cultural Rights, *General Comment No. 14: The Right to the Highest Attainable Standard of Health*, UN Doc. No. E/C.12/2000/4, (Geneva: United Nations, 2000), http://www.unhchr.ch/tbs/doc.nsf/%28symbol%29/E.C.12.2000.4.En. (Accessed June 19, 2014).

[309] Committee on Economic, Social and Cultural Rights, *General Comment No. 14: The Right to the Highest Attainable Standard of Health,* UN Doc. No. E/C.12/2000/4, (Geneva : United Nations, 2000), http://www.unhchr.ch/tbs/doc.nsf/%28symbol%29/E.C.12.2000.4.En. (Accessed June 19, 2014).

[310] Gorik Ooms, Katharine Derderian and David Melody, "Do We Need a World Health Insurance to Realise the Right to Health? *PLoS* 3, no. 12 (December 2006): e530.

year by WHO.[311] Taken into consideration the principle of *ultra posse nemo obligatur*, which implies in this context that no person (or country) can be obligated beyond what he, she or it is able to do, there is then some doubts whether it is reasonable to define core obligations that cannot be afforded in low-income countries. It is perfectly relevant to do so appealing to the Committee on Economic, Social and Cultural Rights which notes that "Each State Party to the present Covenant undertakes to take steps, individually and through international assistance and cooperation, especially economic and technical, to the maximum of its available resources."[312] To consider the ability or inability of low-income countries to realize their core obligations, it is pertinent that one should not simply consider their national resources, but as well resources they obtain from international assistance.[313]

At the May 2000 Committee on Economic, Social and Cultural Rights session Paul Hunt noted: "if the Committee decided to approve the list of core obligations, it would be unfair not to insist also that richer countries fulfill their obligations relating to international cooperation under article 2, paragraph 1, of the Covenant. The two sets of obligations should be seen as two halves of a package."[314] It is important to note that if the right to health is considered meaningless without the achievement of at the minimum its core content, and if some countries lack resources required to achieve the core content of the right to health, then the right to health itself cannot exist in the absence of international obligations to provide assistance.[315] Ooms and Hammonds argue that "Without international obligations to provide assistance – without global responsibility, that is – the right to health is not a right but a privilege reserved for those who are born outside of the world's poorest countries."[316] However, it would not be misunderstood that such global responsibility does not imply that low-income countries have an unconditional and unlimited claim to international assistance in order to achieve the core content of the right to health.[317] Philip Alston succinctly writes, "The correlative obligation would, of course, be confined to situations in which a developing country had demonstrated its best efforts to meet the (Millennium Development) Goals and its inability to do so

---

[311] Carrin, Evans, and Ku, "Designing Health Financing Policy towards Universal Coverage," 652.

[312] Committee on Economic, Social and Cultural Rights, *General Comment No. 14: The Right to the Highest Attainable Standard of Health*, UN Doc. No. E/C.12/2000/4, (Geneva : United Nations, 2000), http://www.unhchr.ch/tbs/doc.nsf/%28symbol%29/E.C.12.2000.4.En. (Accessed June 19, 2014).

[313] Ooms and Hammonds, "Taking Up Daniels' Challenge: The Case for Global Health Justice," 36.

[314] Committee on Economic, Social and Cultural Rights, *Summary Record of the 10th Meeting*, (2000), http://www.unhchr.ch/tbs/doc.nsf/(Symbol)/d8711da53e337f75802568d9003bc930. (Accessed July 15, 2014).

[315] Ooms and Hammonds, "Taking Up Daniels' Challenge: The Case for Global Health Justice," 36

[316] Ooms and Hammonds, "Taking Up Daniels' Challenge: The Case for Global Health Justice," 36.

[317] Ooms and Hammonds, "Taking Up Daniels' Challenge: The Case for Global Health Justice," 36.

because of a lack of financial resources."[318] Although, Alston makes reference to Millennium Development Goals, instead of core content of socioeconomic human rights, there is a contention that the same argument can as well be made for the achievement of the core content of socioeconomic human rights.[319] Therefore, any claim to international assistance would be contingent on countries that show their best efforts.[320]

Ooms and Hammonds further contend that "a claim to international assistance would not only be conditional, but also limited.[321] Article 2, paragraph 1 of the ICESCR endorses both domestic obligations and international obligations of assistance but doesn't show clearly the difference between both. If international obligations of assistance result only when domestic obligations have been entirely realized, actually they would never result because the right to health is a dynamic goal and the fact is that "the highest attainable standard of physical and mental health" would never be fully realized. In this context, high-income countries could incessantly contest international obligations of assistance, alluding to their domestic obligations. The concept of core content actually dictates a hierarchy which stresses that it is more pressing to achieve the minimum essential standard of health for all humans, in the absence of which the right to health itself becomes meaningless, than to strive for the very highest attainable standard of health within the domestic setting. The hierarchy would cease to exist as soon as the minimum essential standard of health were achieved everywhere. At that moment, in consonant with the concept of core content, affluent countries could appeal to the primacy of domestic obligations to contend a change from offering international assistance to give preference alternatively to the highest attainable standard in domestic setting.[322] Ooms and Hammonds argue that "interpreting the core content of the right in this way would also provide for the possibility of a sliding scale of responsibility, one that falls between exclusively national responsibility and wholly global responsibility."[323] The combination of national and global responsibilities in realizing the right to health in poor developing nations establishes the context for a sliding scale of responsibility which is discussed next.

[318] Philip Alston, "Ships Passing in the Night: The Current State of the Human Rights and Development Debate Seen Through the Lens of the Millennium Development Goals," *Human Rights Quarterly* 27, no. 3 (August 2005): 755–829.

[319] Ooms and Hammonds, "Taking Up Daniels' Challenge: The Case for Global Health Justice," 45.

[320] Ooms and Hammonds, "Taking Up Daniels' Challenge: The Case for Global Health Justice," 36.

[321] Ooms and Hammonds, "Taking Up Daniels' Challenge: The Case for Global Health Justice," 36.

[322] Ooms and Hammonds, "Taking Up Daniels' Challenge: The Case for Global Health Justice," 36.

[323] Ooms and Hammonds, "Taking Up Daniels' Challenge: The Case for Global Health Justice," 36.

6.3.1.3 The Sliding Scale of National and Global Responsibilities and the Minimum Level of the Right to Health

A general consensus on making necessary efforts to guarantee for everyone globally the core components of the right to health, makes it imperative to examine the costs involved and how those costs should be allocated between the various nations and the international community, and among those nations that are well positioned to assist.[324] Eide however acknowledges the difficulty associated with the calculation of costs for numerous reasons. There is currently a prevalent recognition of the social determinants of health, which implies that convincing efforts towards global health must take into consideration factors that are frequently not incorporated in the calculation of health expenditure.[325]

The rough calculation by the Commission on Macroeconomics of Health deduced that the minimum expenditure obligation within the health sector narrowly explained would come to US$ 40. These rough calculations are currently about ten years old and the amount would have to be likely increased. The calculations simply incorporated the expenditures on health within the health sector narrowly defined, and thus did not encompass the costs of guaranteeing access to preconditions of health.[326]

Critics including Ooms indicated that the calculations provided by the Commission on the Macroeconomics of Health were seriously flawed.[327] Ooms however endorses that the figures provided by the Commission may be helpful as a point of departure for reflections on how the costs, narrowly restricted to the health sector, could be allocated as well as a recognition that many developing countries could not realize the core content of right to health for their citizens without foreign assistance.[328] In consonant with this perspective, Ooms and Hammonds proposed the creation of a Global Health Fund, in line with the current Global Fund to fight AIDS, Tuberculosis and Malaria which will be extensively discussed in the next part of the chapter.

As a basis for the sliding scale of national and global responsibilities for the realization of the minimum level of the right to health, Ooms and Hammonds made some critical proposals in relation to the costs of core obligations for global health justice. First, they assume that achieving the content of the right to health would require that governments must be able to spend at least US$40 per person per year on health-related goods recognized by WHO as essential for an "adequate package of healthcare interventions" (taking into account an adjustment for inflation).[329]

---

[324] Eide, *The Health of the World's Poor – A Human Rights Challenge*, 106.

[325] Eide, *The Health of the World's Poor – A Human Rights Challenge*, 106.

[326] Eide, *The Health of the World's Poor – A Human Rights Challenge*, 107.

[327] Gorik Ooms, "The Right to Health and the Sustainability of Healthcare: Why a New Global Health Aid Paradigm is Needed," (PhD diss., University of Ghent, 2008), 149–153.

[328] Ooms, "The Right to Health and the Sustainability of Healthcare: Why a New Global Health Aid Paradigm is Needed," 149–153.

[329] Carrin, Evans, and Ku, "Designing Health Financing Policy towards Universal Coverage," 652.

Second, they also assume that government revenue, with the exclusion of grants could reach the target of 20% of GDP in low-income countries.[330]

The sliding scale paradigm proposed by Ooms and Hammonds requires each country to spend at least 3% of average Gross Domestic Product (GDP) per person on health-related goods distribution. The 3% approach to GDP refers to a benchmark for spending on national health by developing countries that would qualify them to receive global assistance. Each developing country would need to spend 3% of its GDP in order to show that it has exhausted its efforts and resources in realizing the core content of the right to health.[331] This 3% approach to GDP is worth US$40 per person per year as proposed by the World Health Organization. They proposed using GDP per capita as the starting foundation for allocation and then calculating the domestic responsibility as well as global responsibility that would be required to fulfill the target of US$40 per person. For example, an identified country has a GDP per person of US$333 and is supposed to be able to spend 3% of this amount, or US$10 per person per year, on health-related goods distribution. Then, the global responsibility towards the identified country is restricted to guaranteeing that it can realize the distribution of health-related goods valued at US$40 per person per year, presuming that this level of financing is what it entails to achieve the core content of the right to health, or the equivalent of US$30 per person per year.[332] The implication is that if a country can only afford US$10 per person as its national responsibility, then evidently the global responsibility towards the country is US$30.

Based on their calculation, the amount required for funding the cost of global responsibility for the right to health, or the cost of obligations of global health justice is estimated at US$50 billion per year. This is an estimate of the amount of assistance that would be required by about 59 low-income and lower-middle-income countries with a population of 2.5 billion. The 66 countries classified by the World Bank as high-income countries have a collective GDP of US$43 trillion in 2008, and US$26 trillion in 2000. It is projected that they will reach a collective GDP of US$49 trillion in a moment, despite setbacks from global financial crisis. It implies then that to discharge their global responsibility, affluent countries would need to allot just about 0.1% of their GDP to international assistance for health.[333] This is considered "a modest share of their wealth."[334] It is pertinent to note that as a result of the revised assessment by WHO ten years following the Commission's report, from US$40 to US$44, the total amount would be accordingly higher.[335]

---

[330] International Monetary Fund (IMF), *World Economic Outlook Database*, (Accesed July 16, 2014).

[331] Ooms and Hammonds, "Taking Up Daniels' Challenge: The Case for Global Health Justice," 36–37.

[332] Ooms and Hammonds, "Taking Up Daniels' Challenge: The Case for Global Health Justice," 37.

[333] Ooms and Hammonds, "Taking Up Daniels' Challenge: The Case for Global Health Justice," 37.

[334] Eide, *The Health of the World's Poor – A Human Rights Challenge*, 108.

[335] Eide, *The Health of the World's Poor – A Human Rights Challenge*, 108.

Eide contends that the figures demonstrate that it should be feasible for the global community constituting primarily the high-income countries to provide the funds needed to supplement the resources of poor nations, in order that the minimum core of the right to health can be guaranteed globally.[336] He further calls for significant efforts "to establish the institutions and the procedures necessitate determining the scope of contributions, the allocation between donors, and the supervision of compliance both by the home state and by the external contributors."[337] The analysis further deals with an explanation of how a Global Health Fund can be established and operated in order to affirm the feasibility of the Global Health paradigm for improving affordable access to drugs.

6.3.2 Access to Post-trial Drugs in Developing Nations

The critical need for an international agency for the distribution of health-related goods that would rectify the injustice arising from the current global system has been forcefully argued by Ooms and Hammonds.[338] This global basic institution that may govern the distribution of health-related goods in a fair approach could take a number of forms. Ooms and Hammonds suggest two different forms: a Framework Convention on Global Health proposed by Lawrence Gostin[339] and a Global Health Fund proposed by Gorik Ooms and Rachel Hammonds.[340] This implies that some type of conventional global institution is imperative for enforcing "the interactive and practical applications of national and global responsibility."[341] A Global health fund in line with the current Global Fund to fight AIDS, Tuberculosis and Malaria is viewed as one way to effectively handle an agreement between most of the countries on critical parameters regarding realizing the core content of the right to health which involves improved access to essential drugs. A more detailed discussion on the Global Health Fund and affordable access to drugs begins with addressing the traditional concept of a Global Health Fund and its impeding factors.

---

[336] Eide, *The Health of the World's Poor – A Human Rights Challenge*, 108.

[337] Eide, *The Health of the World's Poor – A Human Rights Challenge*, 108.

[338] Ooms and Hammonds, "Taking Up Daniels' Challenge: The Case for Global Health Justice," 40.

[339] Lawrence Gostin, "A Proposal for a Framework Convention on Global Health," *Journal of International Economic Law* 10, no. 4 (2007): 989–1008.

[340] Ooms and Hammonds, "Correcting Globalisation in Health: Transactional Entitlements Versus the Ethical Imperative of Reducing Aid-Dependency," 154–170.

[341] Ooms and Hammonds, "Taking Up Daniels' Challenge: The Case for Global Health Justice," 40.

6.3.2.1 Traditional Concept of Global Health Fund

The term "Global Health" can be viewed from different perspectives. Ooms et. al, discussed the various trends of global health in a recent working paper as what global health has been so far, what it should be, and what it could become. Regarding what global health has been so far, it is a shift from international health to global health explained by an era in which richer countries accept shared responsibility for poor public health in poorer countries and seek shared benefits with such countries. The international health era was marked by richer countries accepting little or no direct responsibility for poor public health in poorer countries, but rather held that those poorer countries would attain economic development in the end to take care of their public health issues.[342] Regarding what global health would be, they build on the requirement of universalization of international human rights law, focusing on the notion of progressive realization, primacy of national responsibility, emphasis on the core content of the right to health, a proportional distribution of health expenditures for essential levels of health, fair trade instead of free trade in international cooperation.[343] Regarding what global health could become, they acknowledged that the likely result will be a compromise between the pursuit of richer countries' enlightened self- interests and the more idealistic or altruistic viewpoint of a human rights-based obligation of all countries to the reduction of the health risks and burdens of the poor. What global health could become also requires a global social contract that gives adequate attention to the priorities of poorer countries.[344]

The traditional concept of Global Health Fund (GHF) invokes the conventional health development paradigm with all its limitations. It emphasizes the aim of financial sustainability. Pertinent to note, is a distinction between national sustainability, which refers to present or future domestic resource availability, and international sustainability which refers to the availability of a mixture of domestic resources and sustained foreign assistance.[345] The post-trial access of drugs to participants and host populations would not fit within the conventional health development paradigm because of its exorbitant cost. Providing post-trial access of interventions in

[342] Gorik Ooms et. al, "Global Health: What it has been so far, what it should be, and what it could become." Studies in Health Services Organisation & Policy, Working Paper Series No. 2 (Antwerpen, Belgium: The Department of Public Health, Institute of Tropical Medicine, 2011), 7. http://www.itg.be/WPshsop (Accessed December 3, 2016).

[343] Gorik Ooms et. al, "Global Health: What it has been so far, what it should be, and what it could become." Studies in Health Services Organisation & Policy, Working Paper Series No. 2 (Antwerpen, Belgium: The Department of Public Health, Institute of Tropical Medicine, 2011), 40–52. http://www.itg.be/WPshsop (Accessed December 3, 2016).

[344] Gorik Ooms et. al, "Global Health: What it has been so far, what it should be, and what it could become." Studies in Health Services Organisation & Policy, Working Paper Series No. 2 (Antwerpen, Belgium: The Department of Public Health, Institute of Tropical Medicine, 2011), 53. http://www.itg.be/WPshsop (Accessed December 3, 2016).

[345] Ooms, "The Right to Health and the Sustainability of Healthcare: Why a New Global Health Aid Paradigm is Needed," 25.

poor health systems or in poor health budgets, will only be feasible if traditional notions of sustainability are abandoned in favor of that of sustained international funding.[346] The implication is that, "sustainable health care – narrowly defined as independent from international aid – is illusionary in the world's poorest countries."[347] Public health budgets in these countries need to be expanded which in essence entail greater financial commitments to national and international resources. Ooms forcefully articulates, "Rejecting concerns about sustainability might be the best way to defeat the illusion of sustainability and paradoxically, to promote sustainability at a different level: the sustainability of international assistance."[348]

Mindsets definitely need to change at all levels including that of the Executive Director of UNICEF, who continues to maintain that effective interventions to reduce infant mortality should be "phased in according to the ability of both the health system to deliver them at scale, and of governments to afford them and to sustain them in the longer term."[349] In the context of traditional concept of GHF, several effective health interventions are identified as "unsustainable" in poor countries because they cannot be paid for by national health budgets.[350] The prospect of increasing national health budgets considerably and permanently through international assistance is hardly taken into account. The establishment of the Global Fund to Fight AIDS, Tuberculosis and Malaria proves the advantage of a more robust and ambitious perspective in which projects such as the provision of antiretroviral treatment to AIDS patients, formerly considered as unsustainable was finally accepted as tenable as the Global Fund offered a long-term funding viewpoint.[351] Other health interventions such as provision of post-trial access to drugs would fit within this strategy.

The novel idea of GHF articulated in this book endorses the legal obligation to provide international assistance for health. Health in this context is considered a human right, a right preserved by international treaties. Two critical problems were underscored as associated with fulfilling the legal obligation to provide international assistance for health – shared responsibility and progressive realization which implies that the obligation is not immediate but can take some time to achieve.[352]

---

[346] Ooms, Derderian and Melody, "Do We Need a World Health Insurance to Realise the Right to Health? Ooms, " e530.

[347] Ooms, "Health Development Versus Medical Relief: The Illusion Versus the Irrelevance of Sustainability," e345.

[348] Ooms, "Health Development Versus Medical Relief: The Illusion Versus the Irrelevance of Sustainability," e345.

[349] A.M. Veneman, "Achieving Millennium Development Goal 4." Lancet 368, no.954 (September, 2006): 1044–1047.

[350] World Health Organization, Core Health Indicators: Malawi, (Geneva: WHO, 2006), http://www3.who.int/whosis/core/core_select_process.cfm?country=mwi&indicators=healthpersonnel&intYear_select=all&language=en (Accessed February 21, 2017).

[351] Ooms, Derderian and Melody, "Do We Need a World Health Insurance to Realise the Right to Health? e530.

[352] K. Roth, "Human Rights and the AIDS Crisis: The Debate over Resource," Canadian HIV/AIDS Policy and Law Review 5, no.4 (February 2000): 93. http://hrw.org/english/docs/2000/07/11/global2195.htm. (Accessed February 21, 2017).

The problem of progressive realization rests on the recognition that all economic, social, and cultural rights cannot be completely fulfilled in a brief period. This offers states the opportunity to assert that they are exhausting or have exhausted all their efforts. The fulfilment of the minimum essential level is seen to be immediate rather than progressive.[353] The obligation to provide assistance for health is avoided by a second problem – shared responsibility. In this context, poor states can blame rich states for not keeping their obligation to provide assistance, resulting that poor states do not have adequate resources to fulfil their core obligations. Furthermore, rich states can blame poor states and each other for not adequately doing what they are supposed to do.[354]

The novel idea of GHF would require: "(1) A willingness to share health risks and the burden of health care between rich and poor states; (2) A mechanism to allocate resources to poor states; and (3) A mechanism to determine the contributions from rich states."[355] The establishment of the Global Fund substantially changed public health policies in low-income countries. The provision of post-trial access to drugs would have been considered unsustainable before the creation of Global Fund. The Global Fund needs every aspect of the intervention to be sustainable, except the funding which it guarantees itself. The Global Fund's notion of sustainability – focusing on sustained international assistance, rather than on present or future self-financing appears to have affected the reasoning of rich states.[356] The United Nations General Assembly adopted the Political Declaration on HIV/AIDS in 2006 which pledges the "provision of funds for (ART) in a sustained manner."[357] This essentially differs from previous pledge to " provide (ART) progressively and in a sustainable manner," as articulated in the "Declaration of Commitment" in the wake of the United Nations General Assembly Special Session on AIDS in June 2001.[358]

The reliability of the funding from Global Fund also helps to surmount other barriers. The World Bank and the International Monetary Fund (IMF) have usually imposed ceilings on public health expenditure.[359] In this context, low-income countries are not permitted to violate the ceiling, or to increase health budgets more

---

[353] Ooms, Derderian and Melody, "Do We Need a World Health Insurance to Realise the Right to Health? e530.

[354] Ooms, Derderian and Melody, "Do We Need a World Health Insurance to Realise the Right to Health? e530.

[355] Ooms, Derderian and Melody, "Do We Need a World Health Insurance to Realise the Right to Health? e530.

[356] Ooms, Derderian and Melody, "Do We Need a World Health Insurance to Realise the Right to Health? e530.

[357] United Nations General Assembly, Political Declaration on HIV/AIDS, (2006), http://data.unaids.org/pub/Report/2006/20060615_HLM_PoliticalDeclaration_ARES60262_en.pdf. (Accessed February 24, 2016).

[358] United Nations General Assembly, Declaration of Commitment on HIV/AIDS, (2001), http://www.un.org/ga/aids/docs/aress262.pdf. (Accessed February 24, 2017).

[359] Gorik Ooms and Ted Schrecker, "Expenditure Ceilings, Multilateral Financial Institutions, and the Health of Poor Populations," Lancet 365, no. 9473 (May 2005):1821–1823.

than the ceiling, even if they receive more Official Development Assistance (ODA) to pay for more expenses.[360] The WHO asserted that "Financial ceilings …may need to be stretched,"[361] while on the contrary, the World Bank articulated: " It is not prudent for countries to commit to permanent expenditures for such items as salaries for nurses and doctors on the basis of uncertain financing flows from development assistance funds."[362] The IMF and World Bank policies significantly hinder the maximum utilization of foreign assistance. These policies constitute one of the critical issues with the health development paradigm as they are viewed as a "financial straitjacket used by the IMF to safeguard national sustainability, as foreign assistance beyond the level of sustainable expenditure – or above fiscal space – is not allowed to be spent."[363] However, because the Global Fund was considered reliable and predictable, the World Bank and the IMF were disposed to make an exception for more ODA coming from it.[364]

The Global Fund struggles with the task of concentrating on the three diseases, as well as strengthening the weak public health systems that are critical for the fight of the diseases. The Global Fund at some point signals more emphasis on the three diseases and to allow the strengthening of health systems and support for the workforce to others.[365] This would result in a "Medicines without Doctors" scenario which implies that medicines to fight AIDS, tuberculosis, and malaria are available whereas the doctors and nurses to prescribe those medicines are not.[366] This trend would be a disaster for the Global Fund as it has a significant advantage to play a pivotal role in supporting health workforces. Many other donors inevitably rely on sustainability in the conventional sense which implies that beneficiary countries should at some point replace international funding with domestic resources. On the contrary, the Global Fund relies on sustained funding by the international community, bolstering sustained commitments to beneficiary countries.[367] The Global Fund supports this novel approach to sustainability. Some authors rightly articulated that

[360] Ooms, Derderian and Melody, "Do We Need a World Health Insurance to Realise the Right to Health? e530.

[361] World Health Organization, Health and the MDGs: Keep the Promise, (2005), http://www.who.int/mdg/publications/MDG_Reporter_08_2005.pdf. (Accessed February 25, 2016).

[362] World Bank, The Millennium Goals for Health: Rising to the Challenges, (2004), http://siteresources.worldbank.org/INTEAPREGTOPHEAUNUT/PublicationsandReports/20306102/296730PAPEROMilentOgoalsOforOhealth.pdf. (Accessed February 25, 2017).

[363] Ooms, "The Right to Health and the Sustainability of Healthcare: Why a New Global Health Aid Paradigm is Needed," 25.

[364] Ooms, Derderian and Melody, "Do We Need a World Health Insurance to Realise the Right to Health? e530.

[365] Gorik Ooms, Wim Van Damme, and Marieen Temmerman, "Medicines Without Doctors: Why the Global Fund Must Fund Salaries of Health Workers to Expand AIDS Treatment," PLos Medicine 4, no. 4 (April 2007): e128.

[366] Ooms, Damme, and Temmerman, "Medicines Without Doctors: Why the Global Fund Must Fund Salaries of Health Workers to Expand AIDS Treatment," e128.

[367] Ooms, Damme, and Temmerman, "Medicines Without Doctors: Why the Global Fund Must Fund Salaries of Health Workers to Expand AIDS Treatment," e128.

"It is the only donor mechanism that benefits from an explicit endorsement from the international community to practice a novel approach to sustainability."[368] Similarly, the Global Health Fund emphasizes a novel approach to sustainability focusing on sustained domestic and international resources to address all diseases and primary health care system of developing nations. The Global Health Fund proposed by the author requires a new global health aid paradigm rooted in a human rights framework.

6.3.2.2 Ethical Issues in Global Health Fund

The international community was pressured by AIDS activists to establish the Global Fund to fight AIDS, Tuberculosis and Malaria, which when examined from a human rights framework is nothing except a tool for conformity to the transnational obligation to realize a critical component of the core content of the right to health. The existence of the Global Fund applied more pressure on governments of countries in need of assistance to develop AIDS intervention plans and on government of countries well positioned to assist to offer the required assistance.[369] It established entitlements for individual countries and to a reasonable extent the "equitable contributions framework" already alluded to, that is, duties for individual countries.[370]

It is almost an impossible mission to realize progress in the fight against a single disease or even three diseases without effectively tackling the weakness of the public health systems in developing countries.[371] Ooms and Hammonds argue that "to address this fundamental problem the world needs a Global Fund to fight poor health, including AIDS treatment but not excluding other essential health care, or a Global Health Fund to realise the core content of the right to health."[372]

The approach utilized by the Global Fund involves a serious limitation. It primarily concentrates on deadly diseases in Sub-Saharan Africa such as AIDS, tuberculosis and Malaria and supports simply interventions to deal with those diseases. This situation does not create any problem for most middle-income countries which can fund their public health systems from domestic resources. Nevertheless, the same situation which is Global Fund's limitation creates a significant problem for low-income countries which results in a two-tier system: the fight against AIDS,

---

368 Ooms, Damme, and Temmerman, "Medicines Without Doctors: Why the Global Fund Must Fund Salaries of Health Workers to Expand AIDS Treatment," e128.

369 Ooms and Hammonds, "Correcting Globalisation in Health: Transactional Entitlements Versus the Ethical Imperative of Reducing Aid-Dependency," 160.

370 Ooms and Hammonds, "Correcting Globalisation in Health: Transactional Entitlements Versus the Ethical Imperative of Reducing Aid-Dependency," 160.

371 Ooms and Hammonds, "Correcting Globalisation in Health: Transactional Entitlements Versus the Ethical Imperative of Reducing Aid-Dependency," 160.

372 Ooms and Hammonds, "Correcting Globalisation in Health: Transactional Entitlements Versus the Ethical Imperative of Reducing Aid-Dependency," 160.

tuberculosis and malaria is not impeded by the limitation of financial self-sufficiency, while the fight for primary health care in general is.[373] Ooms contends that "the result is the current paradox; international health aid to strengthen the backbone of the health systems is much harder to find (because of the financial self-sufficiency restriction) than international health aid for extra muscle to fight AIDS, tuberculosis or malaria."[374]

This two-tier system has engendered an increasing difference between adequately funded muscle to fight AIDS, tuberculosis and malaria, and heavily underfunded backbone of the health systems.[375] Global Fund has been criticized on this ground. Roger England argued that "HIV is receiving relatively too much money, with much of it used inefficiently and sometimes counterproductively."[376] He further raises objections to the cost-effectiveness of HIV interventions and calls for utilizing money intended for HIV interventions for other health needs.[377] He ends his argument with a passionate call for transforming the Global Fund into a Global Health Fund. He writes "A global basket fund is needed to transfer sustainable and predictable funding to countries, avoiding hugely unpredictable aid flows from fickle donors that make planning impossible. The Global Fund to Fight AIDS, Tuberculosis and Malaria could abandon disease dedicated support to become this fund... Improving health systems should form the platform for action and research now, transcending HIV and other disease-specific programmes."[378]

It is pertinent to note that while most critiques essentially fault the Global Fund for having a very restricted mandate, none of them faults the conventional health development approach and its aim of self-sufficiency, which is actually the crux of the problem. These critiques are apparently condemning the Global Fund for the favorable outcomes of its exceptional approach partly because they prefer this exceptional approach to be employed for primary health care generally.[379] Ooms further contends that they should be pressing for expanding the Global Fund mandate rather than preoccupying themselves with condemning its success.[380] The

[373] Ooms, *From the Global AIDS Response towards Global Health?* A Discussion Paper for the Helene De Beir Foundation and the International Civil Society Support Group, (January 2009), 16. http://www.hdbf.org/?pageid=345&lang=en (Accessed June 13, 2014).

[374] Ooms, *From the Global AIDS Response towards Global Health?* A Discussion Paper for the Helene De Beir Foundation and the International Civil Society Support Group, (January 2009), 16. http://www.hdbf.org/?pageid=345&lang=en (Accessed June 13, 2014).

[375] Ooms, *From the Global AIDS Response towards Global Health?* A Discussion Paper for the Helene De Beir Foundation and the International Civil Society Support Group, (January 2009), 16. http://www.hdbf.org/?pageid=345&lang=en (Accessed June 13, 2014).

[376] Roger England, "Are We Spending Too Much on HIV? *BMJ* 334, (February 2007): 344.

[377] England, "Are We Spending Too Much on HIV? 344.

[378] England, "Are We Spending Too Much on HIV? 344.

[379] Ooms, *From the Global AIDS Response towards Global Health?* A Discussion Paper for the Helene De Beir Foundation and the International Civil Society Support Group, (January 2009), 16. http://www.hdbf.org/?pageid=345&lang=en (Accessed June 13, 2014).

[380] Ooms, *From the Global AIDS Response towards Global Health?* A Discussion Paper for the Helene De Beir Foundation and the International Civil Society Support Group, (January 2009), 16. http://www.hdbf.org/?pageid=345&lang=en (Accessed June 13, 2014).

Global Fund is counteracting these critiques by assuming or reestablishing its responsibility in funding the backbone of health systems. However, this would involve more funding, which would invariably just occur if donor countries entirely support an expanded mandate for the Global Fund.[381]

The IMF application of the concept of fiscal space to health financing in developing countries was very detrimental to Global Health Fund. Peter Heller from the IMF defines fiscal space as "room in a government's budget that allows it to provide resources for a desired purpose without jeopardizing the sustainability of its financial position or the stability of the economy."[382] He claims that fiscal space is a more urgent issue in developing countries than in developed countries or advanced economies due to more urgent needs for expenditure currently. Nonetheless, longer-term issues are as well included, still for lower-income countries, because of the need to guarantee that there would be room to deal with unexpected challenges.[383] Thus, he articulates "Countries that receive significant flows of foreign resources for a specific sector (such as health care) may, as a result of the associated expansion of the sector, face additional future spending needs that may essentially preempt a share of the growth of future domestic budgetary resources."[384]

The IMF presumes that aid-driven health sector growth would unavoidably forestall a share of domestic resources. It is not disposed to envisage that international health aid could finance the growth of the health sector ultimately.[385] Hence, it incessantly warns countries against utilizing excessive international aid for broadening the health sector, a warning also reiterated by the World Bank: "Obviously, then, it is not prudent for countries to commit to permanent expenditures for such items as salaries for nurses and doctors on the basis of uncertain financing flows from development assistance funds."[386]

The IMF presses for its message about the unreliability of international solidarity with the objective of scaring recipient governments regarding growing expenditures. It carries out this by continuously reechoing the findings of Ales Bulir and

---

[381] Ooms, *From the Global AIDS Response towards Global Health?* A Discussion Paper for the Helene De Beir Foundation and the International Civil Society Support Group, (January 2009), 16. http://www.hdbf.org/?pageid=345&lang=en (Accessed June 13, 2014).

[382] Peter Heller, "Fiscal Space: What It is and How to Get It," *Finance and Development* 42, no. 2 (June2005): http://www.imf.org/external/pubs/ft/fandd/2005/06/basics.htm (Accessed July 17, 2014).

[383] Heller, "Fiscal Space: What It is and How to Get It," *Finance and Development* 42, no. 2 (June2005): http://www.imf.org/external/pubs/ft/fandd/2005/06/basics.htm (Accessed July 17, 2014).

[384] Heller, "Fiscal Space: What It is and How to Get It," *Finance and Development* 42, no. 2 (June2005): http://www.imf.org/external/pubs/ft/fandd/2005/06/basics.htm (Accessed July 17, 2014).

[385] Ooms, *From the Global AIDS Response towards Global Health?* A Discussion Paper for the Helene De Beir Foundation and the International Civil Society Support Group, (January 2009), 17. http://www.hdbf.org/?pageid=345&lang=en (Accessed June 13, 2014).

[386] Adam Wagstaff and Miriam Claeson, *The Millennium Development Goals for Health: Rising to the Challenges* (Washington, D.C.: World Bank, 2004), http://www.hlfhealthmdgs.org/documents.asp (Accessed July 18, 2014).

A. Javier Hamann who discovered that international aid is "substantially more volatile than domestic revenues."[387] It vehemently refuses to pay attention to another pair of relevant data on the issue by Paul Collier who discovered that international aid is, actually, more reliable than domestic revenue[388] and by Oya Celasun and Jan Walliser who discovered that whereas international aid is to some extent less reliable than domestic revenue, international aid deficits previously did not compel recipient countries to decrease recurrent expenditure, as these deficits were compensated by decreased investments.[389]

A report published in March 2007 by the Independent Evaluation Office of the IMF showed that only 27% of the extra international aid to sub-Saharan Africa from 1999 to 2006 was in fact permitted or authorized to be spent.[390] The remaining 73% was placed in savings. This practice permits the IMF to impose conformity to fiscal space constraints: whenever a country is in danger of going beyond fiscal space, the IMF can program international aid to be saved by the recipient nation rather than being spent, for instance, to broaden health services to vulnerable populations. This obviously falls short of an incentive for donors to grow international aid.[391]

Pertinent to note also is that the global AIDS response discovered a breakthrough strategy at the intersection of the medical relief and health development paradigms. International aid in the type of medical relief is not impeded by fiscal space constraints.[392] This issue may have prompted Peter Piot, the former general director of the United Nations Joint AIDS Programme (UNAIDS), to clearly compare countries affected by AIDS with countries in or emerging from conflict, when he requested for a general exception for AIDS expenditure from fiscal space constraints.[393]

Nevertheless, the unreliability of international aid over a long period of time constitutes a problem. It is hard for health ministries of various low-income coun-

[387] Ales Bulir and A. Javier Hamann, "Aid Volatility: An Empirical Assessment" *IMF Staff Papers* 50, no. 1 (2003): 83.

[388] Paul Collier, "Aid Dependency: A Critique," *Journal of African Economies* 8, no. 4 (1999): 528–545.

[389] Oya Celasun and Jan Walliser, *Predictability of Budget Aid: Experiences in Eight African Countries* (Washington, D.C.: Center for Global Development, 2005), http://www.cgdev.org/doc/event%20docs/Predictability%20of%20Budget%20Aid%20revised.pdf (Accessed July 18, 2014).

[390] Independent Evaluation Office of the International Monetary Fund, *The IMF and Aid to Sub-Saharan Africa* (Washington D.C.: Independent Evaluation Office of the International Monetary Fund, 2007), http://www.imf.org/external/np/ieo/2007/ssa/eng/pdf/report.pdf. (Accessed July 8, 2014)

[391] Ooms, *From the Global AIDS Response towards Global Health?* A Discussion Paper for the Helene De Beir Foundation and the International Civil Society Support Group, (January 2009), 17. http://www.hdbf.org/?pageid=345&lang=en (Accessed June 13, 2014).

[392] Ooms, *From the Global AIDS Response towards Global Health?* A Discussion Paper for the Helene De Beir Foundation and the International Civil Society Support Group, (January 2009), 17. http://www.hdbf.org/?pageid=345&lang=en (Accessed June 13, 2014).

[393] Peter Piot, *AIDS: The Need for An Exceptional Response to An Unprecedented Crisis,* A Presidential Fellows Lecture (Washington, D.C.: World Bank, 2003), http://go.worldbank.org/L1940XMKYO (Accessed July 18, 2014).

tries to commit to long-term salaries for more health workers, for instance, if those commitments are simply supported by short-term international health aid commitments. On the other hand, the practice of fiscal space austerity establishes a vicious circle: it is defended by the unreliability of international aid over a long period of time; it engenders international aid being saved instead of being spent, hence generating frustrations for donors who do not observe the expected outcomes; which results in a feeling that "all that aid is not helping anyhow" and hence increases the unreliability of international aid.[394]

It has been forcefully argued by Ooms that "a Global Health paradigm – in the sense of a globally shared responsibility for the health of all people – would solve this problem, or turn it into a merely technical matter."[395] Some technical solutions are already in place in order to address the issue of the unpredictability and unreliability of international health aid: the "replenishments" of the International Development Association (IDA) – as well known as the soft loan arm of the World Bank – are grounded on the principle of burden-sharing between affluent countries,[396] and at least some of those countries regard them as mandatory.[397] Ooms contends that "if we would copy this practice of burden-sharing and mandatory contributions to the financing of primary health care in low-income countries, we would not need to place limits on increases in recurrent health expenditure in low-income countries that are funded by increased international health aid."[398] Despite these impeding factors, a Global Health Fund that effectively discharges its functions would enhance the realization of the core content of the right to health including affordable access to essential drugs which is discussed next.

---

[394] Ooms, *From the Global AIDS Response towards Global Health?* A Discussion Paper for the Helene De Beir Foundation and the International Civil Society Support Group, (January 2009), 18. http://www.hdbf.org/?pageid=345&lang=en (Accessed June 13, 2014).

[395] Ooms, *From the Global AIDS Response towards Global Health?* A Discussion Paper for the Helene De Beir Foundation and the International Civil Society Support Group, (January 2009), 18. http://www.hdbf.org/?pageid=345&lang=en (Accessed June 13, 2014).

[396] International Development Association, *Report from the Executive Directors of the International Development Association to the Board of Governors, Additions to IDA Resources: Fourteenth Replenishment* (Washington, D.C.: World Bank, 2005), http://siteresources.worldbank.org/IDA/Resources/14th_Replenishment_Final.pdf. (Accessed July 18, 2014).

[397] Directorate-General for Development Cooperation, *DGCD Annual Report 2005* (Brussels: Directorate-General for Development Cooperation, 2006), http://www.dgos.be/documents/en/annual_report/2005/dgdc_annual_report_2005.pdf (Accesed July 18, 2014).

[398] Ooms, *From the Global AIDS Response towards Global Health?* A Discussion Paper for the Helene De Beir Foundation and the International Civil Society Support Group, (January 2009), 18. http://www.hdbf.org/?pageid=345&lang=en (Accessed June 13, 2014).

6.3.2.3 Responsibilities of Global Health Fund

First, a Global Health Fund would establish a convention that extensively outlines the scope of national and global responsibilities for all the nations involved. It would pool and monitor contributions from high-income nations and redistribute to low-income nations that need assistance in realizing the right to health which includes affordable access to drugs.[399]

Furthermore, a Global Health Fund would work out a burden-sharing mechanism between all high-income nations. The level of the contribution for each nation would reflect the nation's capacity based on the relative wealth of its economy. The solidarity of all high-income countries would have to continue, if they accept mutual accountability for the health compacts they assisted in endorsing.[400]

A Global Health fund would have to recognize that the primary responsibility for achieving the right to health lies with the state. It would have to establish a double benchmark for domestic contribution to health care that can be required from developing nations. One benchmark would focus on the amount of domestic resources a developing nation can adequately mobilize for government expenditure. A second benchmark would focus on the amount of domestic resources a developing nation can adequately allot to health care, which includes affordable access to drugs.[401] These benchmarks without much emphasis on the details and the figures represent a method, not exact estimates. The figures simply need to be adequately pragmatic to show the feasibility of a Global Health Fund, as "a method to transpose collective entitlements and duties into individual states' entitlements and duties, and not to provide precise estimates."[402]

To make a pragmatic proposal for the first bench-mark involves examining current levels of government revenue with the exclusion of grants in low-income countries of Sub-Saharan Africa, as roughly calculated by IMF in its October 2007 Regional Economic Outlook report for Sub-Saharan Africa.[403] These countries succeeded in increasing government revenue with the exclusion of grants from 15.6% of GDP in 2003 to 17.8% of GDP in 2008. This is obviously an increase of 0.44% of GDP per year. It is projected that progressing at this rate, by 2015, government revenue with the exclusion of grants might be 20%.[404]

---

[399] Ooms and Hammonds, "Taking Up Daniels' Challenge: The Case for Global Health Justice," 41.

[400] Ooms, *From the Global AIDS Response towards Global Health?* A Discussion Paper for the Helene De Beir Foundation and the International Civil Society Support Group, (January 2009), 37–38. http://www.hdbf.org/?pageid=345&lang=en (Accessed June 13, 2014).

[401] Ooms and Hammonds, "Correcting Globalisation in Health: Transactional Entitlements Versus the Ethical Imperative of Reducing Aid-Dependency," 160.

[402] Ooms and Hammonds, "Correcting Globalisation in Health: Transactional Entitlements Versus the Ethical Imperative of Reducing Aid-Dependency," 160.

[403] International Monetary Fund, *World Economic and Financial Surveys: Regional Economic Outlook*, Sub-Saharan Africa, (2007), http://www.imf.org/external/pubs/ft/reo/2007/AFR/ENG/sreo-0407.pdf (Accessed July 18, 2014).

[404] Ooms and Hammonds, "Correcting Globalisation in Health: Transactional Entitlements Versus the Ethical Imperative of Reducing Aid-Dependency," 160.

Regarding the second benchmark, allocating 15% of government revenue to health care could be established as condition for international aid for developing nations.[405] This idea coincides with the pledge made by African Heads of state and government in the 2001 Abuja Declaration.[406] Ooms remarks that "this idea may sound like "patronising conditionality", but it should be understood as "emancipating conditionality": a human rights approach, considering both national and international responsibilities and duties, or simply mutual accountability."[407] Combining these two benchmarks results in a general benchmark which requires low-income countries to mobilize and allot 3% of their GDP to healthcare in order to demonstrate that they have made their best efforts to achieve the core content of the right to health.[408]

A Global Health Fund would involve civil society as a watchdog to detect and fight corruption and misuse of funding disbursed to developing nations.[409] The success of a Global Health Fund would considerably be contingent on the involvement of civil society. Civil society will play a critical role in pressuring low-income countries' governments to allot the equivalent of 3% of their GDP to health. Civil society will as well play a significant role in guaranteeing that the increased health expenditure is utilized judiciously; that is primarily for broadening essential healthcare to remote rural districts and not simply for additional, exorbitant and non-accessible health services in low-income countries' capitals.[410]

On the other hand, civil society would assume the responsibility of ensuring that the Global Health Fund receives adequate funding to finance the approved proposals. It would launch campaigns to push high-income countries' governments to increase their contributions during every replenishment cycle.[411] Ooms aptly describes this as "a rare example of mutual accountability at the level of civil society: civil society of the "Global North" mobilising to generate the international health aid needed; civil society of the "Global South" mobilising to generate

---

[405] Ooms, *From the Global AIDS Response towards Global Health?* A Discussion Paper for the Helene De Beir Foundation and the International Civil Society Support Group, (January 2009), 38–39. http://www.hdbf.org/?pageid=345&lang=en (Accessed June 13, 2014).

[406] Organization of African Unity, *Abuja Declaration on HIV/AIDS, Tuberculosis and Other Related Diseases,* (Abuja, Nigeria: OAU, 2001), http://www.un.org/ga/aids/pdf/abujadeclaration.pdf (Accessed July 2, 2014).

[407] Ooms, *From the Global AIDS Response towards Global Health?* A Discussion Paper for the Helene De Beir Foundation and the International Civil Society Support Group, (January 2009), 39. http://www.hdbf.org/?pageid=345&lang=en (Accessed June 13, 2014).

[408] Ooms and Hammonds, "Correcting Globalisation in Health: Transactional Entitlements Versus the Ethical Imperative of Reducing Aid-Dependency," 161.

[409] Ooms, *From the Global AIDS Response towards Global Health?* A Discussion Paper for the Helene De Beir Foundation and the International Civil Society Support Group, (January 2009), 39. http://www.hdbf.org/?pageid=345&lang=en (Accessed June 13, 2014).

[410] Ooms, "The Right to Health and the Sustainability of Healthcare: Why a New Global Health Aid Paradigm is Needed," 251.

[411] Ooms, *From the Global AIDS Response towards Global Health?* A Discussion Paper for the Helene De Beir Foundation and the International Civil Society Support Group, (January 2009), 37. http://www.hdbf.org/?pageid=345&lang=en (Accessed June 13, 2014).

increased domestic health financing and to make sure that all health financing is well spent."[412]

A Global Health Fund rooted in the new concept of financial sustainability would result in a considerable increase in international health aid.[413] Ooms argues that "it would also change the nature of international aid for health: it would change from temporary to ongoing and from charity to a collective obligation corresponding to a collective entitlement, or a global dimension to social protection."[414] The sustained domestic and international financial support through a Global Health Fund would help in realizing the right to health including affordable access to anti-retroviral drugs for developing nations.

6.4 Conclusion

Concluding remarks on this chapter which discusses global health responsibility focuses on a summary of a proposed paradigm of global health justice which emphasizes sliding scale of national and global responsibilities in realizing the core content of the right to health, including post-trial access to drugs in developing nations. Essentially, improved global health governance was considered a desideratum for global health. There was a consensus that health is a shared responsibility especially in relation to affordable access to essential health services and collective defense against transnational threats, including communicable diseases. Hence, it was forcefully argued that the responsibility for guaranteeing the right to health for all does not rest exclusively with states and their obligations to their populations, but as well with the global community.

It was argued using Daniels' works that Rawls's and Pogge's approaches for global health were insufficient. Rawls's statist version of relational justice narrowly stresses the national responsibility of each country to realize the right to health of all its citizens and effectively excludes the global responsibility of individual nations. In contrast, Pogge's cosmopolitan approach concentrates too much on global responsibility without adequate focus on the national responsibility of individual nations. Daniels criticizes the two extreme positions of Rawls and Pogge and advocated for a middle ground that resists the pull of the two opposing alternatives. Hence, the proposed paradigm combines these two approaches by espousing their strengths and avoiding their weaknesses. The paradigm refers to sliding scale of

[412] Ooms, *From the Global AIDS Response towards Global Health?* A Discussion Paper for the Helene De Beir Foundation and the International Civil Society Support Group, (January 2009), 37. http://www.hdbf.org/?pageid=345&lang=en (Accessed June 13, 2014).

[413] Ooms, *From the Global AIDS Response towards Global Health?* A Discussion Paper for the Helene De Beir Foundation and the International Civil Society Support Group, (January 2009), 34. http://www.hdbf.org/?pageid=345&lang=en (Accessed June 13, 2014).

[414] Ooms, *From the Global AIDS Response towards Global Health?* A Discussion Paper for the Helene De Beir Foundation and the International Civil Society Support Group, (January 2009), 34. http://www.hdbf.org/?pageid=345&lang=en (Accessed June 13, 2014).

national and global responsibilities concerning the right to health in general and affordable access to drugs in particular. This combined approach emphasizes that global responsibility supplements rather than replaces national responsibility for health.

From this perspective, it was argued that the primary responsibility for realizing the right to health rests with every nation. However, cognizant that most developing countries are too poor and that they could exhaust their domestic resources without still able to realize the right to health of their citizens, affluent countries can intervene to exercise their global responsibility as a secondary responsibility for the realization of the right to health.

The justification of the focus on government programs to address the issue of post-trial access to drugs for participants and host populations rather than requiring multinational pharmaceutical companies to fund the cost of these programs was extensively discussed. The duty to promote global justice justifies broad collaboration of all stakeholders involved in the clinical research, including researchers, sponsors, governments of the host country, governments of affluent countries and international governmental and non-governmental aid agencies to provide drugs to participants and host populations at the conclusion of the clinical research. The reference to available resources through international cooperation and assistance buttressed that developed countries have obligations of international assistance for access to essential drugs to participants and host populations.

The obligations of researchers and sponsors to participants and host populations was argued to be transferable in the form of building healthcare infrastructure, contributing health care and research equipments, training local health care workers, researchers and members of the research ethics committee, and providing basic health services. Three different approaches for sharing benefits of clinical research with participants and host populations were discussed in conjunction with their shortcomings, including the reasonable availability requirement, the fair benefits framework and the human development approach.

The challenges of global health as well as global responsibility and health inequalities between nations. It was further strongly argued that there is a global responsibility for global health and that there are obligations of justice to realize the right to health in other nations due to increasing wealth inequality between nations which has considerable direct impact on health inequity. The global capacity to redress health inequalities was discussed, consisting of the health development paradigm, the medical relief paradigm and the new global health paradigm. The health development paradigm and the medical relief paradigm were argued as not sufficiently designed to effectively redress global inequalities, and thus the need to introduce a global health paradigm was imperative. The global health paradigm evolved from the global AIDS response and is at the intersection of health development paradigm and medical relief paradigm. The new concept of sustainability introduced by the global health paradigm entails sustainability at the international level, relying on sustained international support as well as on domestic resources.

International human rights law was argued as a foundation for global health justice and global health paradigm. It was emphasized that without international obligations to provide assistance as well as without global responsibility, the right to health would simply be a privilege meant for the affluent. Any claim to international assistance would be based on countries that showed their best efforts.

The combination of national and global responsibilities in realizing the right to health in poor developing countries establishes the context for a sliding scale of responsibility. The sliding scale paradigm requires each developing country to spend at least 3% of average Gross Domestic Product (GDP) per person on health related goods distribution. The 3% approach to GDP alludes to a benchmark for spending on national health by developing countries that would qualify them to receive global assistance. This 3% approach to GDP is worth US $40 per person per year as proposed by the World Health Organization. If a country can only afford US $10 per person per year as its national responsibility, then, the global responsibility towards the country is US $30 per person per year. A projected estimate for funding the cost of global responsibility for the right to health, or the cost of obligations of global health justice was US$50 billion per year.

There was a consensus for a crucial need for an international agency such as Global Health Fund for the distribution of health-related goods that would rectify the injustice stemming from the current global system. The existing Global Fund to fight AIDS, tuberculosis and malaria was viewed as a prototype, but required expansion of its current mandate in order to include interventions for other diseases as well as addressing the weakness of the public health systems in developing countries. It was argued that the Global Health Fund would be designed to effectively fight for primary health care and to realize the core content of the right to health, including affordable access to essential drugs. The fiscal space constraints imposed on low-income countries by IMF due to the unreliability of the international aid was seen as detrimental to their public health systems and invariably to the proposed Global Health Fund. The traditional concept of Global Health Fund was discussed with a focus on conventional health development paradigm and its severe limitations. The post-trial access of drugs to participants and host populations could not be addressed within the context of conventional health development paradigm. Providing post-trial access of drugs to participants and host populations would entail abandoning traditional notion of sustainability in favor of a sustained international resources. Several critical responsibilities of the Global Health Fund were discussed, including working out burden-sharing mechanisms, pooling, monitoring and redistributing contributions, establishing a double benchmark for domestic contribution to health care and involving the civil society as a watchdog in order to detect and fight corruption and misuse of funding disbursed to developing countries. A well constituted and efficiently operated Global Health Fund rooted in the new concept of financial sustainability would enhance the realization of right to health and affordable access to drugs in developing countries. A brief summary of the entire debate on the ethical justification of post-trial access to drugs for participants and host populations in developing nations anchored on the Universal Declaration on Bioethics and Human Rights which links human rights and bioethics, and emphasizes the principles of human dignity, human rights and fundamental freedom in its efforts to promote responsible biomedical research and clinical practice is discussed in the final chapter and the conclusion.

Chapter 7
UNESCO Declaration on Bioethics and Human Rights

The concluding reflections of this book is anchored on the UNESCO Universal Declaration on Bioethics and Human Rights, because the aim of the Declaration coincides with the critical aspect of the book which is guaranteeing the protection of the participants and host populations of developing countries in clinical research under the auspices of International Human Rights Law. The Universal Declaration on Bioethics and Human Rights is seen as a logical extension of the principles of the 1948 Universal Declaration of Human Rights that emphasizes the dignity and equality of all human persons. The UNESCO Declaration also extends the rights espoused in the previous UN Declaration to the evolving field of bioethics.[1] The Declaration extensively links human rights and bioethics.[2] The Declaration aims to accomplish the objective of safeguarding the protection of the rights of research participants and host population in developing countries by setting "global minimum standards in biomedical research and clinical practice."[3] The Declaration, ipso facto, aims "to guide the actions of individuals, groups, communities, institutions and corporations, public and private."[4] Most importantly, the Declaration prioritizes the principles of "human dignity, human rights and fundamental freedoms" in its efforts to promote responsible biomedical research and clinical practice.[5]

[1] Edmund D. Pellegrino, "Benefit and Harm" in *The UNESCO Universal Declaration on Bioethics and Human Rights: Background, Principles and Application* ed. Henk A. M. J. ten Have and Michele S. Jean (Paris, France: UNESCO Publishing, 2009), 99.

[2] Roberto Andorno, "Global Bioethics at UNESCO: In Defence of the Universal Declaration on Bioethics and Human Rights," *Journal of Medical Ethics* 33, no. 3 (March 2007): 150.

[3] Roberto Andorno, "Global Bioethics at UNESCO: In Defence of the Universal Declaration on Bioethics and Human Rights," 150.

[4] Michael Kirby, "Aims" in *The UNESCO Universal Declaration on Bioethics and Human Rights: Background, Principles and Application* ed. Henk A. M. J. ten Have and Michele S. Jean (Paris, France: UNESCO Publishing, 2009), 81.

[5] Roberto Andorno, "Human Dignity and Human Rights," in *The UNESCO Universal Declaration on Bioethics and Human Rights: Background, Principles and Application* ed. Henk A. M. J. ten Have and Michele S. Jean (Paris, France: UNESCO Publishing, 2009), 91.

E.C. Obi, *Post-Trial Access to Drugs in Developing Nations*, Advancing Global Bioethics 7, DOI 10.1007/978-3-319-60028-4_7

The author also articulates as in the Declaration that the requirement for voluntary informed consent are grounded on the ethical principles of respect for persons, human dignity, and autonomy.

The Declaration further emphasizes the priority of the individual over science or society.[6] The 1964 version of Declaration of Helsinki cited by the author had earlier acknowledged this fundamental principle. The priority of the human being over science is an explicit result of the principle of respect for human dignity and points to stress two basic notions: "First, that science is not an end in itself but only a means for improving the welfare of individuals and society; second, that people should not be reduced to mere instruments for the benefit of science."[7] This notion of human dignity was further clearly articulated in the well-known Kantian formula asserting that persons should always be treated as an end in themselves and never as a means only.[8] The implication is that human beings have intrinsic value and should never be treated as instruments. The author also argues that exploitation of the research participants and host population would be minimized by responsiveness of research to the needs and priorities of the host population. The Declaration also recognizes the importance of transnational health research to be responsive to the needs of host countries, and to contribute to the mitigation of urgent global health issues.[9] The responsiveness requirement invokes several aspects: address the health problems of the population; ensure that successful products are available at the conclusion of the research (reasonable availability); and link other benefits of research with responsiveness (Fair benefits). The Declaration acknowledges the importance of dealing with the problem of exploitation of vulnerable research participants in developing countries and resource communities.[10] Exploitation in international clinical research is averted by the emphasis on providing a fair level of benefits rather than types of benefits. The benefits would accrue to research participants during the research, to the population during the research, and to the population after the completion of the research. The Declaration's focus on benefit and harm to research participants and host population is evidently intended to safeguard "the inherent dignity and human rights of every human person."[11] The author articulated that safeguarding the rights of participants and host population would be realized by maximizing benefits and minimizing risks. The Declaration echoes the same idea as it articulates that the

---

[6] Roberto Andorno, "Human Dignity and Human Rights," in *The UNESCO Universal Declaration on Bioethics and Human Rights: Background, Principles and Application*, 91.

[7] Roberto Andorno, "Human Dignity and Human Rights," in *The UNESCO Universal Declaration on Bioethics and Human Rights: Background, Principles and Application*, 93.

[8] Immanuel Kant, Fundamental Principles of the Metaphysic of Morals, translated T.K. Abbott (Amherst, New York: Prometheus Books, 1988), 58.

[9] Leonardo D. de Castro, "Transnational Practices" in *The UNESCO Universal Declaration on Bioethics and Human Rights: Background, Principles and Application,* ed. Henk A. M. J. ten Have and Michele S. Jean (Paris, France: UNESCO Publishing, 2009), 283.

[10] Leonardo D. de Castro, "Transnational Practices" in *The UNESCO Universal Declaration on Bioethics and Human Rights: Background, Principles and Application,* 286–287.

[11] Edmund D. Pellegrino, "Benefit and Harm" in *The UNESCO Universal Declaration on Bioethics and Human Rights: Background, Principles and Application,* 108.

application of Article 4 lies on the maximizing benefits and minimizing harms to patients and research participants whose interest and welfare should be prioritized over the interests of science or society.[12]

The author argues that the rights of research participants and host populations in developing countries are protected by the requirement of two safeguards, voluntary informed consent and review of research. The Declaration acknowledges the requirement of the critical role of the respect for the autonomy of persons engaged in clinical research.[13] The autonomy of persons to make decisions in clinical research derives its foundation on the fundamental principle of human dignity.[14] Debate on the principle centered on the balance between autonomy and the responsibility of the individual towards others. The Declaration acknowledges the rights and freedom of each person to make decisions as well as the need to respect the autonomy of others.[15] The autonomy of research participants is respected when adequate information is provided. Absence of adequate provision of information to potential research participants creates circumstances that exclude the possibility of informed consent.[16] Substituted judgements was identified as a way of respecting a person's autonomy when that autonomy is compromised.[17] The Declaration recognizes individual and communal autonomy but stresses that community autonomy does not substitute for individual autonomy.[18]

The autonomy of persons is realized in the requirement of informed consent as a critical element of clinical research. The doctrine of informed consent is an ethical and legal requirement for clinical research that protects the rights to integrity and self-determination of research participants.[19] Several challenges are encountered in the process of implementing the informed consent in different cultural settings because of its foundation in Western thought, philosophy, and culture that emphasizes the autonomy of the individual.[20] The Western culture sees informed consent

[12] Edmund D. Pellegrino, "Benefit and Harm" in *The UNESCO Universal Declaration on Bioethics and Human Rights: Background, Principles and Application,* 108.

[13] Donald Evans, "Autonomy and Individual Responsibility" in *The UNESCO Universal Declaration on Bioethics and Human Rights: Background, Principles and Application,* ed. Henk A. M. J. ten Have and Michele S. Jean (Paris, France: UNESCO Publishing, 2009), 113.

[14] Donald Evans, "Autonomy and Individual Responsibility" in *The UNESCO Universal Declaration on Bioethics and Human Rights: Background, Principles and Application,* 115.

[15] Donald Evans, "Autonomy and Individual Responsibility" in *The UNESCO Universal Declaration on Bioethics and Human Rights: Background, Principles and Application,* 115.

[16] Donald Evans, "Autonomy and Individual Responsibility" in *The UNESCO Universal Declaration on Bioethics and Human Rights: Background, Principles and Application,* 118.

[17] Donald Evans, "Autonomy and Individual Responsibility" in *The UNESCO Universal Declaration on Bioethics and Human Rights: Background, Principles and Application,* 118.

[18] Donald Evans, "Autonomy and Individual Responsibility" in *The UNESCO Universal Declaration on Bioethics and Human Rights: Background, Principles and Application,* 120.

[19] Regine Kollek, "Consent" in *The UNESCO Universal Declaration on Bioethics and Human Rights: Background, Principles and Application* ed. Henk A. M. J. ten Have and Michele S. Jean (Paris, France: UNESCO Publishing, 2009), 126.

[20] Regine Kollek, "Consent" in *The UNESCO Universal Declaration on Bioethics and Human Rights: Background, Principles and Application,* 134.

as an expression of personal autonomy and the locus of decisional capacity as the individual. On the other hand, several cultures in developing countries stress the importance of the relationship of the individual with the family and community. The decisional capacity in this context is expressed socially, rather than individually.[21] The decision of the individual is significantly shaped by both family and community. Community participation in clinical research was highly recommended. However, community consent should not substitute for individual consent. The declaration recognizes the challenges presented by different understandings of informed consent.[22]

The author articulates that the primary responsibility of the ethics review committee is to protect the rights of research participants. The ethics review committee discharges the task of protecting the rights of research participants through oversight, review and approval of research protocols. The Declaration recommends ethics review of research proposals in both sponsor and host countries to regulate the conduct of international research and to clarify their legal and ethical frameworks.[23] The protection of the rights of research participants and host population was also discussed in the Declaration within the context of social responsibility and the sharing of benefits arising from research results.[24] The Declaration reiterates that ethics review of research should guarantee the responsiveness of research to the needs of host population and to other global health problems.[25]

The author argues for post-trial access to drugs in developing countries based on the right to health derived from international human rights law. The right to health creates the national and international obligations to realize the minimum essential level of right to health for poor developing countries. The right to health cannot be realized in poor developing countries without the international obligation to provide assistance from the affluent countries. The Declaration acknowledges and encourages the application of the principle of international cooperation based on solidarity among human beings at the level of international institutions and organizations to address global challenges.[26] The author argues for a new global health paradigm resulting in the creation of Global Health Fund (GHF) supported by the right to health and the national and international obligations stemming from it.

---

[21] Regine Kollek, "Consent" in *The UNESCO Universal Declaration on Bioethics and Human Rights: Background, Principles and Application,* 134.

[22] Regine Kollek, "Consent" in *The UNESCO Universal Declaration on Bioethics and Human Rights: Background, Principles and Application,* 134.

[23] Leonardo D. de Castro, "Transnational Practices" in *The UNESCO Universal Declaration on Bioethics and Human Rights: Background, Principles and Application,* 287.

[24] Leonardo D. de Castro, "Transnational Practices" in *The UNESCO Universal Declaration on Bioethics and Human Rights: Background, Principles and Application,* 288.

[25] Leonardo D. de Castro, "Transnational Practices" in *The UNESCO Universal Declaration on Bioethics and Human Rights: Background, Principles and Application,* 288–289.

[26] Alphonse Elungu, "Solidarity and Cooperation" in *The UNESCO Universal Declaration on Bioethics and Human Rights: Background, Principles and Application* ed. Henk A. M. J. ten Have and Michele S. Jean (Paris, France: UNESCO Publishing, 2009), 216.

Several authors in previous research also grappled with the critical issue of post-trial access to drugs in developing countries from different perspectives. Christine Grady discussed the challenge of assuring continued post-trial access to beneficial treatment. She acknowledges some moral reasons for assuring continued treatment appealing to principles of non-maleficence and beneficence, justice as reciprocity and fiduciary responsibility of duty of care by researchers. She contends that terminating a beneficial treatment for a participant can be harmful. She also argues that certain things are owed research participants because they bore some risk for the good of the society and the progress of science. She further agrees that researchers have a corresponding responsibility to care for participants who have entrusted some aspects of their health to them.[27] Grady acknowledges the compelling nature of the foundation of these obligations but identifies the contentious nature of the debate about "the extent of the obligation—or what exactly is owed to research participants and by whom."[28] She further discusses some possible strategies for providing continued post-trial access to beneficial treatment. Collaboration among several stakeholders to a research—investigators, sponsors, communities, national health systems, and international organizations was identified as a possible way of providing continued post-trial access to beneficial treatment.[29] She cited an example with HIV Netherlands, Australia, Thailand Research Collaboration (HIV-NAT) co-payment and sliding scale drug fund program.

Joseph Millum discusses post-trial access to antiretroviral drugs with emphasis on justifications for providing post-trial anti-retroviral drugs coinciding with mechanisms proposed for supplying it. He identifies the moral principle underlying the justification, the duty bearer and to whom the duty is owed based on the principle.[30] He articulates six justifications for providing post-trial treatment to participants and host population—harm to participants, fiduciary relationship, reciprocity, duty of rescue, imperfect duty of beneficence, and global justice. Justifications based on duty owed to participants including harm to participants, fiduciary relationship, and reciprocity put obligation on researchers, sponsors, and sometimes governments who facilitate research. Justifications based on duty owed to people in urgent need, people in need, and people in unjust situations as in duty of rescue, imperfect duty of beneficence, and global justice put obligation on all those who can help. Seema Shah et al. discusses planning for posttrial access to antiretroviral treatment for research participants in developing countries with emphasis on National Institutes of Health funded trials. The authors based on NIH guidance document encourages investigators and contractors to collaborate with ministries of health of host countries and other stakeholders including ART access programs to formulate plans in advance to provide access to ART. No long-term mechanisms for posttrial access by sponsors was guaranteed.[31] Richard Ashcroft also discusses researchers and sponsors

[27] Grady, "The Challenge of Assuring Continued Post-Trial Access to Beneficial Treatment," 430.

[28] Grady, "The Challenge of Assuring Continued Post-Trial Access to Beneficial Treatment," 430.

[29] Grady, "The Challenge of Assuring Continued Post-Trial Access to Beneficial Treatment," 433.

[30] Millum, "Post-Trial Access to Antiretrovirals: Who Owes What to Whom?" 145–154.

[31] Seema, Elmer and Grady, "Planning for Posttrial Access to Antiretroviral Treatment for Research Participants in Developing Countries," 1556–1562.

obligations to provide care to participants and host communities at the end of the trial with emphasis on fairness and responsibility. He advocates for a middle ground position between minimal and maximal obligations of researchers with major constraints to shape any solution as: rationality, limited responsibility, and non-exploitation.[32]

Gorik Ooms and Rachel Hammonds make a strong case for global health justice, arguing that there is a global responsibility for global health and that there are obligations of justice to assist poor countries to fulfill their citizens' right to health. They contend that international human rights law is the basis of obligations of global health justice and national and global responsibilities for fulfilling the core obligations of the right to health and tackling global health inequities. They argue that international assistance and cooperation is imperative to help poor countries fulfill the rights to health of their citizens. They also argue for a new global health paradigm based on the global AIDS response and the creation of Global Fund to fight AIDS, Tuberculosis and Malaria.[33]

On the other hand, the book discusses post-trial access to drugs in developing nations, arguing for obligation of global justice. The author draws from the strengths of Rawls's statist and Pogge's cosmopolitan theories and on the international human rights law to argue for a new paradigm of global health justice that involves a sliding scale of national and global responsibilities for the realization of post-trial access to drugs. He argues that a legal obligation for international assistance and cooperation is imperative to help poor countries realize the minimum essential level of the right to health including post-trial access to drugs. He argues for a new health global health fund based on right to health and rooted in the concept of financial sustainability, i.e. sustained international funding to realize post-trial access to drugs. He advocates for expanding the mandate for current Global Fund to fight AIDS, Tuberculosis and Malaria to include interventions for other diseases and to address the weakness of the public health systems in developing nations.

The practical realization of the Global Health Fund would require specific roles of global and local stakeholders. The collaboration of both global and local stakeholders is critical to the practical realization of the Global Health Fund. The Global Health Fund like the Global Fund should be "an inclusive partnership with emphasis on country ownership, demand-driven program development, and performance-based funding."[34] The Global Health Fund would inherit the current structure of the Global Fund but would expand the mandate to include other diseases and

[32] Richard Ashcroft, "After the Trial is Over: What are the Sponsor's Obligations? *Policy Briefs*, May 1, 2005. http://www.scidev.net/en/policy-briefs/after-the-trial-is-over-what-are-the-sponsor-s-obl... (Accessed June 21, 2014).

[33] Ooms and Hammonds, "Taking Up Daniels' Challenge: The Case for Global Health Justice," 29–46.

[34] Rifat Atun and Michael Kazatchkine, "Promoting Country Ownership and Stewardship of Health Programs: The Global Fund Experience," J Acquir Immune Defic Syndr 52, sup. 1 (November 2009): S67-S68.

strengthening of the primary health care system in developing countries. The countries would be at the foundation of the partnership.[35] The Country Coordinating Mechanism (CCM)[36] as in Global Fund would be critical to Global Health Fund's efforts to ensure local ownership and participation in decision-making. The CCM would comprise of a wide range of global and local stakeholders from the public and private sectors including, the Ministry of Health of the government, multilateral and bilateral agencies, civil society, international and national non-governmental organizations, United Nations agencies, faith-based organizations, academic institutions, private for-profit organizations like pharmaceutical companies, associations of health services users, affected communities, marginalized groups and people living with the disease.

The Global Health Fund's approach to address health issues including post-trial access to drugs in developing countries is broad-based and inclusive as it involves "multiple levels of non-state actors in the design, implementation, monitoring and oversight"[37] of the health care programs. The Global Health Fund would emphasize building the capacity of local health leadership to improve governance of health care programs. The Global Health Fund would finance the country's health needs and priorities and in that case, foster "development of local capacity to generate quality demand through the Country Coordinating Mechanisms—whose inclusive nature has ensured that this demand reflects priorities of broad set of stakeholders."[38] The planning capacity in countries are strengthened in this way.

The Global Health Fund would expand the number and capacity of nongovernmental actors involved in health care programs and in that case, assist to develop broad country leadership. The Global Health Fund can also replicate the Global Fund's dual-track financing that allows countries to nominate one government and one non-governmental principal recipient to head program implementation.[39] In this case, the Global Health Fund would enable "civil society organizations, nongovernmental institutions, faith-based entities, and community-based organizations to develop and play a critical role in the design, implementation, and oversight"[40] of health care programs.

The Global Health Fund would invest in local leadership that is important to generate community demand for services. The Global Health Fund's investments to

[35] Atun and Kazatchkine, "Promoting Country Ownership and Stewardship of Health Programs: The Global Fund Experience," S67.

[36] Gorik Ooms, *From the Global AIDS Response towards Global Health?* A Discussion Paper for the Helene De Beir Foundation and the International Civil Society Support Group, (January 2009), http://www.hdbf.org/?pageid=345&lang=en (Accessed June 13, 2014).

[37] Atun and Kazatchkine, "Promoting Country Ownership and Stewardship of Health Programs: The Global Fund Experience," S68.

[38] Atun and Kazatchkine, "Promoting Country Ownership and Stewardship of Health Programs: The Global Fund Experience," S68.

[39] Atun and Kazatchkine, "Promoting Country Ownership and Stewardship of Health Programs: The Global Fund Experience," S68.

[40] Atun and Kazatchkine, "Promoting Country Ownership and Stewardship of Health Programs: The Global Fund Experience," S68.

establish stronger community systems would foster greater participation of community leaders to "effectively mobilize demand for services and to successfully scale up programs to reach the poor and the vulnerable."[41] For example, a broad partnership of government, civil society, and faith-based organization in Zambia shows how greater involvement of the civil society, the community and government actors has facilitated addressing AIDS epidemic effectively by expanding access to prevention, treatment, and care at community level.[42]

The significant involvement of civil society is also critical for the practical realization of the Global Health Fund. Civil society would play a critical role in pressing the governments of low-income countries to allocate the equivalent of 3% of their GDP to health.[43] Civil society will also play an essential role to guarantee that funds disbursed to low-income countries for their national health priorities are appropriately utilized. Civil society will serve as a watchdog to detect corruption and misuse of funding.[44] Furthermore, civil society of high-income countries would pressure their governments to increase their contributions to the Global Health Fund.[45]

The book extensively discussed an ethical justification for post-trial access to drugs for participants and host populations in developing countries within the context of global justice, emphasizing the combination of national and global responsibilities in realizing this objective. Drawing on the strengths of Rawls' statist and Pogge's cosmopolitan theories and on the International Human Rights Law, the work argued for a paradigm of Global Health Justice involving sliding scale of national and global responsibilities for the realization of the right to health in general and access to drugs in particular.

The book began the discourse by identifying the ethical issues in global health inequalities and global health justice which established the context of the debate. It highlighted the alarming ethical issue of global health inequalities between developed and developing countries which was exacerbated by the advent and calamitous impact of HIV/AIDS in developing countries, where many people living with HIV/AIDS lacked access to anti-retroviral drugs.

The work also acknowledged the critical role of two major ethical issues, including distributive justice and responsiveness of research to the needs and priorities of

[41] Atun and Kazatchkine, "Promoting Country Ownership and Stewardship of Health Programs: The Global Fund Experience," S68.

[42] Atun and Kazatchkine, "Promoting Country Ownership and Stewardship of Health Programs: The Global Fund Experience," S68.

[43] Ooms, "The Right to Health and the Sustainability of Healthcare: Why a New Global Health Aid Paradigm is Needed," 251.

[44] Gorik Ooms, *From the Global AIDS Response towards Global Health?* A Discussion Paper for the Helene De Beir Foundation and the International Civil Society Support Group, (January 2009), http://www.hdbf.org/?pageid=345&lang=en (Accessed June 13, 2014).

[45] Gorik Ooms, *From the Global AIDS Response towards Global Health?* A Discussion Paper for the Helene De Beir Foundation and the International Civil Society Support Group, (January 2009), http://www.hdbf.org/?pageid=345&lang=en (Accessed June 13, 2014).

host populations in evaluating the ethical justification of any global health research especially in developing countries.

The book also discussed the global clinical research which emphasizes the priority of safeguarding the rights of research participants and host populations in the design and implementation of research protocols. Obtaining voluntary informed consent from research participants and thorough review of research protocols by well constituted and competent ethics review committee were considered desiderata in conducting clinical trials in developing countries.

Cultural and language barriers were acknowledged as critical challenges in conducting clinical research in developing countries especially regarding complying with substantive ethical standard of voluntary informed consent, but were not justified as grounds for deviating from it. On the other hand, the Declaration acknowledges the significance of "cultural diversity" and "pluralism" but emphasizes that "such considerations are not to be invoked to infringe upon human dignity, human rights and fundamental freedoms ...."[46] Researchers and sponsors from developed countries were encouraged in the work to respect the local culture and values of research participants and host populations and to adapt standards of informed consent to the cultural norms and practices of developing countries.

The work as well discussed the post-trial access to drugs in developing countries with emphasis on the compelling need and urgency for development of cheaper generic versions of anti-retroviral drugs for fighting HIV/AIDS pandemic to illustrate for the general analysis. The post-trial access to drugs in developing countries was discussed within the broader contexts of intellectual property law, international trade agreements and non-patent factors. The tension between enforcement of strict patent protection and affordable access to essential drugs for the poor people in developing countries was recognized. A discussion about the analysis of the issue of post-trial access to drugs established that both patent and non-patent factors adversely impede access to affordable drugs in developing countries.

The severe impact of international trade agreements currently negotiated by United States, which further exacerbate impeded access to essential drugs for people in developing countries was clearly acknowledged. Two key aspects of TRIPs agreement were discussed, including strict patent protection to promote incentives for innovation and promotion of public health interests, and maintaining a delicate balance between them was considered imperative. Compulsory licensing and parallel importation that authorize countries to manufacture and import generic versions of patented drugs respectively to address national health crises like HIV/AIDS were encouraged.

Some current strategies for addressing the issue of affordable access to drugs for developing countries were discussed, including private donations, price reductions and differential pricing, international collaborative initiatives and public-private

[46] Michel Revel, "Respect for Cultural Diversity and Pluralism," in *The UNESCO Universal Declaration on Bioethics and Human Rights: Background, Principles and Application* ed. Henk A. M. J. ten Have and Michele S. Jean (Paris, France: UNESCO Publishing, 2009), 199.

partnerships and compulsory licensing. Private donations and price reductions were viewed as improvised solutions that are not effective and sustainable.

The book also discussed national responsibility and humanitarian assistance with emphasis on Rawls's statist approach in dealing with the issue of post-trial access to drugs in developing countries especially for participants and host populations. Rawls's two different approaches to justice both in the domestic and in the international arena were discussed. His account of domestic justice is known as justice as fairness emphasizing the idea that fundamental agreements of the parties to the original position are fair. The idea of the basic structure of the society which applies to a well-ordered society was considered fundamental because it is the primary subject of justice. The principles of justice as fairness were discussed, including equal liberty principle, difference principle and fair equality of opportunity principle.

Daniels's extension of Rawls's theory of justice to health care in the domestic society was discussed. He argued that health care institutions should be regulated by fair equality of opportunity principle, since meeting health care needs has a significant impact on the distribution of opportunity. He further contended that fulfilling health care needs protects people's normal opportunity range and helps them to maintain or restore normal species-typical functioning.

Rawls' statist approach discussed the two aspects of Rawls's account of international relations and international justice, including ideal theory and non-ideal theory. The ideal theory focuses on how the laws of peoples should regulate the political relations among liberal and decent hierarchical peoples. The non-ideal theory focuses on burdened societies that lack basic resources to become well-ordered. Rawls argued that well-ordered societies have a duty of assistance to burdened societies to help them attain required level of economic and political developments to become well-ordered. His account of global justice lacks a commitment to principles of distributive justice, because according to him such principles would be redundant and would produce unacceptable results. He argued further that they often lacked a clear target and a cutoff point.

The book explored three arguments in analyzing post-trial access to drugs within the context of national responsibility and humanitarian assistance, including argument based on defense of human rights, argument based on redress for the unjustified distributive effects of cooperative organizations and argument based on access to drugs as a transition strategy supporting the establishment of politically well-ordered nations. The argument based on defense of human rights did not go through because of the minimal and ad hoc account of human rights. Concerning the argument based on redressing the unjustified distributive effects of cooperative organizations, there was a consensus that interventions for providing affordable access to essential drugs for developing countries would constitute logical approaches for redress. The argument based on access to drugs as a transition strategy that supports the establishment of politically well-ordered nations was justified. More so, it was argued that improving access to drugs would help as an effective transition strategy that would enable burdened societies to become politically well-ordered.

The book in contrast also discussed international responsibility and the health impact fund focusing on Pogge's cosmopolitan approach which is a more robust and expansive international perspective to global justice for providing affordable access to drugs for participants and host populations in developing countries. He forcefully argued for a stronger interpretation of global responsibilities grounded on human rights for providing affordable access to drugs for participants and host populations in developing countries.

Pogge's challenge of Rawls's thesis of explanatory nationalism which appeals to domestic factors as engendering persistent severe poverty and global inequality was discussed. He insisted that global factors perpetuate severe poverty and global inequality and urged moral responsibility on the part of affluent countries to redress this alarming situation. He contended that unjust global institutional order imposed on the global poor perpetuates harm and consequently violates their negative rights. He further argued that citizens of affluent countries have not merely positive duties to assist, but also more stringent negative duties not to harm the global poor.

The present rules for incentivizing pharmaceutical research were discussed. The current global patent system was argued as an institutional failure, because the TRIPS agreement and the TRIPS-plus provisions significantly impede affordable access to essential drugs for the global poor in developing countries. Seven major problems of the current global patent system were identified and discussed, comprising high prices, neglect of diseases concentrated among the poor, bias towards maintenance drugs, wastefulness, counterfeiting, excessive marketing and last-mile problem.

The book also discussed Pogge's proposed new rules for reforming and incentivizing pharmaceutical research in response to the problems engendered by the current system. He proposed two basic reform strategies for addressing the monopoly pricing issues and for improving access to essential drugs, including the differential pricing strategy and the public good strategy. Differential pricing and compulsory licensing were argued as inadequate in resolving the disadvantages of the current patent system. The public good strategy was defended as more promising in engendering a reform plan that would prevent the major disadvantages of the current monopoly patent regime, while at the same time, preserving most of its important benefits. The three critical components of public good strategy were discussed, including open access, alternative incentives, and funding.

The HIF was discussed as a global institution for the effective implementation of Pogge's new plan for reforming and incentivizing research and innovation. The HIF's significant structural reform of the current global patent system was discussed. Additionally, the HIF's systematic solution to seven disadvantages of the current global scheme was discussed. Some key objections by some authors were also discussed.

The book finally discussed global health responsibility for health inequalities and affordable access to drugs which emphasizes sliding scale of national and global responsibilities in realizing the core content of the right to health, including affordable access to drugs. Improved global health governance was essentially considered a desideratum for global health. There was a consensus that health is a shared responsibility especially in relation to affordable access to essential health services and collective defense against transnational threats, including communicable diseases. Hence, it was forcefully argued that the responsibility for guaranteeing the right to health for all does not rest exclusively with the states and their obligations to their populations, but as well with the global community.

It was argued using Daniels' works that Rawls's and Pogge's approaches for global health justice were insufficient. Rawls's statist version of relational justice narrowly stresses the national responsibility of each country to realize the right to health of all its citizens and effectively excludes the global responsibility of individual nations. Conversely, Pogge's cosmopolitan approach focuses too much on global responsibility without adequate attention to the national responsibility of individual nations. A middle ground that resists the pull of the two opposing alternatives of Rawls's and Pogge's extreme positions was advocated for by Daniels. Therefore, the proposed paradigm combines these two approaches by espousing their strengths and avoiding their weaknesses. The paradigm refers to a sliding scale of national and global responsibilities concerning the right to health, including affordable access to drugs. The sliding scale paradigm requires each developing country to spend 3% of its GDP on national health in order to qualify to receive global assistance. This combined approach emphasizes that global responsibility supplements rather than replaces national responsibility for health.

The book also discussed justification of the focus on government programs to address the issue of affordable access to drugs for participants and host populations. The obligations of researchers and sponsors to participants and host populations were argued to be transferable. The global capacity to redress health inequalities was discussed, comprising the health development paradigm, the medical relief paradigm and the new global health paradigm.

International human rights law was argued as a foundation for global health justice and global health paradigm. International human rights law was also defended as providing a theoretical framework for national and global responsibilities for realizing the core obligations that arise from socio-economic human rights and for addressing global health inequalities. The critical role of the principle of core obligation and the obligation to provide international assistance in realizing the core content of the right to health and consequently engendering obligation of global health justice was emphasized.

A consensus for a crucial need for an international agency such as Global Health Fund for the distribution of health-related goods that would rectify the injustice stemming from the current global system was discussed. Expanding the mandate of the current Global Fund in order to include interventions for other diseases as well as addressing the weakness of the public health systems in developing countries was argued as a good start for establishing the Global Health Fund. An effective Global Health Fund rooted in the new concept of financial sustainability would significantly improve the realization of right to health and affordable access to drugs in developing countries.

Bibliography

Adebamowo, Clement. 2007. The Impact of Community Dynamics on Conducting Scientific Research in Sub-Saharan Africa. In *Ethical Challenges in Study Design and Informed Consent for Health Research in Resource-Poor Settings*, ed. Patricia A. Marshall, 44–45. Geneva: Social, Economic and Behavioral Research.

African Civil Society. 2010. Letter to July 2010 African Union Summit on Upholding African Health and Social Development Commitments. *EQUINET (Regional Network on Equity in Health in Southern Africa).* http://www.equinetafrica.org/newsletter/index.php?id=7411. Accessed 2 July 2014.

AIDS Vaccine Advocacy Coalition (AVAC). 2005. Vaccine Trials: Leaving Communities Better Off. In *AIDS Vaccine Handbook: Global Perspectives*, ed. P. Kahn, 137–144. New York: AVAC.

Allen, William T. 1992. Our Schizophrenic Conception of the Business Corporation. *Cardozo Law Review* 14: 261–281.

Alston, Philip. 2005. Ships Passing in the Night: The Current State of the Human Rights and Development Debate Seen Through the Lens of the Millennium Development Goals. *Human Rights Quarterly* 27 (3, August): 755–729.

Ananworanich, Jintanat, et al. 2004. Creation of A Drug Fund for Post-Clinical Access to Antiretrovirals. *Lancet* 364 (9428, July): 101–102.

Andorno, Roberto. 2007. Global Bioethics at UNESCO: In Defence of the Universal Declaration on Bioethics and Human Rights. *Journal of Medical Ethics* 33 (3, March): 150–154.

———. 2009. Human Dignity and Human Rights. In *The UNESCO Universal Declaration Bioethics and Human Rights: Background, Principles and Application*, ed. A.M.J. Henk and Michele S. Jean ten Have, 91–98. Paris: UNESCO Publishing.

Angell, Marcia. 2000. Investigators' Responsibilities for Human Subjects in Developing Countries. *The New England Journal of Medicine* 342 (13, March 30): 967–969.

Annas, George J. 2009. Globalized Clinical Trials and Informed Consent. *The New England Journal of Medicine* 360 (20, May): 2050–2053.

Appelbaum, Paul S., and Charles W. Lidz. 2008. Therapeutic Misconception. In *The Oxford Textbook of Clinical Research Ethics*, ed. Ezekiel J. Emmanuel et al., 633–644. New York: Oxford University.

Appelbaum, Paul S., Loren H. Roth, and Lidz Charles. 1982. The Therapeutic Misconception: Informed Consent in Psychiatric Research. *International Journal of Law and Psychiatry* 5 (3–4): 319–329.

Archer, Robert. 2003. *Duties Sans Frontieres: Human Rights and Global Social Justice*, 1–95. Versoix: International Council on Human Rights Policy.

Arras, John D., and Elizabeth M. Fenton. 2009. Bioethics and Human Rights: Access to Health-Related Goods. *Hastings Center Report* 39 (5): 27–38.

E.C. Obi, *Post-Trial Access to Drugs in Developing Nations*, Advancing Global Bioethics 7, DOI 10.1007/978-3-319-60028-4

Arrow, Kenneth J. 1963, December. Uncertainty and the Welfare Economics of Medical Care. *American Economic Review* 53 (5): 941–973.

———. 1973. Some Ordinalist-Utilitarian Notes on Rawls's Theory of Justice. *Journal of Philosophy* 70 (9, May): 245–263.

Arthur, John. 1996. Rights and the Duty to Bring Aid. In *World Hunger and Morality*, ed. W. Aiken and H. LaFollette, 39–50. Englewood Cliffs: Prentice Hall.

Ashcroft, Richard. 2005. After the Trial is Over: What are the Sponsor's Obligations? *Policy Briefs* 1 (May).

Attaran, Amir. 2004. How Do Patents and Economic Policies Affect Access to Essential Medicines in Developing Countries? *Health Affairs* 23 (3, May): 155–166.

Attaran, Amir, and Gillespie-White Lee. 2001. Do Patents for Antiretroviral Drugs Constrain Access to AIDS Treatment in Africa? *JAMA* 286 (15, October): 1886–1892.

Atun, Rifat, and Michael Kazatchkine. 2009. Promoting Country Ownership and Stewardship of Health Programs: The Global Fund Experience. *J Acquir Immune Defic Syndr* 52 (suppl 1, November): S67–S68.

Bale, H.E. 2001. *Patents and Public Health: A Good or Bad Mix.* Pfizer forum, Pfizer forum.

Barnes, D.M., et al. 1998. Infromed Consent in a Multicultural Cancer Patient Population: Implications for Nursing Practice. *Nursing Ethics* 5 (5): 412–423.

Barry, Brian. 1982. Humanity and Justice in Global Perspective. In *Ethics, Economics and the Law*, ed. Roland J. Pennock and John W. Chapman, vol. Nomos XXIV, 219–252. New York: New York University Press.

———. 1991. Justice as Reciprocity. In *Liberty and Justice: Essays in Political Theory*, ed. Brian Barry, vol. 2, 211–241. Oxford: Clarendon.

———. 2008. Humanity and Justice in Global Perspective. In *Global Justice: Seminal Essays*, ed. Thomas Pogge and Darrel Moellendorf, Global Responsibilities, vol. 1, 179–209. St. Paul: Paragon House.

Barsdorf, N., and D. Wassenar. 2005. Racial Differences in Public Perceptions of Voluntariness of Medical Research Participants in South Africa. *Social Science & Medicine* 60 (6, March): 1087–1098.

Barsdorf, Nichola, et al. 2010. Access to Treatment in HIV Prevention Trials: Perspectives from A South African Community. *Developing World Bioethics* 10 (2, August): 78–87.

Bartlett, John A., Ashwini Shewade, and Rukmini Rajagopalan. 2009. Obstacles and Proposed Solutions to Effective Antiretroviral Therapy in Resource-Limited Settings. *Journal of the International Association of Physicians in AIDS Care* 8 (4, July/August): 253–268.

Beauchamp, Tom L., and James F. Childress. 2001. *Principles of Biomedical Ethics*. 5th ed. New York: Oxford University Press.

Beitz, Charles. 1983. Cosmopolitan Ideals and National Sentiment. *The Journal of Philosophy* 80 (10, October): 591–600.

Beitz, Charles R. 1999a. *Political Theory and International Relations*. Princeton: Princeton University Press.

———. 1999b. Social and Cosmopolitan Liberalism. *International Affairs* 75 (3, July): 515–529.

———. 1999c. International Liberalism and Distributive Justice: A Survey of Recent Thought. *World Politics* 51 (2, January): 269–292.

Belsky, Leah, and Henry S. Richardson. 2004. Medical Researchers' Ancillary Clinical Care Responsibilities. *BMJ* 328 (7454, June): 1494–1496.

Benatar, Solomon R. 1998. Global Disparities in Health and Human Rights: A Critical Commentary. *American Journal of Public Health* 88 (3, February): 295–300.

———. 2001. Distributive Justice and Clinical Trials in the Thrid World. *Theoretical Medicine and Bioethics* 22 (3, June): 169–176.

———. 2002. Reflections and Recommendations on Research Ethics in Developing Countries. *Social Science & Medicine* 54 (7, April): 1137–1138.

Benatar, Solomon R., and Peter A. Singer. 2000. A New Look at International Research Ethics. *BMJ* 321 (September): 824–826.

Benatar, Solomon R., Abdallah S. Daar, and Peter A. Singer. 2003. Global Health Ethics: The Rationale for Mutual Caring. *International Affairs* 79 (1, January): 107–138.
Berle, Adolf A., Jr. 1931. Corporate Powers as Powers in Trust. *Harvard Law Review* 44 (7, May): 1049–1074.
Bernard, D. 2002. In the High Court of South Africa, Case No. 4138/98: The Global Politics of Access to Low-Cost AIDS Drugs in Poor Countries. *Kennedy Institute of Ethics Journal* 12 (2, June): 159–174.
Bhutta, Zulfiquar Ahmed. 2002. Ethics in International Health Research: A Perspective from the Developing World. *Bulletin of the World Organization* 80 (2, January): 114–115.
Bird, Robert C. 2009. Developing Nations and the Compulsory License: Maximizing Access to Essential Medicines While Minimizing Investment Side Effects. *Journal of Law, Medicine & Ethics* 37 (2, Summer): 209–221.
Birdsall, Nancy. 2008. Seven Deadly Sins: Reflections on Donor Failings. In *Reinventing Foreign Aid*, ed. William Easterly, 515–551. Cambridge, MA: MIT Press.
Birn, Anne-Emmanuelle. 2011. Addressing the Societal Determinants of Health: The Key Global Health Ethics Imperatives of Our Times. In *Global Health and Global Health Ethics*, ed. Solomon Benatar and Gillian Brock, 37–52. Cambridge: Cambridge University Press.
Black, Fischer, and Myron Scholes. 1973. The Pricing of Options and Corporate Liabilities. *The Journal of Political Economy* 81 (3, May–June): 637–654.
Black, Douglas, et al. 1988. *Inequalities in Health: The Black Report: The Health Divide*. London: Penguin Group.
Blackmer, Jeff, and Henry Haddad. 2005. The Declaration of Helsinki: An Update on Paragraph 30. *Canadian Medical Association Journal (CMAJ)* 179 (9, October): 1052–1053.
Boelaert, M. 2002. Letter to the Editor: Do Patents Prevent Access to Drugs for HIV in Developing Countries? *JAMA* 287 (7, February): 840–841.
Brock, Dan W. 2001. Some Questions About the Moral Responsibilities of Drug Comapnies in Developing Countries. *Developing World Bioethics* 1 (1, May): 33–37.
Bulir, Ales, and A. Javier Hamann. 2003. Aid Volatility: An Empirical Assessment. *IMF Staff Papers* 50: 64–89.
Buvinic, Myra, et al. 2006. Gender Differentials in Health. In *Disease Control Priorities in Developing Countries*, ed. Dean T. Jamison et al., 195–210. Washington, DC: World Bank.
Cage, S. 2005, December 6. *WTO Oks Measures to Improve Drug Access*. New York: Associated Press.
Caney, Simon. 2005. *Justice Beyond Borders: A Global Political Theory*. New York: Oxford University Press.
Carrin, Guy, David Evans, and Ku. Ke. 2007. Designing Health Financing Policy towards Universal Coverage. *WHO Bulletin* 85 (9, September): 652.
Case, Anne, Angela Fertig, and Christina Paxson. 2005. The Lasting Impact of Childhood Health and Circumstance. *Journal of Health Economics* 24 (2, March): 365–389.
Case, Anne, Darren Lubotsky, and Christina Paxson. 2002. Economic Status and Health in Childhood: The Origins of the Gradient. *American Economic Review* 92 (5, December): 1308–1334.
Cassileth, B.R., et al. 1982. Attitudes toward Clinical Trials among Patients and the Public. *JAMA* 248 (8, August): 968–970.
Chaudhuri, S., P.K. Goldberg, and P. Jia. 2003. *The Effects of Extending Intellectual Property Rights Protection to Developing Countries: A Case Study of the Indian Pharmaceutical Market*, Working Paper. Cambridge, MA: National Bureau of Economic Research.
Clemente, C.L. 2001. *Intellectual Property: The Patent on Prosperity*. Pfizer forum, Pfizer forum.
Cohen, Jillian Clare, and Patricia Illingworth. 2003. The Dilemma of Intellectual Property Rights for Pharmaceuticals: The Tension Between Ensuring Access of the Poor to Medicines and Committing International Agreements. *Developing World Bioethiccs* 3 (1, November): 27–48.
Cohen, Joshua, and Charles Sabel. 2006. Extra Rempublicam Nulla Justitia? *Philosophy and Public Affairs* 34 (2, March): 147–175.

Colgan, Ann-Louise. 2002. Hazardous to Health: The World Bank and IMF in Africa. *Africa Action* 1 (2, April).

Collier, Paul. 1999. Aid Dependency: A Critique. *Journal of African Economies* 8 (4): 528–545.

Commission on HIV/AIDS and Governance in Africa. n.d. *AFrica: The Socio-Economic Impact of HIV/AIDS*. Addis Ababa: Economic Commission for Africa.

Commission on Intellectual Property Rights. 2002. *Integrating Intellectual Property Rights and Development Policy*, Commission Report, 1–178. London: Commission on Intellectual Property Rights.

Commission on Macroeconomics and Health. 2001. *Investing in Health for Economic Development*, CMH Report. Geneva: World Health Organization.

Committee on Economic, Social and Cultural Rights. 1990. *General Comment No. 3: The Nature of States Parties' Obligations*, UN Document. Geneva: United Nations.

———. 2000a. *General Comment No. 14: The Right to the Highest Attainable Standard of Health*, UN Report. Geneva: United Nations.

———. 2000b. *Summary Record of the 10th Meeting*. Geneva.

Cook, Kori, Jeremy Snyder, and John Calvert. 2016. Attitudes Toward Post-Trial Access to Medical Interventions: A Review of Academic Literature, Legislation, and International Guidelines. *Developing World Bioethics* 16 (2, August): 70–79.

Cookson, J.B. April 1992. Auditing a Research Ethics Committee. *Journal of the Royal College of Physicians of London* 26 (2): 181–183.

Cooley, Dennis R. 2001. Distributive Justice and Clinical Trials in the Third World. *Theoretical Medicine and Bioethics* 22 (3, June): 151–167.

Council for International Organizations of Medical Sciences. 1993. *International Ethical Guidelines for Biomedical Research Involving Human Subjects*. Geneva: CIOMS.

———. 2002. *International Ethical Guidelines for Biomedical Research Involving Human Subjects*, International Research Guidelines, 1–112. Geneva: CIOMS.

Cox, K. 2002. Informed Consent and Decision-making: Patients' Experiences of the Process of Recruitment to Phase I and II Anti-Cancer Drug Trials. *Patient Education and Counseling* 46 (1, January): 31–38.

Crigger, N.J., L. Holcomb, and J. Weiss. 2001. Fundamentalism, Multiculturalism and Problems of Conducting Research with Populations in Developing Nations. *Nursing Ethics* 8 (5, September): 459–468.

Crossette, Barbara. 2001, April 30. A Wider War on AIDS in Africa and Asia. *New York Times*, A6.

Crouch, Robert A., and John D. Arras. 1998. AZT Trials and Tribulations. *Hastings Center Report* 28 (6, November): 26–34.

Cuddington, John T. 1993. Modelling the Macroeconomic Effects of AIDS, with An Application to Tanzania. *World Bank Economic Review* 7 (2, May): 172–189.

Cutler, David M., and Adriana Lleras-Muney. 2007. *Understanding Differences in Health Behaviors by Education*. Mimeo: Princeton University.

Cutler, David M., Adriana Lleras-Muney, and Tom Vogl. 2008, September. *Socioeconomic Status and Health: Dimensions and Mechanisms*. Cambridge, MA: National Bureau of Economic Research.

Daar, Abdallah, et al. 2002. Top Ten Biotechnologies for Improving Health in Developing Countries. *Nature Genetics* 32 (2, October): 229–232.

Dahos, Peter. 2001. BITS and BIPS: Bilateralism in Intellectual Property. *Journal of World Intellectual Property* 4 (6, November): 791–808.

Daniels, Norman. 1985. *Just Health Care*. New York: Cambridge University Press.

———. 1990. Equality of What: Welfare, Resources, or Capabilities? *Philosophy and Phenomenological Research* 50 (Suppl, Autumn): 273–296.

———. 2001. Social Responsibility and Global Pharmaceutical Companies. *Developing World Bioethics* 1 (1, May): 38–41.

———. 2008. *Just Health: Meeting Health Needs Fairly*. Cambridge: Cambridge University Press.

Daniels, Norman, Bruce P. Kennedy, and Ichiro Kawachi. 1999. Why Justice is Good for our Health: The Social Determinants of Health Inequalities. *Deadalus* 128 (4, Fall): 215–251.

———. 2000. Justice is Good for our Health. *Boston Review* 25 (1, February–March).

Dawson, L., and N.E. Kass. 2005. Views of U.S. Researchers about Infromed Consent in International Collaborative Research. *Social Science and Medicine* 61 (6, Sepember): 1211–1222.

Deaton, Angus. 2003. Health, Inequality and Economic Development. *Journal of Economic Literature (XLI)* 41 (1, March): 113–158.

de Castro, Leonardo D. 2009. Transnational Practices. In *The UNESCO Universal Declaration on Bioethics and Human Rights: Background, Principles and Application*, ed. Henk A.M.J. ten Have and Michele S. Jean, 283–291. Paris: UNESCO Publishing.

Del Rio, C. 1998. Is Ethical Research Feasible in Developed and Developing Countries? *Bioethics* 12 (4, October): 328–330.

Demeka, M. 1993, January. The Potential Impact of HIV/AIDS on the Rural Sector of Ethiopia. Unpublished Manuscript.

Dickert, Neal, and Grady Christine. 1999. What's the Price of a Research Subject? Approaches to Payment for Research Participation. *New England Journal of Medicine* 341 (3, July): 198–203.

———. 2008. Incentives for Research Participants. In *The Oxford Textbook of Clinical Research Ethics*, ed. Ezekiel J. Emmanuel et al., 386–396. New York: Oxford University.

Directorate-General for Development Cooperation. 2006. *DGCD Annual Report 2005*. Brussels: Directorate-General for Development Cooperation.

Dixon, Simon, Scott McDonald, and Jennifer Roberts. 2002. The Impact of HIV and AIDS on Africa's Economic Development. *BMJ* 324 (7331, January): 232–234.

Dodd, Merrick E. 1932. For Whom are Corporate Managers Trustees? *Harvard Law Review* 45 (7, May): 1145–1163.

Donald, Anna. 1999. The Political Economy of Technology Transfer. *BMJ* 319 (7220, November): 1298.

Dorrington, R., et al. 2002. Some Implications of HIV/AIDS on Adult Mortality in South Africa. *AIDS Analysis Africa* 12 (5).

Doval, Dinesh Chandra, Shirali Rashmi, and Sinha Rupal. 2015. Post-trial Access to Treatment for Patients Participating in Clinical Trials. *Perspectives in Clinical Research* 6 (2, April–June): 82–85.

Dunkle, Kristin L., et al. 2004. Gender-Based Violence, Relationship Power, and Risk of HIV Infection in Women Attending Antenatal Clinics in South Africa. *Lancet* 363 (9419, May): 1415–1421.

Eide, Asbjorn. 2011. *The Health of the World's Poor – A Human Rights Challenge*, 1–139. Oslo: Norwegian Directorate of Health.

Elungu, Alphonse. 2009. Solidarity and Cooperation. In *The UNESCO Universal Declaration on Bioethics and Human Rights: Background, Principles and Application*, ed. Henk A.M.J. ten Have and Michele S. Jean, 211–217. Paris: UNESCO Publishing.

Emmanuel, Ezekiel J. 2008. Benefits to Host Countries. In *The Oxford Textbook of Clinical Research Ethics*, ed. Ezekiel J. Emmanuel et al., 719–728. New York: Oxford University Press.

Emmanuel, Ezekiel J., et al. 2004. What Makes Clinical Research in Developing Countries Ethical? The Benchmarks of ethical Research. *Journal of Infectious Diseases* 189 (5, March): 930–937.

———. 2008. Addressing Exploitation: Reasonable Availability and Fair Benefits. In *Exploitation and Developing Countries: The Ethics of Clinical Research*, ed. Jennifer S. Hawkins and Ezekiel J. Emmanuel, 286–313. Princeton: Princeton University.

Engelhardt, Tristram H. 1996. *The Foundations of Bioethics*. New York: Oxford University Press.

Evans, Donald. 2009. Autonomy and Individual Responsibility. In *The UNESCO Universal Declaration on Bioethics and Human Rights: Background, Principles and Application*, ed. Henk A.M.J. ten Have and Michele S. Jean, 111–121. Paris: UNESCO Publishing.

Faden, Ruth, and Tom L. Beauchamp. 1986. *A History and Theory of Informed Consent*. New York: Oxford University Press.

Faden, Ruth R., and Tom L. Beauchamp. 2003. The Concept of Informed Consent. In *Contemporary Issues in Bioethics*, ed. Tom L. Beauchamp and LeRoy Walters, 145–148. Belmont: Wadworth-Thomson Learning.

Ferreira, Lissett. 2002. Access to Affordable HIV/AIDS Drugs: The Human Rights Obligations of Multinational Pharmaceutical Corporations. *Fordham Law Review* 71 (3, December): 1133–1179.

Fink, Carsten, and Patrick Reichenmiller. 2005. *Tightening TRIPS: Intellectual Property Provisions of U.S. Free Trade Agreements*, World Bank Report, 1–11. Washington, DC: World Bank.

Fitzgerald, D.W., et al. 2002. Comprehension During Consent in a Less Developed Country. *Lancet* 360 (9342, October): 1301–1302.

Ford, Nathan, et al. 2004. The Role of Civil Society in Protecting Public Health Over Commercial Interests: Lessons from Thailand. *Lancet* 363 (9408, February): 560–563.

Forgy, Larry. 1993. Mitigating AIDS: The Economic Impacts of AIDS in Zambia and Measures to Counter Them. *REDSO/ESA*.

Forman, Lisa. 2007. Trade Rules, International Property, and the Right to Health. *Ethics and International Affairs* 21 (3, Fall): 337–357.

———. 2009. Trading Health for Profit: The Impact of Bilateral and Rigional Free Trade Agreements on Domestic Intellectual Property Rules on Pharmaceuticals. In *The Power of Pills: Social, Ethical and Legal Issues in Drug Development, Marketing and Pricing*, ed. Jillian Clare, Patricia Illingworth, and Udo Schuklenk Cohen, 190–199. London: Pluto Press.

Freeman, Samuel, ed. 1999. *John Rawls: Collected Papers*. Cambridge, MA: Harvard University Press.

Friedman, Milton. 1970. The Social Responsibility of Business is to Increase its Profits. *The New York Times Magazine*, September 13.

Gairdner, Franklin Tennant. 2009. A Defence of Thomas Pogge's Argument for a Minimally Just Institutional Order. MA thesis, Queens's University.

Galjaard, Hans. 2009. Sharing Benefits. In *The UNESCO Universal Declaration on Bioethics and Human Rights: Background, Principles and Application*, ed. A.M.J. Henk and Michele S. Jean ten Have, 231–241. Paris: UNESCO Publishing.

Gathii, James T. 2002. The Legal Status of The Doha Declaration on TRIPS and Public Health Under the Vienna Convention on the Law of Treaties. *Harvard Journal of Law and Technology* 15 (2, Spring): 291–317.

———. 2005. Third World Perspectives on Global Pharmaceutical Access. In *Ethics and the Pharmaceutical Industry*, ed. Michael A. Santoro and Thomas M. Gorrie, 336–351. Cambridge: Cambridge University Press.

Gelling, Leslie. 1999. Role of the Research Ethics Committee. *Nurse Education Today* 19 (7, October): 564–569.

Gellman, Barton. 2000, July 5. The Belated Global Response to AIDS in Africa. *Washington Post*, A1.

Gilbert, Alan. 1992. *An Unequal World: The Links Between Rich and Poor Nations*. 2nd ed. Nelson: London.

Glantz, L.H., et al. 1998. Research in Developing Countries: Taking Benefit Seriously. *Hastings Center Report* 28 (6, November–December): 38–42.

Global Forum for Health Research. 2004. *The 10/90 Report on Health Research 2003–2004*, GFHR Health Research Report. Geneva: GFHR.

Global Health Council. 2007. *Child Mortality*. http://www.globalhealth.ord/child_health/child_mortality. Accessed 15 Oct 2007.

———. *The Importance of Child Mortality*. http://www.globalhealth.org/childhealth. Accessed 21 Aug 2007.

Gomberg, Paul. 1990. Patriotism is like Racism. *Ethics* 101 (1, October): 144–150.

Goodin, Robert E. 1985. *Protecting the Vulnerable: A Reanalysis of our Social Responsibities*. Chicago: University of Chicago Press.

———. 2008. What Is So Special about Our Fellow Countrymen. In *Global Justice: Seminal Essays*, Global Responsibilities, ed. Thomas Pogge and Darrel Moellendorf, vol. 1, 255–284. St. Paul: Paragon House.

Gostin, Lawrence. 2007. A Proposal for a Framework Convention on Global Health. *Journal of International Economic Law* 10 (4): 989–1008.

Gostin, Lawrence O. 2008. Meeting Basic Survival Needs of the World's Least Healthy People: Toward a Framework Convention on Global Health. *The Georgetown Law Journal* 96: 331–392.

Gostin, Lawrence O., et al. 2010. National and Global Responsibilities for Health. *Bulletin of the World Health Organization* 88: 719.

———. 2011. The Joint Action and Learning Initiative: Towards a Global Agreement on National and Global Responsibilities for Health. *PLoS Medicine* 8 (5, May): e1001031.

Gottret, Pablo, and Georges Schieber. 2006. *Health Financing Revisited: A Practitioner's Guide*, World Bank Report, 1–310. Washington, DC: The World Bank.

Grady, Christie. 2006, April. Ethics of International Research: What Does Responsiveness Mean? *Virtual Mentor* 8 (4): 235–240.

Grady, Christine. 2001. Money for Research Participation: Does it Jeopardize Informed Consent? *American Journal of Bioethics* 1 (2, Spring): 40–44.

———. 2005. The Challenge of Assuring Continued Post-Trial Access to Beneficial Treatment. *Yale Journal of Health Policy, Law and Ethics* 5 (1, Winter): 425–435.

Grady, Christine, et al. 2008. Research Benefits for Hypothetical HIV Vaccine Trials: The Views of Ugandans in the Rakai District. *IRB: Ethics & Human Research* 30 (2, March–April): 1–7.

Griffiths, Sian, and Xiao-Nong Zhou. 2012. Why Research Infectious Diseases of Poverty? In *Global Report for Research on Infectious Diseases of Poverty*, ed. Margaret Harris and Julie N. Reza, 10–43. Geneva: World Health Organization.

Gruskin, Sofia, and Daniel Tarantola. 2000. *Health and Human Rights*, Working Paper. Boston: Francois-Xavier Bagnoud Center for Health and Human Rights, Harvard School of Public Health.

Guillemin, Marilys, et al. 2012. Human Research Ethics Committees: Examining Their Roles and Practices. *Journal of Empirical Research on Human Research Ethics* 7 (3, July): 38–49.

Gwatkin, Davison R., and Michel Guillot. 2000. *The Burden of Disease Among the Global Poor: Current Situation, Future Trends, and Implications for Strategy*, World Bank Report, 1–44. Washington, DC: The World Bank.

Hammonds, Rachel, and Gorik Ooms. 2009. Global Solidarity for Health. *Global Future* 2: 6–7.

Hancock, John, et al. 1996. *The Macroeconomic Impacts of AIDS*. Washington, DC: AIDS in Kenya, Family Health International.

Haupt, A., and T.T. Kane. 2004. *Population Handbook*. Washington, DC: Population Reference Bureau.

Held, David. 1995. *Democracy and the Global Order: From the Modern State to Cosmopolitan Governance*. Cambridge: Polity Press.

Heller, Peter. 2005. Fiscal Space: What It Is and How to Get It. *Finance and Development* 42 (2, June).

Henderson, Gail E., et al. 2007. Clinical Trials and Medical Care: Defining the Therapeutic Misconseption. *PLoS* (4, 11, November): 1735–1738.

HIV Prevention Trials Network. 2006. Community Involvement Tool Box. http://www.hptn.org/community_program/community_involvement_toolbox.htm. Accessed 20 Nov 2013.

Hobcraft, John. 1993. Women's Education, Child Welfare and Child Survival: A Review of the Evidence. *Health Transition Review* 3 (2, October): 159–175.

Hoen, Ellen't et al. 2011. Driving a Decade of Change: HIV/AIDS, Patents and Access to Medicines for All. *Journal* of *the International AIDS Society* 14, no. 15, March. https://msfaccess.org/sites/default/files/MSF_assets/HIV_AIDS/Docs/HIV_MedJourn_JournIntnlAIDS_Decadeofchange_ENG_2011.pdf. Accessed 2 Apr 2017.

Hogerzeil, Hans V., et al. 2006. Is Acces to Essential Medicines as Part of the Fulfillment of the Right to Health Enforceable through the Courts. *The Lancet* 368 (9532, July): 305–311.

Hollis, Aidan. 2005. *An Efficient Reward System for Pharmaceutical Innovation*, Working Paper, Department of Economics. Calgary: University of Calgary.

———. 2007. *Incentive Mechanisms for Innovation*, IAPR Technical Paper, Institute for Advanced Policy Research, 1–42. Calgary: University of Calgary.

Hollis, Aidan, and Thomas Pogge. 2008. *The Health Impact Fund: Making New Medicines Accessible for All*. New Haven: Incentives for Global Health.

Horng, Sam, and Grady Christine. 2003. Misunderstanding in Clinical Research: Distinguishing Therapeutic Misconception, Therapeutic Misestimation, and Therapeutic Optimism. *IRB: Ethics and Human Research* 25 (1, January–February): 11–16.

Horton, Richard. 2007. Health as an Instrument of Foreign Policy. *Lancet* 369 (9564, March): 806–807.

Hotez, Peter. 2007. A New Voice for the Poor. *PLoS Neglected Tropical Diseases* 1 (1, October): e77.

Hughes, J.W., M.J. Moore, and E.A. Snyder. 2002. *Napsterizing Pharmaceuticals Access, Innovation and Consumer Welfare*, Working Paper. Cambridge, MA: National Bureau of Economic Research.

Hyder, A.A., et al. 2004. Ethical Review of Health Research: A Perspective from Developing Country Researchers. *Journal of Medical Ethics* 30 (1, February): 68–72.

Independent Evaluation Office of the International Monetary Fund. 2007. *The IMF and Aid to Sub-Saharan Africa*, IMF Report. Washington, DC: Independent Evaluation Office of the International Monetary Fund.

International Development Association. 2005. *Report from the Executive Directors of the International Development Association to the Board of Governors. Additions to IDA Resources: Fourteenth Replenishment*. Washington, DC: World Bank.

International Federation of Red Cross and Red Crescent Societies. 2008. *World Disasters Report 2008 – Focus on HIV and AIDS*, WDR Report, 1–248. Geneva: International Federation of Red Cross and Red Crescent Societies.

International Intellectual Property Institute. 2000a. *Patent Protection and Access to HIV/AIDS Pharmaceuticals in Sub-Saharan Africa*, IIPI Report for WIPO, 1–100. Washington, DC: International Intellectual Property Institute.

———. 2000b. *Patent Protection and Access to HIV/AIDS Pharmaceuticals in Sub-Saharan Africa*, Report Prepared for WIPO. Washington, DC: International Intellectual Property Institute.

International Monetary Fund. 2007, October. World Economic and Financial Surveys: Regional Economic Outlook, Sub-Saharan Africa. *International Monetary Fund*. http://www.imf.org/external/pubs/ft/reo/2007/AFR/ENG/sreo-0407.pdf. Accessed 18 July 2014.

———. 2009. World Economic. *World Economic Outlook Databse*. http://www.imf.org/external/pubs/ft/weo/2009/01/weodata/index.aspx. Accessed 16 July 2014.

Jefferson, Thomas. 1972. Letter to Isaac McPherson, 1813. In *The Life and Selected Writings of Thomas Jefferson*, ed. A. Knock and W. Peden. New York: Modern Library.

Joffe, S., et al. 2001. Quality of Informed Consent in Cancer Clinical Trials: A Cross-Sectional Survey. *Lancet* 358 (9295, November): 1772–1777.

John, Locke. 1983. *The Second Treatise of Government, 1764*, ed. P. Laslett. Cambridge: Cambridge University Press.

Johnston, Josephine, and Angela A. Wasunna. 2007. Patents, Biomedical Research and Treatments: Examining Concerns, Canvassing Solutions. *The Hastings Center Report* 37 (1, January–Febuary): 1–36.

Johri, Mira et al. 2006. *Sharing the Benefits of Medical Innovation: Ensuring Fair Access to Essential Medicines*. Ensuring Fair Access to Medicines Report.

Joint United Nations Programme on HIV/AIDS (UNAIDS). 2000. *Ethical Considerations in HIV Preventive Vaccine Research: UNAIDS Guidance Document*, UNAIDS Report, 1–48. Geneva: UNAIDS.

Joint U.N. Programme on HIV/AIDS. 2004. *Report on the Global AIDS Epidemic*, Global AIDS Report, 1–231. Geneva: United Nations.

———. 2006. *Global Facts and Figures 06*. http://data.unaids.org/pub/epireport/2006/20061121_EPI_FS_Globalfacts_en.pdf. Accessed 3 May 2013.

Jones, Charles. 1999. *Global Justice: Defending Cosmopolitanism*. New York: Oxford University Press.

Jones-Pauly, Christina. 2001. Loosening the Bounds of Human Rights: Global Justice and the Theory of Justice. *Human Rights and Human Welfare* 1 (3, July): 15–20.

Kant, Immanuel. 1998. Fundamental Principles of the Metaphysic of Morals (Trans. T.K. Abbott). Amherst/New York: Prometheus Books.

Kar, Dev, and Devon Cartwright-Smith. 2009. *Illicit Financial Flows from Development Countries: 2002–2006*, Global Financial Integrity Report. Washington, DC: Center for International Policy.

Karapinar, Baris, and Michelangelo Temmerman. 2008. Benefiting from Biotechnology: Pro-Poor Intellectual Property Rights and Public-Private Partnerships. *Biotechnology Law Report* 27 (3, June): 189–202.

Kass, Nancy, and Adnan A. Hyder. 2001. Attitudes and Experiences of U.S. and Developing Country Investigators Regarding U.S. Human Subjects Regulations. In *Ethical and Policy Issues in International Research: Clinical Trials in Developing Countries, Commissioned Papers and Staf Analysis*, vol. II, B1–220. Bethesda: National Bioethics Advisory Commission.

Kassa, Alain, et al. 2005. *Access to Health Care, Mortality, and Violence, in Democratic Republic of the Congo*, MSF Report, 1–86. Brussels: Medecins Sans Frontieres.

Kaufert, J.M., and J.D. O'Neil. 1990. Biomedical Rituals and Informed Consent: Native Canadians and the Negotiation of Clinical Trust. In *Social Science Perspectives on Medical Ethics*, ed. G. Weisz, 41–64. Philadelphia: University of Pennsylvania Press.

Kaufert, J.M., R.W. Putsch, and M. Lavallee. 1998. Exprience of Aboriginal Health Interpreters in Mediation of Conflicting Values in End of Life Decision Making. *International Journal of Circumpolar Health* 57: 43–48.

Kazatchkine, Michel. 2008. Closing Speech at the XVII International AIDS Conference in Mexico. In *XVII International AIDS Conference*. Geneva: Global Fund to Fight AIDS, Tuberculosis and Malaria.

Keller, Simon. April 2005. Patriotism as Bad Faith. *Ethics* 115 (3): 563–592.

Kent, Gerry. June 1997. The Views of Members of Local Research Ethics Committees, Researchers, and Members of the Public towards the Roles and Functions of LRECs. *Journal of Medical Ethics* 23 (3): 186–190.

Khor, Martin. 2005. Impasse on Talks on TRIPS and Health "Permanent Solution". *Third World Network,* Geneva, October 26.

Kiefer, Christie W. 1992. Militarism and World Health. *Social Science Medicine* 34 (7, April): 719–724.

Kirby, Michael. 2009. Aims. In *The UNESCO Universal Declaration on Bioethics and Human Rights: Background, Principles and Application*, ed. A.M.J. Henk and Michele S. Jean ten Have, 81–89. Paris: UNESCO Publishing.

Kitagawa, Evelyn M., and Philip M. Hauser. 1973. *Differential Mortality in the United States: A Study in Socioeconomic Epidemiology*. Cambridge, MA: Havard University Press.

Kmietowicz, Zosia. 2002. Failure to Tackle AIDS Puts Millions at Risk of Starvation. *BMJ* 325 (7375, November): 1257.

Kollek, Regine. 2009. Consent. In *The UNESCO Universal Declaration on Bioethics and Human Rights: Background, Principles and Application*, ed. Henk A.M.J. ten Have and Michele S. Jean, 123–138. Paris: UNESCO Publishing.

Kunitz, Stephen J. 2000. Globalization, States and the Health of Indigenous Peoples. *Am. J. Public Health* 90 (10, October): 1531–1539.

Kymlicka, Will. 1990. *Contemporary Political Philosophy: An Introduction*. Oxford: Oxford University Press.

Lallemant, M., and V. Vithayasai. 1995. *A Short ZDV Course to Prevent Perinatal HIV in Thailand*, Harvard Study Proposal. Boston: Harvard School of Public Health.

Lambsdorff, Johann Graf. 2008. *Corruption Perception Index (CPI): Transparency International 2008.*

Lavery, James V. 2004. Putting International Research Ethics Guidelines to Work for the Benefit of Developing Countries. *Yale Journal of Health Policy, Law and Ethics* 4 (2, Summer): 319–336.

Lawn, Joy E., et al. 2006. Newborn Survival. In *Disease Control Priorities in Developing Countries*, ed. Dean T. Jamison et al., 531–549. Washington, DC: World Bank.

Leckie, Scott. 1998. Another Step Towards Indivisibility: Identifying the Key Features of Violations of Economic, Social and Cultural Rights. *Human Rights Quarterly* 20 (1, February): 81–124.

Lee, Stacey B. 2010. Informed Consent: Enforcing Pharmaceutical Companies' Obligations Abroad. *Health and Human Rights* 12 (1, June): 15–28.

Lema, V.M., M. Mbondo, and E.M. Kamau. March 2009. Informed Consent for Clinical Trials: A Review. *East African Medical Journal* 85 (3): 133–142.

Lerberghe, Wim Van. 2008. *The World Health Report 2008: Primary Health Care - Now More Than Ever*, World Health Report, 1–119. Geneva: World Health Organization.

Lewis, J.J.C., et al. 2004. The Population Impact of HIV on Fertility in Sub-Saharan Africa. *AIDS* 18 (Suppl 2, June): S35–S43.

Liddell, Kathleen. 2010. The Health Impact Fund: A Critique. In *Incentives for Global Public Health: Patent Law and Access to Essential Medicines*, ed. Matthew Rimmer Thomas and Kim Rubenstein Pogge, 155–180. New York: Cambridge University Press.

Lidz, Charles W., and Paul S. Appelbaum. 2002. The Therapeutic Misconception: Problems and Solutions. *Medical Care* 40 (9, September): V55–V63.

Lie, Reidar K. 2000. *Justice and International Research*, ed. S. Gorowitz, J. Gallagher, and R.J. Levin, 27–40. Genevea: CIOMS.

Lie, R.K. 2017. Ethical Issues in Clinical Trial Collaborations with Developing Countries – With Special Reference to Preventive HIV Vaccine Trials with Secondary Endpoints, Bergen, Norway. https://enterics.tghn.org/site_media/media/articles/trialprotocoltool.../Ethicals.pdf. Accessed 25 March 2017.

Lo, Bernard. 2010. *Ethical Issues in Clinical Research: A Practical Guide*. Philadelphia: Lippincott Williams & Wilkins.

Lo, Bernard, Nancy Padian, and Mark Barnes. 2007. The Obligation to Provide Antiretroviral Treatment in HIV Prevention Trials. *AIDS* 21 (10, June): 1229–1231.

Lockwood, Christopher Lea. 2009. Biotechnology Industry Organization v. District of Columbia: A Preemptive Strike Against State Price Restrictions on Prescription Pharmaceuticals. *Albany Law Journal of Science and Technology* 19 (1, April): 148–149.

Loff, B. 2002. No Agreement Reached in Talks on Access to Cheap Drugs. *Lancet* 360 (9349, December): 1951.

London, Alex John. 2005. Justice and the Human Development Approach to International Research. *Hastings Center Report* 35 (1, January–February): 24–37.

———. 2008. Responsiveness to Host Community Health Needs. In *The Oxford Textbook of Clinical Research Ethics*, ed. Ezekiel J. Emmanuel et al., 737–744. New York: Oxford University.

Loue, Sana, David Okello, and Medi Kawuma. 1996. Research Bioethics in the Ugandan Context: A Program Summary. *Journal of Law, Medicine and Ethics* 24 (1, March): 47–53.

Love, James, and Tim Hubbard. 2007. The Big Idea: Prizes to Stimulate R&D for New Medicines. *Chicago-Kent Law Review* 82 (3, November): 1519–1554.

Lu, Chunling, et al. 2010. Public Financing of Health in Developing Countries: A Cross-National Systematic Analysis. *The Lancet* 375 (9723, April): 1375–1387.

Lurie, Nicole, and Tamara Dubowitz. 2007. Health Disparities and Access to Health. *JAMA* 297 (10, March): 1118–1121.

Macklin, Ruth. 1981. "Due" and "Undue" Inducements: On Paying Money to Subjects. *IRB: A Review of Human Subjects* 3 (5): 1–6.

———. 2001. After Helsinki: Unresolved Issues in International Research. *Kennedy Institute of Ethics Journal* 11 (1, March): 17–36.

———. 2004. *Double Standards in Medical Research in Developing Countries*. Cambridge: Cambridge University Press.

———. 2006. Changing *the* Presumption: Providing ART to Vaccine Research Participants. *American Journal of Bioethics* 6 (1): W1–W5.

Macpherson, Cheryl Cox. 2001. Ethics Committees. Research Ethics: Beyond the Guidelines. *Developing World Bioethics* 1 (1, May): 57–68.

———. 2001. Research Ethics Committee: Getting Started. *West Indian Medical Journal* 50 (3, September): 186–188.

Malan, R. 2004. Africa isn't Dying of AIDS. *Spectator*, December 2004.

Mandle, Jon. 2006. *Global Justice*. Malden: Polity Press.

Marchand, Sarah, Daniel Wikler, and Bruce Landesman. 1998. Class, Health, and Justice. *The Milbank Quarterly* 76 (3, September): 449–467.

Marmot, Michael. 2005. Social Determinants of Health Inequalities. *Lancet* 365 (9464, March): 1099–1104.

Marmot, Michael G., et al. 1978. Employment Grade and Coronary Heart Disease in British Civil Servants. *Journal of Epidemiology and Community Health* 32 (4, December): 244–249.

Marshall, Patricia A. 2001. The Relevance of Culture for Informed Consent in US-Funded International Health Research. In *Ethical and Policy Issues in International Research: Clinical Trials in Developing Countries, Vol. II, Commissioned Papers and Staff Analysis*, C1–38. Bethesda: National Bioethics Advisory Commission.

———. 2004. The Individual and the Community in International Genetic Research. *Journal of Clinical Ethics* 15 (1, Spring): 76–86.

———. 2005. Human Rights, Cultural Pluralism, and International Health Research. *Theoretical Medicine and Bioethics* 26 (6, December): 529–557.

———. 2007. *Ethical Challenges in Study Design and Informed Consent for Health Research in Resource-Poor Settings*, 1–79. Geneva: Social, Economic and Behavioral Research.

Marshall, P.A., and C. Rotimi. 2001. Ethical Challenges in Community Based Research. *American Journal of Medical Sciences* 322 (5, November): 259–263.

Mastroianni, Anna C., Ruth Faden, and Daniel Federman, eds. 1994. *Women and Health Research: Ethical and Legal Issues of Including Women in Clinical Studies*. Vol. 1. Washington, DC: National Academy Press.

McDonough, P., et al. 1997. Income Dynamics and Adult Mortality in the United States, 1972 through 1989. *American Journal of Public Health 87, no.* 87 (9, September): 1476–1483.

McNeil, Donald G. Jr. 2001, February 13. Oxfam Joins Campaign to Cut Drug Prices for Poor Nations. *New York Times*, A6.

———. 2002, October 16. UN Disease Fund Opens Way to Generics. *New York Times*, A6.

Medecins Sans Frontieres. 2001. *Fatal Imbalance: The Crisis in Research and Development for Drugs for Neglected Diseases*, A Survey of Private Sector Drug Research and Development, 16–20. Geneva: MSF and DND Working Group.

———. 2004. *Access to Medicines at Risk Across the Globe: What to Watch Out for in Free Trade Agreements with the United Sates*, MSF Briefing Note. Geneva: MSF.

———. 2005a, October 24. *MSF to WTO: Rethink Access to Life-Saving Drugs Now*. Geneva: MSF.

———. 2005b. *Will the Lifeline of Affordable Medicines for Poor Countries be Cut? Consequences of Medicines Patenting in India*, External Briefing Document, 1–6. Geneva: MSF.

Medhi, Subhi. 1999. *Health and Family Planning Indicators: Measuring Sustainability*, USAID Report, 1–56. Washington DC: US Agency for International Development, Office of Sustainable Development, Bureau of Africa.

Medical Research Council of South Africa (MRC-SA). 1993. *Guidelines on Ethics for Medical Research*. South Africa: MRC-SA.

Mellino, Marla L. 2010. The TRIPS Agreement: Helping or Hurting Least Developed Countries' Access to Essential Pharmaceuticals? *Fordham Intellectual Property, Media and Entertainment Law Journal* 20 (4, Summer): 1349–1388.

Merton, Robert K. 1968. The Matthew Effect in Science: The Reward and Communication Systems of Science are Considered. *Science* 159 (3810, January): 56–63.

Milanovic, Branko. 2005. *Worlds Apart: Measuring International and Global Inequality*. Princeton: Princeton University Press.

Millum, Joseph. 2011. Post-Trial Access to Antiretrovirals: Who Owes What to Whom? *Bioethics* 25 (3): 145–154.

———. 2012. Global Bioethics and Political Theory. In *Global Justice and Bioethics*, ed. Joseph Millum and Ezekiel J. Emmanuel, 17–42. New York: Oxford University Press.

Ministry of Foreign Affairs of Japan. 2009. *The Trust Fund for Human Security: For the "Human Centered" 21st Century*. Tokyo: Global Issues Cooperation Division, Ministry of Foreign Affairs of Japan.

Moellendorf, Darrel. 2002. *Cosmopolitan Justice*. Boulder: Westview Press.

———. 2002. *Cosmopolitan Justice Reconsidered*. Boulder: Westview Press.

Mok, Emily A. 2010. International Assistance and Cooperation for Access to Essential Medicines. *Health and Human Rights* 12 (1): 73–81.

Molyneux, C.S., N. Peshu, and K. Marsh. 2004. Understanding of Informed Consent in a Low-Income Setting: Three Case Studies from the Kenyan Coast. *Social Science and Medicine* 59 (12, December): 2547–2559.

Murthy, Divya. 2002. The Future of Compulsory Licensing: Deciphering the Doha Declarartion on the TRIPS Agreement and Public Health. *American University International Law Review* 17 (6): 1299–1346.

Myrdal, Gunnar. 1957. *Rich Lands and Poor: The Road to World Prosperity*. New York: Harper and Row.

Mystakidou, Kyriaki, et al. 2009. Ethical and Practical Challenges in Implementing Informed Consent in HIV/AIDS Clinical Trials in Developing or Resource-Limited Countries. *Journal of Social Aspects of HIV/AIDS* 6 (2, September): 46–57.

National Bioethics Advisory Commision (NBAC). 2001. *Ethical and Policy Issues in Internal Research: Clinical Trials in Developing Countries*, NBAC Report, 1–138. Bethesda: NBAC.

National Commision for the Protection of Human Subjects of Biomedical and Behavioral Research. 1979. *The Belmont Report: Ethical Principles and Guidelines for the Protection of Human Subjects of Research*. Washington, DC: U.S. Government Printing Office.

National Consensus Conference on Bioethics and Health Research in Uganda. 1997. *Guidelines for the Conduct of Health Research Involving Human Subjects in Uganda*. Kampala: National Consensus Conference, Section V. Part D. 4.

Novak, Kristine. 2003. The WTO's Balancing Act. *The Journal of Clinical Investigation* 112 (9, November): 1269–1273.

Nuffield Council on Bioethics. 2002. *The Ethics of Research related to Healthcare in Developing Countries*, Nuffield Council on Bioethics Report, 1–205. London: Nuffield Council on Bioethics.

Nunn, Nathan. 2008. The Long Term Effects of Africa's Slave Trades. *Quarterly Journal of Economics* 123 (1): 139–176.

Nussbaum, Martha C. 2004. Beyond the Social Contract: Capabilities and Global Justice. *Oxford Development Studies* 32 (1, March): 1–18.

Odell, John, and Susan Sell. 2006. Reframing the Issue: The WTO Coalition Intellectual Property and Public Health. In *Negotiating Trade: Developing Countries in the WTO and NAFTA*, ed. John Odell. Cambridge: Cambridge University Press.

Office of the United States Global AIDS Coordinator. 2005. *Engendering Bold Leadership, First Annual Report to Congress on the President's Emergency Plan for AIDS Relief*, PEPFAR Annual Report. Washington, DC: United States Department of State.

O'Neill, Onora. 1989. *Constructions of Reason: Exploration of Kant's Practical Philosophy*. New York: St. Martin's Press.

———. 2000. *Bounds of Justice*. New York: Cambridge University Press.

Ooms, Gorik. 2006. Health Development Versus Medical Relief: The Illusion Versus the Irrelevance of Sustainability. *PLoS Medicine* 3 (8, August): e345.

———. 2008. The Right to Health and the Sustainability of Healthcare: Why a New Global Health Aid. PhD Dissertation, University of Ghent.

———. 2009. *From the Global Response Towards Global Health?* A Discussion Paper for the Helene De Beir Foundation and the International Civil Society Support Group, 1–47.

Ooms, Gorik, and Rachel Hammonds. 2008. Correcting Globalisation in Health: Transactional Entitlements Versus the Ethical Imperative of Reducing Aid-Dependency. *Public Health Ethics* 1 (2, January): 154–170.

———. 2010. Taking Up Daniels' Challenge: The Case for Global Health Justice. *Health and Human Rights* 12 (1): 29–46.

Ooms, Gorik, and Ted Schrecker. 2005. Expenditure Ceilings, Multilateral Financial Institutions, and the Health of Poor Populations. *Lancet* 365 (9473, May): 1821–1823.

Ooms, Gorik, Katharine Derderian, and David Melody. 2006. Do We Need a World Health Insurance to Realise the Right to Health? *PLoS Medicine* 3 (12, December): e530.

Ooms, Gorik, Wim Van Damme, and Marieen Temmerman. 2007. Medicines Without Doctors: Why the Global Fund Must Fund Salaries of Health Workers to Expand AIDS Treatment. *PLoS Medicine* 4 (4, April): e128.

Ooms, Gorik et al. 2011. Global Health: What It Has Been So Far, What It Should Be, and What It Could Become. In *Studies in Health Services Organisation & Policy*, Working Paper Series No. 2, 40–52. Antwerpen: The Department of Public Health, Institute of Tropical Medicine. http://www.itg.be/WPshsop. Accessed 3 Dec 2016.

Organization of African Unity. 2001. *Abuja Declaration on HIV/AIDS, Tuberculosis and Other Related Diseases*, OAU Health Report, 1–7. Abuja: OAU.

Orts, Eric W. 1992. Beyond Shareholders: Interpreting Corporate Constituency Statutes. *George Washington Law Review* 61 (1, November): 14–135.

Oxfam International. 2007. Investing for Life: Meeting Poor People's Needs for Access to Medicines through Responsible Business Practices. Oxfam Report, Oxfam.

Page, Alice K. 2002. Prior Agreements in International Clinical Trials: Ensuring the Benefits of Research to Developing Countries. *Health Policy, Law and Ethics* 3 (1, Winter): 35–64.

Parashar, Umesh D., et al. 2003. Global Illness and Deaths Caused by Rotavirus Disease in Children. *Emerging Infectious Diseases* 9 (5, May): 565–572.

Park, Rosalyn S. 2002. The International Drug Industry: What the Future Hold for South Africa's HIV/AIDS Patients. *Minnesota Journal of Global Trade* 11 (1, Winter): 125–154.

Pavignani, Enrico, and Sandra Colombo. 2009. *Analysing Disrupted Health Sectors: A Modular Manual*, WHO Report, 1–484. Geneva: World Health Organization.

Pellegrino, Edmund D. 2009. Benefit and Harm. In *The UNESCO Universal Declaration on Bioethics and Human Rights: Background, Principles and Application*, ed. Henk A.M.J. ten Have and Michele S. Jean, 99–109. Paris: UNESCO Publishing.

Petersen, M.. 2001, May 3. Novartis Agrees to Lower Price of a Medicine Used in Africa. *New York Times*, C1.

Petersen, Melody and Rohter, L.. 2001, March 31. Maker Agrees to Cut Price of 2 AIDS Drugs in Brazil. *New York Times*, A4.

Petersen, W., and R. Petersen. 1985. *Dictionary of Demography: Multilingual Glossary*. Westport: Greenwood Press.

Piot, Peter. 2003. *AIDS: The Need For An Exceptional Response to An Unprecedented Crisis. A Presidential Fellows Lecture*. Washington, DC: World Bank.

Pogge, Thomas W. 1989. *Realizing Rawls*. Ithaca: Cornell University Press.

———. 1994. An Egalitarian Law of Peoples. *Philosophy and Public Affairs* 23 (3, Summer): 195–224.

Pogge, Thomas. 1998. The Bounds of Nationalism. In *Rethinking Nationalism*, ed. J.K. Nielsen and M. Seymour Couture, 463–504. Calgary: University of Calgary Press.

———. 2002a. Eradicating Systematic Poverty: Brief for a Global Resources Dividend. In *World Poverty and Human Rights: Cosmopolitan Responsibilities and Reforms*, ed. Thomas Pogge, 196–215. Cambridge: Polity Press.

———. 2002b. *World Poverty and Human Rights: Cosmopolitan Responsibilities and Reforms*. Second ed. Cambridge: Polity Press.

———. 2004. Assisting the Global Poor. In *The Ethics of Assistance*, ed. K. Chatterjee Deen, 260–288. Cambridge: Cambridge University Press.

———. 2005. Severe Poverty as a Violation of Negative Duties. *Ethics & International Affairs* 19 (1, March): 55–83.

———. 2006. Montreal Statement on the Human Rights to Essential Medicines. *Cambridge Quarterly of Healthcare Ethics* 15 (2): 1–15.

———. 2007. Severe Poverty as a Human Rights Violation. In *Freedom from Poverty as a Human Right: Who Owes What to the Very Poor?* ed. Thomas Pogge, 11–54. New York: Oxford University Press.

———. 2008a. Pharmaceutical Innovation: Must We Exclude the Poor? In *World Poverty and Human Rights: Cosmopolitan Responsibilities and Reforms*, ed. Thomas Pogge, 222–261. Cambridge: Polity Press.

———. 2008b. Access to Medicines. *Public Health Ethics* 1 (2, July): 73–82.

———. 2008c. *World Poverty and Human Rights: Cosmopolitan Responsibilities and Reforms*. 2nd ed. Cambridge: Polity Press.

———. 2011a. The Health Impact Fund: Enduring Innovation Incentives for Cost-Effective Health Gains. *Social Europe Journal* 5 (2, June): 5–9.

———. 2011b. The Health Impact Fund: How to Make New Medicines Accessible to All. In *Global Health and Global Health Ethics*, ed. Solomon Benatar and Gillian Brock, 241–250. New York: Cambridge University Press.

———. 2011c. *The Health Impact Fund: More Justice and Efficiency in Global Health*, Development Policy Center Discussion Paper, Crawford School of Economic and Government. Cranberra: The Australian National University.

Pogge, Thomas, Matthew Rimmer, and Kim Rubenstein. 2010. Access to Essential Medicines: Public Health and International Law. In *Incentives for Global Public Health: Patent Law and Access to Essential Medicines*, ed. Matthew Rimmer and Kim Rubenstein Pogge Thomas, 1–32. New York/Cambridge: Cambridge University Press.

Posse, Mariana, et al. 2008. Barriers to Access to Antiretroviral Treatment in Developing Countries: A Review. *Tropical Medicine and International Health* 13 (7, July): 904–913.

Pratt, Bridget, and Bebe Loff. 2011. Justice in International Clinical Research. *Developing World Bioethics* 11 (2): 75–81.

Preziosi, M.P., et al. 1997. Practical Experiences in Obtaining Informed Consent for a Vaccine Trial in Rural Africa. *New England Journal of Medicine* 336 (5, January): 370–373.

Putsch, R.W. 1985. Cross-Cultural Communication: The Special Case of Interpreters in Health Care. *Journal of the American Medical Association* 254 (23, December): 3344–3348.

Quinn, Sandra Crouse. 2004. Ethics in Public Research: Protecting Human Subjects: The Role of Community Advisory Boards. *American Journal of Public Health* 94 (6, June): 918–922.

Rajaratnam, Julie Knoll, et al. 2010. Neonatal, Postneonatal, Childhood, and Under-5 Mortality for 187 Countries, 1970–2010: A Systematic Analysis of Progress Toward Millennium Development Goal 4. *Lancet* 375 (9730, June): 1988–2008.

Rand, Ayn. 1966. *Capitalism: The Unknown Ideal*. New York: New American Library.

Ravvin, Michael. 2008. Incentivizing Access and Innovations for Essential Medicines: A Survey of the Problem and Proposed Solutions. *Public Health Ethics* 1 (2): 110–123.

Rawls, John. 1971. *A Theory of Justice*, Original. Cambridge, MA: Harvard University Press.

———. 1993. *Political Liberalism*. New York: Columbia University Press.

———. 1999a. *A Theory of Justice*. Revised ed. Cambridge, MA: The Belknap Press of Harvard University Press.

———. 1999b. *The Law of Peoples: With the Idea of Public Reason Revisited*. Cambridge, MA: Harvard University Press.

———. 2001. *Justice as Fairness*. Cambridge, MA: Harvard University Press.

Redshaw, M.E., A. Harris, and J.D. Baum. 1996. Research Ethics Committee Audit: Differences between Committees. *Journal of Medical Ethics* 22 (2, April): 78–82.

Reichman, Jerome H. 2009. Compulsory Licensing of Patented Pharmaceutical Inventions: Evaluating the Options. *The Journal of Law, Medicine & Ethics* 37 (2, Summer): 247–263.

Resnik, David B. 1998. The Ethics of HIV Research in Developing Nations. *Bioethics* 12 (4, October): 286–306.

———. 2001. Developing Drugs for the Developing World: An Economic, Legal, Moral and Political Dilemma. *Developing World Bioethics* 1 (1, May): 1–32.

———. 2003. A Pluralistic Account of Intellectual Property. *Journal of Business Ethics* 46 (4, September): 319–335.

———. 2004. The Distribution of Biomedical Research Resources and International Justice. *Developing World Bioethics* 4 (1, May): 42–57.

Resnik, David. 2005. Access to Affordable Medications in Developing World: Social Responsibility VS. Profit. In *Ethics & AIDS in Africa: The Challenge to Our Thinking*, ed. A. Anton Van and Loretta M. Kopelman Niekerk, 111–126. Walnut Creek: Left Coast Press Inc.

Revel, Michel. 2009. Respect for Cultural Diversity and Pluralism. In *The UNESCO Declaration on Bioethics and Human Rights: Background, Principles and Application*, ed. A.M.J. Henk and Michele S. Jean ten Have, 199–209. Paris: UNESCO Publishing.

Richards, David A.J. 1982. International Distributive Justice. In *Ethics, Economics, and the Law: NOMOS XXIV*, ed. Roland J. Pennock and John W. Chapman, 275–299. New York: New York University Press.

Ridley, David B., Henry G. Grabowski, and Jeffrey L. Moe. 2006. Developing Drugs for Developing Countries. *Health Affairs* 25 (2, March): 313–324.

Rimmer, Matthew. 2008. Race Against Time: The Export of Essential Medicines to Rwanda. *Public Health Ethics* 1 (2, July): 89–103.

Roberts, Matthew, and Bill Rau. 1997. *African Workplace Profiles: Private Sector AIDS Policy*. Arlington: AIDSCAP.

Roe, Mark J. 2001. Symposium Norms and Corporate Law: The Shareholder Wealth Maximization Norm and Industrial Organization. *University of Pennsylvania Law Review* 149 (September): 2063–2065.

Rosenberg, Alex. 2004. On the Priority of Intellectual Property Rights, Especially in Biotechnology. *Politics, Philosophy & Economics* 3 (1, February): 77–95.

Roth, K. 2000. Human Rights and the AIDS Crisis: The Debate over Resource. *Canadian HIV/AIDS Policy and Law Review* 5(4, February), 93. http://hrw.org/english/docs/2000/07/11/global2195.htm. Accessed 21 Feb 2017.

Ruger, Jennifer P. 2006. Ethics and Governance of Global Health Inequalities. *Journal of Epidemiology and Community Health* 60 (11, November): 998–1003.

Ruger, Jennifer P., and Hak-Ju Kim. 2006. Global Health Inequalities: An International Comparison. *Journal of Epidemiology and Community Health* 60 (11, November): 928–936.

Rutkow, Lainie. 2011. Corporations and Health Watch: Tracking the Effects of Corporate Practices on Health. *Corporations and Health.org*, April 6, 2011. http://www.corporationsandhealth.org/2011/04/06/should-corporations-serve-shareholders-or-society-the-origins-of-the-debate/. Accessed 20 Feb 2014.

Schaeffer, Monica H., et al. 1996. The Impact of Disease Severity on the Informed Consent Process in Clinical Research. *The American Journal of Medicine* 100 (3, March): 261–268.

Scherer, F.M., and J. Watal. 2001. *Post-TRIPS Options for Access to Patented Medicines in Developing Countries*. Working Paper no. WG4, Commission on Macroeconomics and Health.

Schuklenk, Udo, and Richard Ashcroft. 2005. Affordable Access to Essential Medication in Developing Countries: Conflicts between Ethical and Economic Imperatives. In *Ethics & AIDS in Africa: The Challenge to Our Thinking*, ed. A. Van, Loretta M. Kopelman, and Niekerk Anton, 127–140. Walnut Creek: Left Coast Press Inc.

Second African Conference on Sanitation and Hygiene. 2008. *The eThekwini Declaration*, Sanitation and Hygiene Conference Report, 1–8. Durban: AfricaSan + 5 Conference.

Selgelid, Michael J. 2008. A Full-Pull Program for the Provision of Pharmaceuticals: Practical Issues. *Public Health Ethics* 1 (2, July): 134–145.

Selgelid, Michael J., and Udo Schuklenk. 2002. Letter to the Editor: Do Patents Prevent Access to Drugs for HIV in Developing Countries? *JAMA* 287 (7, February): 842–843.

Selgelid, Michael J., and Eline M. Sepers. 2006. Patents, Profits, and the Price of Pills: Implications for Access and Availability. In *The Power of Pills: Social, Ethical and Legal Issues in Drug Development, Marketing and Pricing*, ed. Jillian Clare, Patricia Illingworth, and Udo Schuklenk Cohen, 153–163. London: Pluto Press.

Sen, Amartya K. 1992. *Inequality Reexamined.* Cambridge, MA: Harvard University Press.

Sen, Gita, Piroska Ostlin, and Asha George. 2007, September. *Unequal, Unfair, Ineffective and Inefficient Gender Inequality in Health: Why It Exists and How We Can Change It.* Geneva: Women and Gender Equity Knowledge Network.

Shaffer, D.N., et al. 2006. Equitable Treatment for HIV/AIDS Clinical Trials Participants: A Focus Group Study of Patients, Clinician Researchers and Administrators in Western Kenya. *Journal of Medical Ethics* 32 (1, January): 55–60.

Shah, Seema, Stacey Elmer, and Christine Grady. 2009. Planning for Posttrial Access to Antiretroviral Treatment for Research Participants in Developing Countries. *American Journal of Public Health* 99 (9, September): 1556–1562.

Shamoo, Adil E. 2005. Debating Moral Issues in Developing Countries. *Applied Clinical Trials* 14 (6, June): 86–96.

Shamoo, Adil E., and David B. Resnik. 2009. *Responsible Conduct of Research.* 2nd ed. New York: Oxford University Press.

Shapiro, K., and S.R. Benatar. 2004. HIV Prevention Research and Global Inequality: Steps towards Improved Standards of Care. *Journal of Medical Ethics* 31: 39–47.

Shue, Henry. 1996. *Basic Rights: Subsistence, Affluence, and U.S. Foreign Policy.* 2nd ed. Princeton: Princeton University Press.

Simms, Chris, Mike Rowson, and Siobhan Peattie. 2001. *The Bitterest Pill of All: The Collapse of Africa's Health System*, Children Briefing Report, 1–27. London: Save the Children.

Singer, Peter. 1972. Famine, Affluence and Morality. *Philosophy and Public Affairs* 1 (3, Spring): 229–243.

Skolnik, Richard. 2008. *Essentials of Global Health.* Sadbury, MA: Jones and Bartlett Publishers.

Slack, C., et al. 2005. Provision of HIV Treatment in HIV Preventive Vaccine Trials: A Developing Country Perspective. *Social Science & Medicine* 60 (6, March): 1197–1208.

Slote, Michael. 1997. The Morality of Wealth. In *World Hunger and Moral Obligation*, ed. W. Aiken and H. LaFollette, 124–147. Englewood Cliffs: Prentice Hall.

Smith, Lynch J., et al. 2004. Is Income Inequality a Determinant of Population Health? Part 1. A Systemic Review. *Milbank Q* 82 (1, March): 5–99.

Snowdon, C., J. Garcia, and D. Elbourne. 1997. Making Sense of Radomization: Responses of Parents of Critically Ill Babies to Random Allocation of Treatment in a Clinical Trial. *Social Science and Medicine* 45 (9, November): 1337–1355.

Sommer, A. 1999, September 16. *Testimony Before NBAC.* Meeting Transcript, Arlington, VA, 214.

Sonderholm, Jorn. 2010a. *Intellectual Property Rights and the TRIPS Agreement: An Overview of Ethical Problems and Some Proposed Solutions*, Policy Research Working Paper 5228, 1–46. Washington, DC: World Bank and Human Develoment Network.

———. 2010b. A Reform Proposal in Need of Reform: A Critique of Thomas Pogge's Proposal for How to Incentivize Research and Development of Essential Drugs. *Public Health Ethics* 3 (2, January): 167–177.
Speers, Marjorie A. 2008. Evaluating the Effectiveness of Institutional Review Boards. In *The Oxford Textbook of Clinical Research Ethics*, ed. Ezekiel J. Emmanuel et al., 560–568. New York: Oxford University.
Sridhar, Devi. 2008. Improving Access to Essential Medicines: How Health Concerns can be Prioritised in the Global Governance System. *Public Health Ethics* 1 (2, July): 83–88.
Stanecki, Karen A. 2004. *The AIDS Pandemic in the 21st Century*, International Population Reports, 1–29. Washington, DC: US Census Bureau.
Stanford Encyclopedia of Philosophy. 2008. *John Rawls*, March 25.
Steinberg, M., et al. 2002. *Hitting Home: How Household Cope with the Impact of the HIV/AIDS Epidemic. A Survey of Household Affected by HIV/AIDS in South Africa*. Health Systems Trust and The Kaiser Family Foundation: Washington, DC.
Stephens, Joe. 2000, December 17. Where Profits and Lives Hang in the Balance. *Washington Post*, A1.
Stevens, Philip. 2004. *Diseases of Poverty and the 10/90 Gap*. London: International Policy Network.
Stiglitz, Joseph E. 2008. The Future of Global Governance. In *The Washington Consensus Reconsidered: Towards a New Global Governance*, ed. Narcis Serra and Joseph E. Stiglitz, 309–323. New York: Oxford University Press.
Stolberg, S.G. 2002. Funds Falls short of Goal and US is Given Some Blame. *New York Times*. New York, February 13.
Stover, John, and Lori Bollinger. 1999. *The Policy Project: The Economic Impact of AIDS*, 1–14. The Futures Group International: Policy Project.
Stuckler, David, et al. 2008. WHO's Budgetary Allocations and Burden of Disease: A Comparative Analysis. *Lancet* 372 (9649, November): 1563–1569.
Sugarman, Jeremy, et al. 2001. International Perspectives on Protecting Human Research Subjects. In *Ethical and Policy Issues in International Research: Clinical Trials in Developing Countries, Vol. II, Commissioned Papers and Staff Analysis*, E1–30. Bethesda: National Bioethics Advisory Commission.
Sun, Haochen. 2004. The Road to Doha and Beyond: Some Reflections on the TRIPS Agreement and Public Health. *European Journal of International Law* 15 (1): 123–150.
Swarns, Rachel L. 2003, January. African Nations Applaud Bush Plan to Fight AIDS Epidemic. *New York Times*, A19.
Tan, Kok-Chor. 2004. *Justice Without Borders: Cosmopolitanism, Nationalism and Patriotism*. Cambridge: Cambridge University Press.
Tarantola, D., et al. 2007. Meeting Report: Ethical Considerations Related to the Provision of Care and Treatment in Vaccine Trials. *Vaccine* 25: 4863–4874.
ten Have, Henk A.M.J., and Michele S. Jean. 2009. Introduction. In *The UNESCO Universal Declaration on Bioethics and Human Rights: Background, Principles and Application*, ed. A.M.J. Henk and Michele S. Jean ten Have, 17–55. Paris: UNESCO Publishing.
The Department of Economic and Social Affairs of the United Nations. 2009. *Implementing the Millennium Development Goals: Health Inequality and the Role of Global Health Partnerships*, Policy Report, 1–64. New York: United Nations.
The Group of Eight. 2005. Gleneagles Summit Documents, Chair's Summary. Gleneagles, July 8.
The Lancet. 2003. One Standard, Not Two. *Lancet* 362 (November): 1005.
The National Commission for the Protection of Human Subjects of Biomedical and Behavioral Research. 1978. *The Belmont Report: Ethical Principles and Guidelines for the Protection of Human Subjects*. Washington, DC: U.S. Government Printing Office.
Tindana, Paulina O., et al. 2007. Grand Challenges in Global Health: Community Engagement in Research in Developing Countries. *PLoS Medicine* 4 (9, September): e273.

Toebes, Brigit C.A. 1999. *The Right to Health as a Human Right in International Law*. Amsterdam: Hart/Intersentia.
Transparency International. 2006. *Global Corruption Report 2006: Corruption and Health*. London: Pluto Press.
Trench, Pippa, et al. 2007. *Beyond Any Drought*. London: Sahel Working Group.
Trouiller, Patrice, et al. 2001. Drugs for Neglected Diseases: A Failure of the Market and a Public Health Failure? *Tropical Medicine and International Health* 6 (11, November): 945–951.
U.S. Department of Health and Human Services. 1993. *Office of Protection from Research Risks, Protecting Human Research Subjects: IRB Guidebook*, 3–44. Washington, DC: Government Printing Office.
———. 2009. *Code of Federal Regulations Title 45 Public Welfare, Part 46 Protection of Human Subjects*. Washington, DC: U.S. Department of Health and Human Services.
UN Human Rights Committee (HRC). 1989, April 7. *Covenant on Civil and Political Rights (CCPR) General Comment No. 17: Article 24 Rights of the Child.* Geneva.
UNAIDS. 1999. *Acting Early to Prevent AIDS: The Case of Senegal*, UNAIDS Report, 1–23. Geneva: UNAIDS.
UNAIDS and World Bank. 2000, July 11. AIDS Hindering Economic Growth, Worsening Poverty in Hard Hit Countries. http://www.thebody.com/content/art641.htm?ts=pf. Accessed 5 Feb 2013.
UNAIDS. *AIDS Epidemic Update: December 2003*, UNAIDS and WHO Report, ed. WHO. Vol. 2003, 1–13. Geneva: UNAIDS/WHO.
UNAIDS, and WHO. 2004. *Global AIDS Epidemics*. Geneva: UNAIDS.
UNICEF. 2001. *The State of World's Children 2001*, UNICEF Report, 1–116. New York: UNICEF.
———. 2003. *Africa's Orphaned Generation*. New York: UNICEF.
United Nations. 1948. *Universal Declaration of Human Rights*. New York: United Nations.
———. 1966a. *International Covenant on Civil and Political Rights*. New York: United Nations.
———. 1966b. *International Covenant on Economic*. New York: Social and Cultural Rights.
———. 1976. *International Covenant on Civil and Political Rights*. New York: United Nations.
———. 1989, November. *Convention on Rights of the Child*. United Nations Human Rights. http://www.ohchr.org/en/professionalinterest/pages/crc.aspx. Accessed 13 July 2014.
———. 2000, September 8. *Millennium Declaration*. New York: United Nations.
———. 2010a. *Global Strategy for Women's and Children's Health*, Health Report. United Nations: New York.
———. 2010b. *Millennium Development Goals*. New York: United Nations.
United Nationa Millennium Project. 2005. The 0.7% Target: An In-depth Look. Millennium Project. . http://www.unmillenniumproject.org/press/07.htm. Accessed 2 July 2014.
United Nations Children Fund (UNICEF). 2006. The State of the World's Children 2007. http://www.unicef.org/sowc07/docs/sowc07.pdf. Accessed 3 May 2013.
United Nations General Assembly. 1986, December. *Declaration on the Right to Development, G.A. Res.41/128, UN GAOR*. New York.
———. 2001. *Declaration of Committment on HIV/AIDS*, UNAIDS Report, 1–47. Geneva: Joint UN Programme on HIV/AIDS (UNAIDS).
———. 2006. *Political Declarartion on HIV/AIDS*. Geneva.
United Nations Office for the Coordination of Humanitarian Affairs. 2005. *Humanitarian Strategic Framework for Southern Africa 2005*, OCHA Report. Johannesburg: United Nations Office for the Coordination of Humanitarian Affairs.
Upvall, M., and S. Hashwani. 2001. Negotiating the Informed Consent Process in Developing Countries: A Comparison of Swaziland and Pakistan. *International Nursing Review* 48 (3, September): 188–192.
Urban Morgan Institute for Human Rights. 1998. The Maastricht Guidelines on Violations of Economic, Social and Cultural Rights. *Human Rights Quarterly* 20 (3, August): 691–704.

Vallely, Andrew, et al. 2010. How Informed is Consent in Vulnerable Populations? Experience Using a Continuous Consent Process during the MDP301 Vaginal Microbicide Trial in Mwanza, Tanzania. *BMC Medical Ethics* 11 (10, June): 1–15.

Van Niekerk, Anton A. 2005. Principles of Global Distributive Justice and the HIV/AIDS Pandemic: Moving Beyond Rawls and Buchanan. In *Ethics & AIDS in Africa: The Challenge to Our Thinking*, ed. Anton A. Van Niekerk and Loretta M. Kopelman, 84–110. Walnut Creek: Left Coast Press.

Veneman, A.M. 2006. Achieving Millennium Development Goal 4. *Lancet* 368 (954, September): 1044–1047.

Wagstaff, Adam. 2002. Poverty and Health Sector Inequalities. *Bulletin of the World Health Organization* 80 (2, January): 97–105.

Wagstaff, Adam, and Miriam Claeson. 2004. *The Millennium Development Goals for Health: Rising to the Challenges*. Washington, DC: World Bank.

Werhane, Patricia H., and Michael E. Gorman. 2005. Intellectual Property Rights, Access to Life-Enhancing Drugs, and Corporate Moral Responsibilities. In *Ethics and the Pharmaceutical Industry*, ed. Michael A. Santoro and Thomas M. Gorrie, 260–281. Cambridge: Cambridge University Press.

Wertheimer, Alan. 1987. *Coercion*. Princeton: Princeton University.

———. 2008. Exploitation in Clinical Research. In *Exploitation and Developing Countries: The Ethics of Clinical Research*, ed. Jennifer S. Hawkins and Ezekiel J. Emmnanuel, 63–104. Princeton: Princeton University.

Westerhaus, Michael, and Arachu Castro. 2006. How Do Intellectual Property Law and International Trade Agreements Affect Access to Antiretroviral Therapy. *PLoS Medicine* 3 (8, August): 1230–1236.

White, Mary Terrell. 2007. A Right to Benefit from International Research: A New Approach to Capacity Building in Less-Developed Countries. *Accountability in Research: Policies and. Quality Assurance* 14 (2): 73–92.

Whiteside, Alan. 2005. AIDS in Africa: Facts, Figures and the Extent of the Problem. In *Ethics & AIDS In Africa: The Challenge to Our Thinking*, ed. Anton A. Van Niekerk and Loretta M. Kopelman, 1–14. Walnut Creek: Left Coast Press Inc.

Whiteside, Alan, and Amy Whalley. 2007. *Reviewing "Emergencies" for Swaziland: Shifting the Paradigm in a New Era*, 1–59. Durban: HEARD.

Whiteside, Alan, et al. 2005. Through a Glass, Darkly: Data and Uncertainty in the AIDS Debate. In *Ethics & AIDS In Africa: The Challenge to Our Thinking*, ed. A. Anton Van and Loretta M. Kopelman Nierkerk, 15–38. Walnut Creek: Left Coast Press Inc.

WHO, and UNAIDS. 2007. *Ethical Considerations in Biomedical HIV Prevention Trials*. Geneva: UNAIDS.

WHO, and WTO. 2002. *WTO Agreements & Public Health: A Joint Study by the WHO and the WTO Secretariat*, Public Health Report, 1–171. Geneva: WHO and WTO.

WHO Secretariat. 2001. *More Equitable Pricing for Essential Drugs: What Do We Mean and What Are the Issues?* Hosbjor: WHO.

WHO/UNICEF/UNFPA. 2006. Maternal Mortality in 2000. http://childinfo.org/areas/maternal-mortality/countrydata.php. Accessed 3 May 2013.

Wilkinson, Richard G. 1990. Income Distribution and Mortality: A Natural Experiment. *Sociology of Health and Illness* 12 (4, December): 391–412.

———. 1992. Income Distribution and Life Expectancy. *British Medical Journal* 304 (6820, January): 165–168.

Wilkinson, Richard. 1996. *Unhealthy Societies: The Afflictions of Inequality*. London/New York: Routledge.

Woodsong, Cynthia, and Quarraisha Abdool Karim. 2005. A Model Designed to Enhance Informed Consent: Experiences from the HIV Prevention Trials Network. *American Journal of Public Health* 95 (3, March): 412–419.

World Bank. 2004. The Millennium Goals for Health: Rising to the Challenges. http://siteresources.worldbank.org/INTEAPREGTOPHEAUNUT/PublicationsandReports/20306102/296730PAPEROMilentOgoalsOforOhealth.pdf. Accessed 25 Feb 2017.

———. 1999.

———. 2008. *Data, Country Classification*. http://go.worldbank.org/K2CKM78CCO. Accessed 9 July 2014.

———. n.d. *WDI Data Query*. http://devdata.worldbank.org/data-query/. Accessed 1 July 2006.

World Health Assembly. 2005. *Social Health Insurance: Sustainable Health Financing, Universal Coverage and Social Health Insurance: Report by the Secretariat*. Geneva, April 7.

———. 2010. *Primary Health Care, Including Health System Strengthening: WHA62.12*, Health Report. Geneva: WHO.

World Health Organization (WHO). 2000. *Operational Guidelines for Ethics Committees That Review Biomedical Research*. Geneva: WHO, para. 6.2.6.6.

World Health Organization. 2003. *World Health Report 2003: Shaping the Future*, Health Report. Geneva: World Health Organization.

World Health Organization (WHO). 2004a. *Medicines and the Idea of Essential Drugs (EDM)*, WHO Report of Essential Medicines. Geneva: WHO.

———. 2004b. *Recruitment of Health Workers from Developing World*. Geneva: WHO.

———. 2005. Health and the MDGs: Keep the Promise, http://www.who.int/mdg/publications/MDG_Reporter_08_2005.pdf. (Accessed February 25, 2016).

———. 2006a. *Core Health Indicators: Malawi*. Geneva: WHO. http://www3.who.int/whosis/core/core_select_process.cfm?country=mwi&indicators=healthpersonnel&intYear_select=all&language=en. Accessed 21 Feb 2017.

———. 2006b. *Health Status Statistics: Mortality*. http://www.who.int/healthinfo/statistic/indneonatalmortality/en. Accessed 25 June 2006.

World Health Organization. 2007. *World Health Report: Health Systems Financing: the Path to Universal Coverage*. World Health Report. *World Health Statistics*. http://who.int/whosis/whostat2007.pdf. Accessed 3 May 2013.

———. 2008. *More than Five Million People Receiving HIV Treatment*. Geneva: World Health Organization, . *National Health Accounts, Country Information*. 2008. http://www.who.int.nha/country/nha_ratios_and_percapita_levels_2001–2005.xls. Accessed 8 July 2014.

World Health Organization (WHO). 2010a. *About WHO*. http://www.who.int/about/en/ Accessed 8 June 2014.

———. 2010b, July 19. *More than Five Million People Receiving HIV Treatment*. Geneva: World Health Organization.

World Health Organization. 2010c. *World Health Report: Health Systems Financing: the Path to Universal Coverage*, World Health Report, 1–106. Geneva: WHO.

———. 2010d. *World Health Statistics 2010*, World Health Report. Geneva: World Health Organization.

———. 2012. *Meeting Report: World Conference on Social Determinants of Health 2011*, WHO Report, 1–66. Geneva: WHO.

World Medical Association. 2000. *Declaration of Helsinki, Ethical Principles for Medical Research Involving Human Subjects*. Edinburgh: World Medical Association.

———. 2001. Declaration of Helsinki: Ethical Principles for Medical Research Involving Human Subjects. *Bulletin of the World Health Organization*: 373–374.

———. 2004. *Workgroup* Report on the Revision of Paragraph 30 of the Declaration of Helsinki. http://www.wma.net/e/ethicsunit/pdf/wgdohjan2004.pdf. Accessed 12 Nov 2016.

———. 2008. *Declaration of Helsink: Ethical Principles for Medical Research Involving Human Subjects*, International Research Guidelines, 1–5. Edinburgh: World Medical Association.

———. 2013. *Declaration of Helsinki: Ethical Principles for Medical Research Involving Human Subjects*, International Research Guidelines. Edinburgh: World Medical Association.

World Trade Organization. 1994. *TRIPS Agreement on Trade-Related Aspects of Intellectual Property Rights*.

World Trade Organization General Council. 2003. *Implementation of Paragraph 6 of the Doha Declaration on the TRIPS Agreement and Public Health: Decision of the General Council of* 30 August 2003. Geneva, September 1.

Zainol, Zinatul A. 2011. Pharmaceutical Patents and Access to Essential Medicines in sub-Saharan Africa. *African Journal of Biotechnology* 10 (58, September): 12376–12388.

Zakus, David J., and Catherine L. Lysack. 1998. Revisiting Community Participation. *Health Policy and Planning* 13 (1, March): 1–12.

Zion, Deborah. 2004. HIV/AIDS Clinical research and the Claims of Beneficence, Justice and Integrity. *Cambridge Quarterly of Healthcare Ethics* 13 (4, October): 404–413.

Zong, Zhiyong. 2008. Should Post-Trial Provision of Beneficial Experimental Interventions be Mandatory in Developing Countries. *Journal of Medical Ethics* 34 (3): 188–192.

Zulueta, Paquita De. 2001. Randomised Placebo-Controlled Trials and HIV-Infected Pregnant Women in Developing Countries. Ethical Imperialism or Unethical Exploitation. *Bioethics* 15 (4, August): 294.

The manufacturer's authorised representative in the EU is Springer Nature Customer Service Centre GmbH, Europaplatz 3, 69115 Heidelberg, Germany. If you have any concerns regarding our products, please contact ProductSafety@springernature.com

Printed and bound by CPI Group (UK) Ltd, Croydon, CR0 4YY
01/10/2026
02225861-0015